METHODS IN MOLECULAR BIOLOGY

Series Editor
John M. Walker
School of Life and Medical Sciences
University of Hertfordshire
Hatfield, Hertfordshire, UK

For further volumes:
http://www.springer.com/series/7651

For over 35 years, biological scientists have come to rely on the research protocols and methodologies in the critically acclaimed *Methods in Molecular Biology* series. The series was the first to introduce the step-by-step protocols approach that has become the standard in all biomedical protocol publishing. Each protocol is provided in readily-reproducible step-by-step fashion, opening with an introductory overview, a list of the materials and reagents needed to complete the experiment, and followed by a detailed procedure that is supported with a helpful notes section offering tips and tricks of the trade as well as troubleshooting advice. These hallmark features were introduced by series editor Dr. John Walker and constitute the key ingredient in each and every volume of the *Methods in Molecular Biology* series. Tested and trusted, comprehensive and reliable, all protocols from the series are indexed in PubMed.

G Protein-Coupled Receptor Screening Assays

Methods and Protocols

Second Edition

Edited by

Sofia Aires M. Martins

INESC Microsistemas e Nanotecnologias, Lisbon, Portugal

Duarte Miguel F. Prazeres

IBB—Institute for Bioengineering and Biosciences, Lisbon, Portugal; Department of Bioengineering, Instituto Superior Técnico, Universidade de Lisboa, Lisbon, Portugal

 Humana Press

Editors
Sofia Aires M. Martins
INESC Microsistemas
e Nanotecnologias
Lisbon, Portugal

Duarte Miguel F. Prazeres
IBB—Institute for Bioengineering and Biosciences
Lisbon, Portugal

Department of Bioengineering
Instituto Superior Técnico
Universidade de Lisboa
Lisbon, Portugal

ISSN 1064-3745   ISSN 1940-6029   (electronic)
Methods in Molecular Biology
ISBN 978-1-0716-1220-0   ISBN 978-1-0716-1221-7   (eBook)
https://doi.org/10.1007/978-1-0716-1221-7

Cover Illustration Caption: RAW 264.7 cells undergoing directional migration as a result of localized optical activation of lamprey parapinopsin, a Gi protein coupled receptor. Cells are expressing Venus-Parapinopsin (green) and PIP3 sensor PH-Akt-mCh (red). Rectangles represent regions illuminated with blue light. Photoactivation of parapinopsin with blue light triggers activation of native Gi signaling pathways in the cell, as evidenced by increased PIP3 and extending lamellipodia in the photoactivated region. Authors: N. Gautman, X. Meshik.

This Humana imprint is published by the registered company Springer Science+Business Media, LLC, part of Springer Nature.
The registered company address is: 1 New York Plaza, New York, NY 10004, U.S.A.

Preface

The world is tackling an unprecedented crisis, perhaps the greatest challenge humanity has faced since the Second World War. A novel coronavirus, unknown to our immune system emerged in Asia by the end of 2019 and rapidly spread throughout the entire planet. The COVID-19 pandemic caught the world unprepared and forced societies into a global lockdown that disrupted the normal practice of economy and governance, as well as our way of life.

As never before, the quest for new therapeutics received a central role in the world's agenda. In an attempt to cope with the devastating consequences of COVID-19—currently well beyond a health crisis, but rather expanding toward an economical emergency—scientific communities and industries are joining efforts worldwide to develop vaccines or pharmaceutical compounds capable of improving the disease's management and outcomes.

The second edition of the book *G Protein-Coupled Receptor Screening Assays: Methods and Protocols* appears within this context. GPCRs are the largest family of membrane receptors, the specialized proteins that convert extracellular stimulus into cellular information. A multitude of ligands are known to bind GPCRs, including chemokines, neurotransmitters, metals, peptides, amino acids, steroids, or photons. Ligand binding to a GPCR causes a change in the receptor conformation and activates a G-protein. The active form of the G-protein dissociates then into its α and β/γ subunits, activating specific effectors with the release of second messengers and producing a signaling cascade that culminates in a cellular response.

Although GPCRs are not directly involved in viral infection, their role in the so-called comorbidities or exacerbated immune responses, which are likely related with poor clinical outcomes in COVID-19 patients, is unquestionable. In fact, GPCRs regulate a plethora of physiological states, in which modulation of the immune response, metabolism, and homeostasis represent only limited examples. Their importance can be emphasized by the fact that more than one third of the current drugs in the market are GPCR-acting compounds. However, no ligand has yet been identified for over 100 of the approximately 800 known human GPCRs. Hence, screening assays are critical tools to discover new GPCR targets and explore new opportunities in the pharmacology arena.

Key Outcomes of This Edition

This second edition features over 20 chapters dedicated to the study of GPCRs and related screening assays that aim at a better understanding of the receptors' mode of action or at the identification of potential drug candidates. In the book, different methodologies for the expression and purification of a multitude of GPCRs are described, ranging from the isolation of messenger RNA, to the development of recombinant constructs targeting reporter detection systems based on fluorescence and luminescence read-outs. In Chapter 1, Tiu et al. propose a method for the isolation of lipid rafts—localized domains in the plasmatic membranes of defined lipidic composition—that serve as anchors for the organization and interaction dynamics of biomolecules, including GPCRs. Embedded receptors and respective signaling mechanisms can then be studied in their natural milieu and cellular environment. Alternatively, GPCRs can be directly isolated from primary cells. However, functionality is most often compromised due to losses in their 3-D structure. In addition, GPCR protein levels are sparse and validated antibodies to guide specific isolation are often not available. Molecular techniques that comprise the isolation and validation of messenger RNA (mRNA) and cloning in heterologous expression systems constitute two important methodologies to support subsequent functional studies. In Chapter 2, Sriram et al. describe a method for the isolation of total mRNA and validation of GPCR-related sequences by RNA-seq and GPCR Taqman assays. In Chapter 3, Reeves proposes a method for the stable expression of bovine rhodopsin in the HEK293 cell line. The author takes advantage of both cytomegalovirus promoter and control elements from the $Tn10$ tetracycline operon present in an $E.\ coli$ transposon, to generate an inducible gene expression system. In this view, a stable and strong protein production can be achieved at higher cell concentrations. In Chapter 4, Yeliseev describes the expression and purification of type II human cannabinoid receptor CB_2, in both $Escherichia\ coli$ and suspension cultures of mammalian cells. To achieve an efficient purification, the CB_2 molecular construct comprises an His_{10}-Tag and two identical Streptag sequences for the subsequent affinity chromatography purification. The purified receptors are suitable for further applications including functional studies.

Functional assays are defined as in vivo experiments designed to evaluate the activity of the partners involved in a particular cellular pathway. In here, the direct detection of the ligand-binding event, the activation of G-proteins or other effector enzymes, the release of second messengers, as well as receptor internalization can be monitored to unveil the GPCR activation mechanism and identify potential targets and respective modes of action. Reporter systems are in the spotlight of this book's edition, with fluorescence and bioluminescence assays, resonance energy transfer (FRET/BRET), opsins and enzyme complementation assays gaining increased maturity to monitor nearly all individual steps of the GPCR signaling cascade. In Chapter 5, Yasi and Peralta-Yahya use $Saccharomyces\ cerevisiae$ as the recombinant host to express the human serotonin receptor 4 (5-HTR_4). Whereas the correct expression of 5-HTR_4 in yeasts could be confirmed by recombinant constructs with green fluorescent protein ($5\text{-HTR}_4\text{-GFP}$), a luminescent assay based on the NanoLuc luciferase reporter system is used for the fast screening of serotonin receptor-acting compounds. In Chapter 6, Vidic and Hou describe a method for the purification of olfactory (OR) GPCRs from yeast cells. OR respond to odor molecules and are mainly expressed in the membrane of olfactory sensory neurons. Their intrinsically high sensitivity and selectivity makes them ideal candidates for the development of biosensors for the detection of volatile compounds (e.g., electronic noses).

Cell-free (CF) expression systems offer a new technological platform for protein expression by mobilizing the cellular translation machinery present in cell lysates for the in vitro production of the target proteins. In Chapter 7, Bernhard et al. applied the technology for the purification of the human endothelin B receptor (ETB) and the turkey β1-adrenergic receptor (β1AR). For further functional studies, however, cell-free systems are particularly challenging for membrane proteins because detergents are necessary to disrupt the cellular membrane and extract the translation machinery. The authors overtook this limitation by including nanodiscs (ND) in the cell-free reaction mixtures. The GPCR/ND particles are then purified by conventional affinity chromatography and ready for subsequent functional assays.

In Chapter 8, Veiksina et al. describe a method based on fluorescence anisotropy (FA) to characterize the ligand-binding event—the first step of the GPCR signaling cascade—based on emitting light polarization differences. The authors used budded baculoviruses as a gene expression tool to express high levels of proteins in insect cells, and in this way, overcome the limited expression of receptors in native cells. Schihada et al. describe a BRET platform in Chapter 9 that is compatible with HTS to monitor ligand binding to the GPCRs of interest. The latter are modified at the full-length or at the truncated C-terminus with Nanoluciferase—NLuc—(donor) and the HaloTag® NanoBret™ Ligand in the third intracellular loop (acceptor). Upon ligand binding, the receptor conformation changes and concomitantly modifies the relative distance of the energy partners, which in turn produces a change in BRET.

Laschet and Hanson address the interaction of GPCRs with their cognate G-proteins by developing a fragment complementation assay based also on nanoluciferase enzyme. Taking advantage of the enzyme subunits SmBiI and LgBiT, the authors further modified the natural peptide regions of the enzyme to increase the affinity between the enzyme units, which is important to measure the fast and transient interactions of GPCR/G-protein (*see* Chapter 10).

Bordes et al. (Chapter 11) describe a protocol to construct either transient or stable cell lines that enable FRET-based bioassays. The authors address the $G_{12/13}$ activation pathway by using constructs with the sensitive FRET pairs, paving the way for an expedite screening of Rho GTPases activating G proteins. The latter are most often monitored by following phenotypic responses (e.g., cytoskeleton rearrangements) that occur later, after stimulus activation. Protocols for cell transduction of EPAC sensors and monitoring of Gi/o-mediated cAMP response are also described. Lavogina et al. further detail the use of EPAC sensors to monitor Gs-dependent cAMP cellular release. In their Chapter 12, the authors use recombinant constructs of EPAC proteins, combined with sensitive FRET pairs. The constructs are amplified in BacMam expression system (a baculovirus vector for gene delivery to mammalian cells) and used for the transduction of Madin-Darby Canine Kidney cells expressing FSH receptors. Upon ligand binding, the cAMP increase is monitored by measuring the fluorescence Sensitized Emission (SE) and calculating the acceptor/donor emission ratio.

The fluorescence detection of intracellular calcium following activation of $G\alpha q/11$-coupled receptors constitutes a hallmark in HTS assays and a related protocol is described in Chapter 13. Typically, recombinant cell lines are cultured on microtiter plates (96–1536 wells) and stained with calcium sensitive dyes that exhibit enhanced fluorescence upon calcium binding. Cell permeable Ca^{2+} dye formulations are available from several companies, including no-wash formulations, making these assays compatible with fluorescence imaging plate readers (FLIPR). Woszczek et al. describe a methodology for Ca^{2+} that allows for the simultaneous concentration-dependent analysis of several receptor agonists and antagonists.

Protein redistribution assays focus on the later steps of the GPCR signaling cascade by following receptor internalization and recruitment of β-arrestins or by monitoring protein translocations across the plasma membrane. In Chapter 14, Meshik and Gautman describe the use of optically activatable GPCRs—opsins—to detect the translocation of Gβγ sub-units. The methodology comprises the transient transfection of eukaryotic cells with opsins, which are then optically activated at specific wavelengths. The resulting cellular response is then monitored through live cell imaging. Using this approach, it is possible to detect with high spatial and temporal resolution the activation of different G proteins' sub-types as well as several effectors, 2nd messengers' release and cell migration. Alonso-Gardón and Estévez describe a method to transfect HeLa cells with alternative constructs targeting the detection of sphingosine 1-phosphate receptor activity by the Split-TEV method. The latter is based on tobacco Etch Virus (TEV) protease fragment complementation assay. The authors used the GV transcription factor (fused to the N-TEV fragment) to activate a luminescent-based reporter gene system based on *Gaussia* luciferase (*see* Chapter 15). Ma et al. propose a method, compatible with microplate readers, to evaluate the interaction between Histamine-1 receptor (H_1R) and β-arrestin via BRET by taking advantage of nanoluciferase (Nluc) activity. Histamine-induced β-arrestin2 recruitment to the H_1R brings the two Nluc fragments in proximity, allowing functional reconstitution of Nluc activity (*see* Chapter 16). Dijon et al. explore the same principle of NanoBiT for the analysis of agonists and antagonist's pharmacology in Chapter 17. The authors carefully describe important considerations for the development of the molecular constructs that enable both endpoint and kinetic analysis. By collecting timecourse data from the ligand's concentration response curves and by applying an operational model for the analysis of the kinetic parameters, the method can infer on potential ligand bias. The latter refers to a process in which different ligands stabilize different receptor conformations (in contrast, for example, with the endogenous ligands), thereby activating different signaling pathways.

GPCRs microtiter-based assays or cell imaging techniques are well implemented and suitable to high-throughput screening assays (HTS). However, cells in their physiological environment are likely to be exposed to ligand gradients and temporal profiles rather than being in contact with discrete concentrations of particular compounds. In this view, the emergence of microfluidic technology has prompted the development of cell-based assays where fluid flow and shear stress can be fine-tuned to mimic physiological conditions. Gradient microfluidics are described by Suzuki et al. in Chapter 18. Using the peptide mating pheromone α-factor in yeasts as a model GPCR, the authors describe the fabrication and operation of a PDMS microfluidic device with a gradient flow generator or pulsating flow to monitor chemotropic guided cell elongation upon ligand addition. In Chapter 19, Martins et al. describe a Ca^{2+} mobilization assay that is performed in a microfluidic system. In here, muscarinic M1 agonist (carbachol) and antagonist (pirenzepine) efficacy can be determined with similar accuracy as in standard microscopy cell assays.

Finally, the recent disclosure of several GPCR structures has paved the way for the rational design of GPCR acting compounds, and in vitro drug design is gaining ground in drug discovery programs. The last chapter of this edition is dedicated to in silico tools with a focus on homology modeling. This methodology uses available 3D structures to build a model of a target sequence (e.g., binding molecule) based on a given sequence alignment between the template and target. Thus, in Chapter 20, Miszta et al. introduce the GPCRM web service and detail stepwise instructions on how to use the software to obtain high-quality GPCR models.

Editors' Final Remarks

This edition of *G Protein-Coupled Receptor Screening Assays* was planned considering not only the assays directly involved in the discovery of GPCR-active compounds but also those involved in cell-based experiments designed to study physiological responses. Whether coming from academia or industry, or being an experienced researcher or a newcomer to the field, the reader will find a comprehensive list of methods and protocols that cover the latest developments on receptor purification, molecular biology, recombinant engineering, and analytical techniques that enable the real-time monitoring of the complex GPCR signaling cascade and identification of potential drug targets.

We sincerely hope that the content provided here can somehow contribute to advancing GPCR research and discovery and ultimately lead to the availability of innovative and more efficient drugs.

Acknowledgments

The guest editors are grateful to the authors that contributed with their work to this book edition and under such unsettled times.

Lisbon, Portugal   *Sofia Aires M. Martins*
                                                                 Duarte Miguel F. Prazeres

Contents

Contributors

PETER ABDELMASEEH • *Einstein Medical Center Philadelphia, Philadelphia, PA, USA*

ANNI ALLIKALT • *Institute of Chemistry, University of Tartu, Tartu, Estonia*

MARTA ALONSO-GARDÓN • *Unitat de Fisiologia, Departament de Ciències Fisiològiques, Genes Disease and Therapy Program, IDIBELL-Institute of Neurosciences, Universitat de Barcelona, L'Hospitalet de Llobregat, Spain; Centro de Investigación en red de enfermedades raras (CIBERER), ISCIII, Instituto de Salud Carlos III, Madrid, Spain*

LAUREANO D. ASICO • *Division of Kidney Diseases & Hypertension, Department of Medicine, School of Medicine & Health Sciences, The George Washington University, Washington, DC, USA*

FRANK BERNHARD • *Centre for Biomolecular Magnetic Resonance, Institute for Biophysical Chemistry, Goethe-University of Frankfurt/Main, Frankfurt/Main, Germany*

LUCA BORDES • *Section Molecular Cytology, Swammerdam Institute for Life Sciences, University of Amsterdam, Amsterdam, The Netherlands*

SERGEI CHAVEZ-ABIEGA • *Section Molecular Cytology, Swammerdam Institute for Life Sciences, University of Amsterdam, Amsterdam, The Netherlands; Section Systems Bioinformatics, Amsterdam Institute for Molecules, Medicines and Systems, VU University, Amsterdam, The Netherlands*

VIRGINIA CHU • *INESC Microsistemas e Nanotecnologias, Lisbon, Portugal*

JOÃO P. CONDE • *INESC Microsistemas e Nanotecnologias, Lisbon, Portugal; Department of Bioengineering, Instituto Superior Técnico, Universidade de Lisboa, Lisbon, Portugal*

NICOLA C. DIJON • *School of Life Sciences, The Medical School, Queen's Medical Centre, University of Nottingham, Nottingham, UK; Centre of Membrane Proteins and Receptors, University of Birmingham and University of Nottingham, UK*

ANNA DI NARDO • *Department of Dermatology, University of California San Diego, La Jolla, CA, USA*

HENRIK G. DOHLMAN • *Curriculum in Bioinformatics and Computational Biology, UNC School of Medicine, Chapel Hill, NC, USA; Department of Pharmacology, UNC School of Medicine, Chapel Hill, NC, USA*

VOLKER DÖTSCH • *Centre for Biomolecular Magnetic Resonance, Institute for Biophysical Chemistry, Goethe-University of Frankfurt/Main, Frankfurt/Main, Germany*

TIMOTHY C. ELSTON • *Curriculum in Bioinformatics and Computational Biology, UNC School of Medicine, Chapel Hill, NC, USA; Computational Medicine Program, UNC School of Medicine, Chapel Hill, NC, USA; Department of Pharmacology, UNC School of Medicine, Chapel Hill, NC, USA*

RAÚL ESTÉVEZ • *Unitat de Fisiologia, Departament de Ciències Fisiològiques, Genes Disease and Therapy Program, IDIBELL-Institute of Neurosciences, Universitat de Barcelona, L'Hospitalet de Llobregat, Spain; Centro de Investigación en red de enfermedades raras (CIBERER), ISCIII, Instituto de Salud Carlos III, Madrid, Spain*

SŁAWOMIR FILIPEK • *Faculty of Chemistry, Biological and Chemical Research Centre, University of Warsaw, Warsaw, Poland*

ELISABETH FUERST • *School of Immunology & Microbial Sciences, King's College London, Guy's Hospital, London, UK; MRC & Asthma UK Centre in Allergic Mechanisms of Asthma, London, UK*

NARASIMHAN GAUTAM • *Department of Anesthesiology, Washington University School of Medicine, St. Louis, MO, USA; Department of Genetics, Washington University School of Medicine, St. Louis, MO, USA*

JOACHIM GOEDHART • *Section Molecular Cytology, Swammerdam Institute for Life Sciences, University of Amsterdam, Amsterdam, The Netherlands*

JULIEN HANSON • *Laboratory of Molecular Pharmacology, GIGA-Molecular Biology of Diseases, University of Liège, Liège, Belgium; Laboratory of Medicinal Chemistry, Centre for Interdisciplinary Research on Medicines (CIRM), University of Liège, Liège, Belgium*

NICHOLAS D. HOLLIDAY • *School of Life Sciences, The Medical School, Queen's Medical Centre, University of Nottingham, Nottingham, UK; Centre of Membrane Proteins and Receptors, University of Birmingham and University of Nottingham, UK; Excellerate Bioscience, Biocity, Nottingham, UK*

YANXIA HOU • *University Grenoble Alpes, CEA, CNRS, IRIG-SyMMES, Grenoble, France*

PAUL A. INSEL • *Department of Pharmacology, University of California San Diego, La Jolla, CA, USA; Department of Medicine, University of California San Diego, La Jolla, CA, USA*

PEDRO A. JOSE • *Division of Kidney Diseases & Hypertension, Department of Medicine, School of Medicine & Health Sciences, The George Washington University, Washington, DC, USA; Department of Pharmacology-Physiology, School of Medicine & Health Sciences, The George Washington University, Washington, DC, USA*

JOSHUA B. KELLEY • *Molecular and Biomedical Sciences Department, University of Maine, Orono, ME, USA*

ZOE KÖCK • *Centre for Biomolecular Magnetic Resonance, Institute for Biophysical Chemistry, Goethe-University of Frankfurt/Main, Frankfurt/Main, Germany*

SERGEI KOPANCHUK • *Institute of Chemistry, University of Tartu, Tartu, Estonia*

OLGA KUKK • *Institute of Chemistry, University of Tartu, Tartu, Estonia; Competence Centre on Reproductive Medicine & Biology, Tartu, Estonia*

TÕNIS LAASFELD • *Institute of Chemistry, University of Tartu, Tartu, Estonia*

CÉLINE LASCHET • *Laboratory of Molecular Pharmacology, GIGA-Molecular Biology of Diseases, University of Liège, Liège, Belgium*

DARJA LAVOGINA • *University of Tartu, Institute of Chemistry, Tartu, Estonia; University of Tartu, Institute of Clinical Medicine, Clinic of Hematology and Oncology, Tartu, Estonia; Competence Centre on Reproductive Medicine & Biology, Tartu, Estonia*

ROB LEURS • *Amsterdam Institute for Molecules, Medicines and Systems, Division of Medicinal Chemistry, Faculty of Science, Vrije Universiteit Amsterdam, Amsterdam, The Netherlands*

REET LINK • *Institute of Chemistry, University of Tartu, Tartu, Estonia*

MARTIN J. LOHSE • *Institute of Pharmacology and Toxicology, University of Wuerzburg, Wuerzburg, Germany; Max-Delbrueck-Center for Molecular Medicine, Berlin, Germany; ISAR Bioscience, Planegg, Germany*

XIAOYUAN MA • *Amsterdam Institute for Molecules, Medicines and Systems, Division of Medicinal Chemistry, Faculty of Science, Vrije Universiteit Amsterdam, Amsterdam, The Netherlands*

THOMAS J. A. MAGUIRE • *School of Immunology & Microbial Sciences, King's College London, Guy's Hospital, London, UK; MRC & Asthma UK Centre in Allergic Mechanisms of Asthma, London, UK*

ISABELLA MAIELLARO • *School of Life Sciences, Queen's Medical Centre, University of Nottingham, Nottingham, UK*

SOFIA AIRES M. MARTINS • *INESC Microsistemas e Nanotecnologias, Lisbon, Portugal*

XENIA MESHIK • *Department of Anesthesiology, Washington University School of Medicine, St. Louis, MO, USA*

PRZEMYSŁAW MISZTA • *Faculty of Chemistry, Biological and Chemical Research Centre, University of Warsaw, Warsaw, Poland*

KATARINA NEMEC • *Max-Delbrueck-Center for Molecular Medicine, Berlin, Germany*

DESISLAVA N. NESHEVA • *School of Life Sciences, The Medical School, Queen's Medical Centre, University of Nottingham, Nottingham, UK; Centre of Membrane Proteins and Receptors, University of Birmingham and University of Nottingham, UK*

SZYMON NIEWIECZERZAŁ • *Faculty of Chemistry, Biological and Chemical Research Centre, University of Warsaw, Warsaw, Poland*

PAWEŁ PASZNIK • *Faculty of Chemistry, Biological and Chemical Research Centre, University of Warsaw, Warsaw, Poland*

PAMELA PERALTA-YAHYA • *School of Chemistry and Biochemistry, Georgia Institute of Technology, Atlanta, GA, USA; School of Chemical and Biomolecular Engineering, Georgia Institute of Technology, Atlanta, GA, USA*

DUARTE MIGUEL F. PRAZERES • *IBB—Institute for Bioengineering and Biosciences, Lisbon, Portugal; Department of Bioengineering, Instituto Superior Técnico, Universidade de Lisboa, Lisbon, Portugal*

PHILIP J. REEVES • *School of Life Sciences, University of Essex, Colchester, Essex, UK*

AGO RINKEN • *Institute of Chemistry, University of Tartu, Tartu, Estonia*

SELIM ROZYYEV • *Division of Kidney Diseases & Hypertension, Department of Medicine, School of Medicine & Health Sciences, The George Washington University, Washington, DC, USA; Sheikh Zayed Institute for Pediatric Surgical Innovation, Children's National Health System, Washington, DC, USA*

CRISTINA SALMERÓN • *Department of Pharmacology, University of California San Diego, La Jolla, CA, USA*

HANNES SCHIHADA • *Section of Receptor Biology & Signaling, Department of Physiology & Pharmacology, Karolinska Institutet, Stockholm, Sweden; Institute of Pharmacology and Toxicology, University of Wuerzburg, Wuerzburg, Germany*

KRISHNA SRIRAM • *Department of Pharmacology, University of California San Diego, La Jolla, CA, USA*

SARA KIMIKO SUZUKI • *Curriculum in Bioinformatics and Computational Biology, UNC School of Medicine, Chapel Hill, NC, USA*

MARIS-JOHANNA TAHK • *Institute of Chemistry, University of Tartu, Tartu, Estonia*

ANDREW C. TIU • *Einstein Medical Center Philadelphia, Philadelphia, PA, USA; Division of Kidney Diseases & Hypertension, Department of Medicine, School of Medicine & Health Sciences, The George Washington University, Washington, DC, USA*

SANTA VEIKSINA • *Institute of Chemistry, University of Tartu, Tartu, Estonia*

JASMINA VIDIC • *INRAE, AgroParisTech, Micalis Institute, Université Paris-Saclay, Jouy-en-Josas, France*

VAN ANTHONY M. VILLAR • *Division of Kidney Diseases & Hypertension, Department of Medicine, School of Medicine & Health Sciences, The George Washington University, Washington, DC, USA*

HENRY F. VISCHER • *Amsterdam Institute for Molecules, Medicines and Systems, Division of Medicinal Chemistry, Faculty of Science, Vrije Universiteit Amsterdam, Amsterdam, The Netherlands*

GRZEGORZ WOSZCZEK • *School of Immunology & Microbial Sciences, King's College London, Guy's Hospital, London, UK; MRC & Asthma UK Centre in Allergic Mechanisms of Asthma, London, UK*

EMILY A. YASI • *School of Chemistry and Biochemistry, Georgia Institute of Technology, Atlanta, GA, USA*

ALEXEI A. YELISEEV • *National Institute on Alcohol Abuse and Alcoholism, National Institutes of Health, Bethesda, MD, USA*

Chapter 1

Isolation of Lipid Rafts by the Detergent-Based and Non-detergent-Based Methods for Localization of GPCRs with Immunoblotting and Laser Scanning Confocal Microscopy

Peter Abdelmaseeh, Andrew C. Tiu, Selim Rozyyev, Laureano D. Asico, Pedro A. Jose, and Van Anthony M. Villar

Abstract

The understanding of how biological membranes are organized and how they function has constantly been evolving over the past decades. Instead of just serving as a medium in which specific proteins are located, certain parts of the lipid bilayer contribute to platforms that assemble signaling complexes by providing a microenvironment that facilitates effective protein–protein interactions. G protein-coupled receptors (GPCRs) and relevant signaling molecules, including the heterotrimeric G proteins, key enzymes such as kinases and phosphatases, trafficking proteins, and secondary messengers, preferentially partition to these highly organized cell membrane microdomains, called lipid rafts. Lipid rafts are essential for the trafficking and signaling of GPCRs. The study of GPCR biology in the context of lipid rafts involves the localization of the GPCR of interest in lipid rafts, at the basal state and upon receptor agonism, and the evaluation of the biological functions of the GPCR in appropriate cell lines. The lack of standardized methodologies to study lipid rafts, in general, and of the workings of GPCRs in lipid rafts, in particular, and the inescapable drawbacks of current methods have hampered the complete understanding of the underlying molecular mechanisms. Newer methodologies that allow the study of GPCRs in their native form are needed. The use of complementary approaches that produce mutually supportive results appears to be the best way for drawing conclusions with regard to the distribution and activity of GPCRs in lipid rafts.

Key words α-Cyclodextrin, Caveolin-1, Lipid raft, Methyl-β-cyclodextrin, Sucrose gradient ultracentrifugation

1   Introduction

The plasma membrane is a semipermeable, biological membrane that separates the intracellular domain from the extracellular environment. Amphipathic lipids, such as phospholipids and sphingolipids, are the building blocks of these bilipid membranes because of their aggregative properties, i.e., their hydrophobic tails associate together, while their hydrophilic heads interact with both

Sofia Aires M. Martins and Duarte Miguel F. Prazeres (eds.), *G Protein-Coupled Receptor Screening Assays: Methods and Protocols*, Methods in Molecular Biology, vol. 2268, https://doi.org/10.1007/978-1-0716-1221-7_1,

extra- and intracellular aqueous environments [1]. The fluidity of the fatty acyl groups of phospholipids at 37 °C enables the membranes to act as a medium in which dissolved membrane proteins are afforded ample lateral mobility, especially in response to environmental cues. Since the first description of an "organization of the lipid components of membranes into domains" [2] and the elaboration of the "lipid raft hypothesis" by Simons and van Meer [3–5], the existence of lipid rafts is now established. However, the direct imaging of these membrane domains in vivo needs improvement [6]. It has also recently been described that the lipid mixtures have the ability of separating into two different phases, namely the lipid ordered phase (Lo) and the lipid disordered phase (Ld). These phases were recognized in the study of giant plasma membrane vesicles (GPMVs) [7]. Recognition of these phases, as well as critical fluctuations, that is the reformation of the domains into smaller assemblies when exposed to elevated temperatures, is one step closer to proving the presence of lipid rafts, despite the challenges persisting with direct visualization of these structures in living cells [6, 8].

Lipid rafts (Fig. 1) are tightly packed, highly organized plasma membrane microdomains that are enriched in phospholipids, glycosphingolipids, and cholesterol and serve as a platform for the organization and dynamic interaction of biomolecules involved in various biological processes.

The cholesterol has an essential role in the structure and rigidity of lipid rafts, by attaching to the hydrophobic gaps between the phospholipid acyl chains. Certain structural proteins abound in lipid rafts to serve as scaffold or anchor for other proteins, including caveolins [9–12], flotillins [9, 13], tetraspanins [14], and glycosylphosphatidylinositol-linked (GPI-linked) proteins [15]. The spatial concentration and organization of specific sets of membrane proteins allow greater efficiency and specificity of signal transduction by facilitating protein–protein interactions and by preventing crosstalk between competing pathways. The nonhomogeneous lateral distribution of membrane components helps explain the differences in composition between apical and basolateral membrane domains of polarized epithelial cells [1].

The best-characterized lipid raft microdomains are the caveolae, which were first described by Palade and Yamada in the 1950s [16, 17]. These are small (60–80 nm) invaginations of the plasma membrane formed by the polymerization of caveolins with cholesterol [18]. Caveolae have been implicated in a variety of cellular processes, including signal transduction, endocytosis, transcytosis, and cholesterol trafficking [19]. Lipid rafts are located in the apical plasma membrane in polarized epithelial cells and in axonal membranes in neurons. Basolateral and dendritic membranes contain lipid rafts but in more limited quantities [3]. Interestingly, caveolae are found mostly at the basolateral membrane that faces the blood supply and is more active during signal transduction [20].

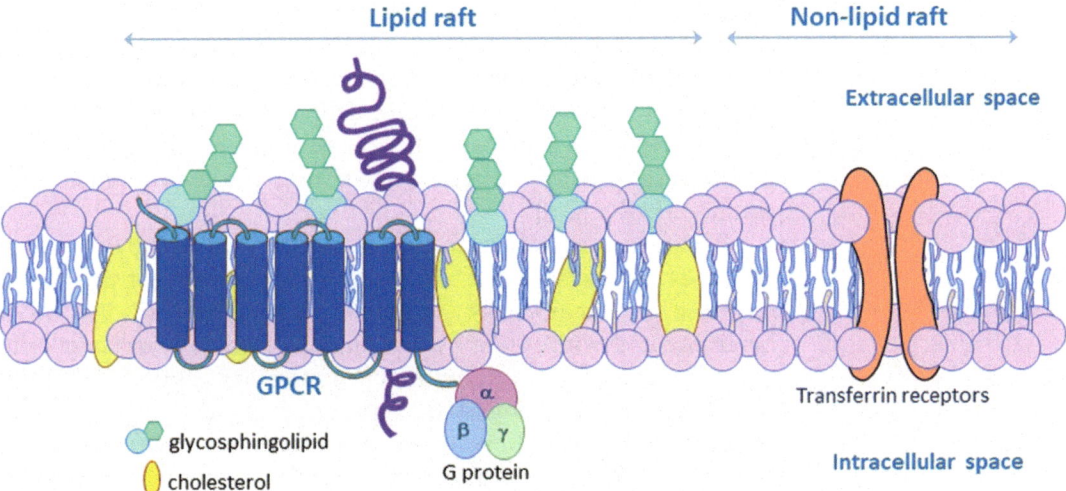

Fig. 1 A lipid raft membrane microdomain. Lipid rafts are highly organized plasma membrane microdomains enriched in phospholipids, glycosphingolipids, and cholesterol, and serve as a matrix for receptors, such as G protein-coupled receptors (GPCRs), and other signaling molecules

Lipid rafts are mostly found at the plasma membrane; however, they may also be found in intracellular membranes involved in the biosynthetic and endocytic pathways. Lipid raft microdomains play a crucial role in cellular processes such as membrane sorting, receptor trafficking, signal transduction, and cell adhesion.

1.1 GPCR Signaling and Trafficking

G protein-coupled receptors (GPCRs) constitute the largest superfamily of seven transmembrane proteins that respond to a myriad of environmental stimuli that are transduced intracellularly as meaningful signals through secondary messengers. Agonist stimulation of a GPCR leads to a conformational change that promotes the exchange of GDP for GTP on the G_α subunit of the G protein, resulting in the uncoupling of the G protein from the GPCR and the dissociation of G_α and $G\beta\gamma$ subunits. The G_α subunit either activates ($G_{\alpha S}$) or inhibits ($G_{\alpha i}$) intracellular signaling pathways depending on the receptor subtype, while the $G\beta\gamma$ subunit recruits G protein-coupled receptor kinases (GRKs), which selectively phosphorylate serine and threonine residues localized within the third intracellular loop and carboxyl-terminal tail domains of the receptor to promote the binding of cytosolic cofactor proteins called arrestins [21]. The β-arrestins play important roles in the uncoupling process and sequestration and internalization of GPCRs through a dynamin-dependent, clathrin-mediated endocytosis. Once internalized, the GPCRs, in vesicles termed as early endosomes, are sorted by sorting nexins and follow divergent pathways [22]. The receptors are sorted into recycling endosomes for their return to the cell membrane (recycling and resensitization), accumulate in late endosomes that target the lysosomes for their subsequent degradation, or are transported initially to the perinuclear endosomes (trans-Golgi network) and then to the late endosomes for eventual lysosomal degradation. Additional proteolytic mechanisms, such as proteasomes or cell-associated endopeptidases, are also implicated in mediating the downregulation of certain GPCRs [23].

The signal transduction that follows ligand occupation of the GPCR is highly regulated to ensure the specificity of the cellular response, both temporally and spatially. The signal transduction can be attenuated with relatively fast kinetics through a process called desensitization or through a much slower process of downregulation following prolonged or repeated exposure to an agonist. Desensitization, or the waning of a receptor's responsiveness to agonist with time, is an inherent molecular "feedback" mechanism that prevents receptor overstimulation and helps in creating an integrated and meaningful signal by filtering out information from weaker GPCR-mediated signals [24].

Desensitization is accomplished through two complementary mechanisms, i.e., the functional uncoupling of GPCRs from their cognate G proteins, which occurs without any detectable change in the number of cell surface receptors, and GPCR phosphorylation, sequestration, and internalization/endocytosis. GPCR resensitization protects the cells from prolonged desensitization and is carried out via dephosphorylation by phosphatases as the GPCR traffics through the endosomal pathway. GPCR activity is the outcome of a fine-tuned balance between receptor desensitization and resensitization.

It is now established that lipid rafts serve as dynamic platforms for GPCRs and pertinent signaling molecules such as G proteins, enzymes, and adaptors [25, 26]. However, understanding the molecular mechanisms involved has been thwarted by the lack of standardized methodology to study lipid rafts, in general, and of the workings of GPCRs in lipid rafts, in particular. The minute size of lipid rafts has made lipid rafts difficult to resolve by standard light microscopy, unless the lipid raft components are crosslinked with antibodies or lectins [20]. Studying how GPCR works in lipid rafts may be accomplished by determining if the GPCR of interest localizes to the lipid rafts and by evaluating if GPCR signaling and activity are lost when lipid rafts are disrupted.

There are many established protocols available that allow the study of GPCR activity, per se, using commercially available kits or, less commonly, proprietary materials. Studying the activity of GPCRs in the context of their residency in lipid rafts often requires additional steps that would disrupt the integrity of the lipid raft microdomain or dissociate the protein of interest from the rafts. Most of the current strategies to disrupt lipid raft involves either perturbation of the raft stability or modifying the cholesterol content of the lipid rafts. Most of these treatments are performed on cells prior to agonist/antagonist treatment and functional assays, such as cAMP production, sodium transport, and NADPH oxidase activity [9, 27–29].

1.2 Perturbation of Raft Stability

Lipid rafts are dynamic assemblies of phospholipids and glycosphingolipids that contain mostly saturated hydrocarbon chains, which allow cholesterol to intercalate between the fatty acyl chains. The surrounding membrane has greater fluidity because of the preponderance of phospholipids with unsaturated acyl groups. The addition of exogenous gangliosides [30] and polyunsaturated fatty acids [31], such as docosahexaenoic acid or DHA [32], in the growth medium, results in a change in the lipid raft composition and the dissociation of proteins from the lipid raft. Inhibition of the biosynthesis of glycosphingolipids and sphingomyelins using the fungal metabolite fumonisin B1 [33, 34] may also perturb the integrity of lipid rafts. Supplementation with 7-ketocholesterol, which differs from cholesterol by the additional ketone group that protrudes perpendicularly to the cyclopentano-perhydro-phenanthrene ring, decreases lipid raft order and increases membrane polarity [35, 36]. Interestingly, the nonsteroidal, anti-inflammatory drug aspirin has a high affinity for phospholipid membranes and partitions into the lipid head groups. This interaction impairs the molecular organization brought about by cholesterol, and thus leads to increased mobility in a lipid raft model [37, 38]. The use of short-chain ceramides, i.e., C2-ceramide and C6-ceramide, decreases the plasma membrane lipid order and disrupts the lipid rafts, as indicated by a reduction in the extent of fluorescence resonance energy transfer (FRET) between lipid raft markers [39].

1.3 Changing the Cholesterol Content

Cholesterol is an integral component of lipid rafts in mammalian cell membranes, and membrane cholesterol levels are crucial in determining the stability and organization of lipid rafts [40]. Thus, modifying the content of cholesterol in the plasma membrane is another option to disrupt the lipid raft and evaluate the function of GPCRs. The antifungal polyene antibiotics filipin [41, 42], nystatin [43], and amphotericin B [44] disrupt lipid rafts by binding and sequestering cholesterol within the plasma membrane. Pore-forming agents such as saponin [45, 46], digitonin [47], and streptolysin O [48] may also be used. β-MCD is one of the most frequently used agents to deplete the endogenous cholesterol content of lipid rafts [28, 10, 29]. One advantage in using β-MCD is the availability of a control for its use, i.e., α-CD [49]. Inhibitors of the rate-limiting enzyme for cholesterol synthesis, the HMG-CoA reductase, may be used also to inhibit endogenous cholesterol biosynthesis. These include drugs such as simvastatin [42] and lovastatin [50]. A summary of protocols using these approaches is found in Table 1.

Table 1
Common agents used to disrupt the lipid raft

Strategies	Comments
A. Disruption of raft stability	
D-PDMP	5–10 µM for 20–30 min [higher concentrations and longer incubation period resulted in cell death [90]] 10–20 µM, overnight [91],
High EC glucose	28 mM glucose for 72 h [92]
Gangliosides	10–100 µM for 1 h [31]
PUFAs	50 µM, overnight [30]
4F	50 µg/10^6 cells for 7 days [93]
B. Changing the cholesterol content	
5,22-cholestadien-3β-ol	1 µg/mL for 24 h [94]
AY 9944	5 µM for 24 h [94]
Amphotericin B	10 µg/mL for 1 h [44]
β-MCD	5–20 mM for 30 min [42] 2% for 1 h [28, 29] 10 mM for 60 min [50] 5 mM for 15 min [47] 10 mM for 1 h [31] 1–10 mM for 30 min [91]
Digitonin	0.003% for 30 min on ice [47]
Filipin	2.5–5 µg/mL for 15 min [stock solution: 5 mg/mL in ethanol] [42]
Fumonisin B1	250 µM for 24 h [95]
Lovastatin	1 mg/mL for 20 h [50]
Haloperidol	50 µM for 24 h [94]
Nystatin	20–50 µg/mL for 1 h [43]
Saponin	0.5% in 20 mM phosphate buffer, pH 7.4, at 4 °C for 10 min followed by extraction in 0.5% triton-X100 at 4 °C [46] 1% for 2 h [48]
Simvastatin	5 µg/mL for 12 h [42]
Streptolysin O	500 ng/mL for 2 h [48]
SKF 104976	3 µM for 24 h [94]
Triparanol	3 µM for 24 h [94]

D-PDMP D-threo-1-phenyl-2-decanoylamino-3-morpholino-1-propanol, *β-MCD* methyl-β-cyclodextrin, *EC* extracellular, *PUFAs* polyunsaturated fatty acids

1.4 Fluorescence Imaging

The advent of FRET and BRET (bioluminescence resonance energy transfer) biosensors has allowed the study of GPCR activity in lipid rafts in living cells. N-Way FRET microscopy can quantify interacting and noninteracting FRET pairs in live cells [51]. The freely diffusible FRET sensor Epac2-camps has been used to measure global cAMP responses of lipid raft-associated receptors since it responds to changes in cAMP occurring throughout the cytosolic compartment of cells [52]. Moreover, versions of the Epac2-camps probe allow the selective targeting to lipid raft (Epac2-MyrPalm) and non-raft (Epac2-CAAX) domains, which are useful in monitoring local cAMP production near the plasma membrane [52]. Photoactivated localization microscopy (PALM), as indicated above, and stochastic optical reconstruction microscopy (dSTORM) have also been used to track the reorganization of lipid rafts [53, 54]. Movement of single molecules in living cells could also be tracked (Single Molecule Tracking) [55]. In addition, fluorescent nanosensors that measure sodium in real time are reversible and completely selective over other cations [56]. Real-time monitoring of sodium transport in response to stimulation or inhibition of GPCRs in intact or disrupted lipid rafts has become feasible.

Current biochemical and biophysical techniques for studying GPCRs in lipid rafts, while helpful in many instances, are still rife with methodological drawbacks and limitations. These include the requirement for cell membrane disruption, the reliance on antibodies that are specific for the GPCR of interest, the inability to study native proteins, and the use of exogenous, often tagged, proteins. Newer methodologies that allow the study of GPCRs in their native form in intact cells are needed, such as the FRET biosensors for cAMP monitoring. Meanwhile, the use of complementary approaches that yield mutually supportive results may be the most judicious way for drawing conclusions regarding the distribution and activity of GPCRs in lipid rafts.

The ganglioside GM1 may be labeled with single quantum dots to measure the lateral mobility and extent of movement of the lipid rafts [57]. Recently, GPI-anchored proteins that segregate into lipid rafts have been visualized using a novel method called enzyme-mediated activation of radical sources (EMARS) [58]. Probes that target the lipid content of lipid rafts have also been used to visualize these membrane microdomains. Laurdan (6-dodecanoyl-2-(dimethylamino)-naphthalene) and C-laurdan (6-dodecanoyl-2-[N-methyl-N-(carboxymethyl)amino]-naphthalene), which are membrane probes that are sensitive to membrane polarity, allow the observation of lipid rafts via two-photon microscopy [59–61]. A fluorophore-tagged domain D4 of perfringolysin O, a cholesterol-binding cytolysin produced by *Clostridium perfringens*, has been used as a probe to study membrane cholesterol [62].

In this chapter, we will describe the methods on how to isolate the lipid rafts using the detergent-based and non-detergent-based methods, with details on how to localize them using immunoblotting and laser scanning confocal microscopy.

2 Materials

2.1 Detergent-Based Method

1. 15-cm Dishes.
2. Methyl-β-cyclodextrin (β-MCD) 2%, dissolved in cell culture media (*see* **Note 1**).
3. Cell culture medium: DMEM/F-12.
4. Cholestane-3,5,6-triol (0.2 mM) + β-MCD (10 mM) dissolved in the cell culture media.
5. 1X Phosphate Buffered Saline (PBS): 137 mM NaCl, 10 mM Phosphate, 2.7 mM KCl, pH 7.4.
6. Halt™ Protease and Phosphatase Inhibitor Single-Use Cocktail 1X.
7. FOCUS™ SubCell kit.
8. OptiPrep™ Density Gradient Medium.

2.2 Detergent-Free Method

1. 2-N-morpholino ethanesulfonic acid (Mes), 250 mM, pH = 6.8.
2. Mes-buffered solution (MBS): 25 mM Mes, 150 mM NaCl.
3. Sodium citrate, 500 mM, pH ~ 11 add protease inhibitor cocktail diluted 1:100.
4. Sucrose solution: 5%, 35%, 80% in MBS. Add protease inhibitor cocktail.
5. Methyl-β-cyclodextrin (β-MCD), 2% dissolved in cell culture media.
6. Cholesterol (0.2 mM) + β-MCD (10 mM) dissolved in cell culture media (*see* **Note 2**).

2.3 Cell Handling

1. HEK-293 cells (heterologously expressing human D_1 receptor (HEK-hD_1)), which has previously been identified to harbor lipid raft-based dopamine receptors.
2. Dounce homogenizer.
3. Centrifuge equipped with a swinging bucket rotor.

2.4 Localization of GPCRs in Lipid Rafts

1. 6X Laemmli sample buffer.
2. 12-mm Coverslips.
3. Glass slides.
4. Mounting medium.

5. 24-Well tissue culture plate.

6. Primary antibody against the GPCR of interest.

7. Labeled secondary antibody.

8. Lipid raft markers, for example, caveolin-1, flotillin-1, CD55, alkaline phosphatase, and pore-forming toxins, such as cholera toxin subunit B (CTxB), equinatoxin II, or perfringolysin. When immunoblotting, tagging with fluorophores is not necessary.

9. Bovine Serum Albumin (BSA) solution: 10% in H_2O.

10. 4% Paraformaldehyde fixing solution.

11. 0.5% Triton X-100 prepared in deionized water.

12. Laser confocal microscopy.

3 Methods

3.1 Isolation of Lipid Rafts

Lipid rafts are characterized by their relative insolubility in nonionic detergents at 4 °C and in light buoyant density on sucrose gradient [63]. The isolation of lipid rafts can be performed using either detergent-based or detergent-free methods [10], with the latter generating a greater fraction of inner leaflet membrane rafts and producing more replicable results [64]. Schnitzer et al. employed a detergent-free method to isolate lipid rafts using cationic colloidal silica particles, which is appropriate for non-cell culture studies [63]. Lipid rafts may be extracted from total cell membranes [65] or just from surface plasma membranes [66]. Detergent insolubility results from the segregation of membrane-associated proteins into the lipid rafts, which are abundant in cholesterol and glycosphingolipids. Nonionic detergents such as Triton X-100, β-octyl glucoside, CHAPS, deoxycholate, Lubrol WX, Lubrol PX, Brij 58, Brij 96, and Brij 98 have been used to prepare lipid raft fractions [67], resulting in varying yields of proteins. Samples obtained by detergent-based methods are termed detergent-resistant membranes (DRMs) or detergent-insoluble fractions. Different detergents may yield different lipid raft components because of the varying degrees of resistance by the proteins to extraction using different reagents. The methods detailed below are based on Yu et al. [10].

1. Grow cells in 15-cm dishes until 90% confluent at 37 °C in 95% air and 5% CO_2 (*see* **Notes 3** and **4**).

2. Prepare two 15-cm dishes for every experimental group (*see* **Note 5**).

3. Serum-starve the cells for 1 h prior to any treatments.

4. Incubate with methyl-β-cyclodextrin solution (β-MCD) for 1 h at 37 °C to deplete the cholesterol while utilizing methyl-β-cyclodextrin (β-MCD) as control [49].

5. Use cholesterol/β-MCD solution for 1 h at 37 °C for cholesterol repletion while utilizing cholestane-3,5,6-triol/ β-MCD, an inactive analog of cholesterol, as a control for the use of exogenous cholesterol [68].

6. Wash the cells with PBS, then scrape and pellet in 5 ml tubes.

7. Centrifuge the cells at ~800 × g for 1 min at 4 °C. Discard the supernatant.

8. Mix Halt™ Protease and Phosphatase Inhibitor Single-Use Cocktail 1X to 10 mL of Subcell Buffer I from FOCUS™ SubCell kit.

9. Add 1 mL of the mixture to the cell sample.

10. Vortex the cells and incubate on ice for 10 min.

11. Lyse the cells using a Dounce homogenizer with 20 strokes per sample.

12. Rinse the homogenizer with 200 µL of Subcell buffer I and pool it with the sample.

13. Transfer the samples to 1.5 mL tubes.

14. Add 500 µL of Subcell Buffer II (3X) to the sample.

15. Perform differential centrifugation as follows:

   (a) Centrifuge cells at 700 × g for 10 min at 4 °C to pellet the nuclei (*see* **Note 6**). Transfer the supernatant to a new tube.

   (b) Centrifuge the supernatant at 12,000 × g for 15 min at 4 °C to pellet the mitochondria (*see* **Note 7**). Transfer the supernatant to a new tube.

   (c) Centrifuge the supernatant at 100,000 × g for 60 min at 4 °C in a swinging bucket rotor to pellet the plasma membranes; confirm with protein markers (e.g., CD40) (*see* **Note 8**).

   (d) Transfer the supernatant to a new tube and confirm that it is the cytoplasm with protein markers (e.g., GAPDH) and the absence of contamination.

3.2 Isolation of Lipid Rafts Using the Detergent-Based Method

1. Precool on ice 10 mL ultracentrifuge tubes and the buffers from OptiPrep™ Density Gradient Medium.

2. Precool the ultracentrifuge to 4 °C.

3. Transfer the pelleted cell membranes to a 10 mL tube for ultracentrifugation.

4. Prepare five solutions that will form the OptiPrep gradient layers according to Table 2 and Fig. 2.

Table 2
Preparation of Optiprep gradient solutions

Gradient layer	Final OptiPrep (%)	Cell lysate (mL)	Lysis buffer (mL)	OptiPrep (mL)	Total volume (mL)
1 (bottom)	35	0.84	0	1.16	2
2	30	–	1	1	2
3	25	–	1.16	0.84	2
4	20	–	1.3	0.7	2
5 (top)	0	–	1	0	1

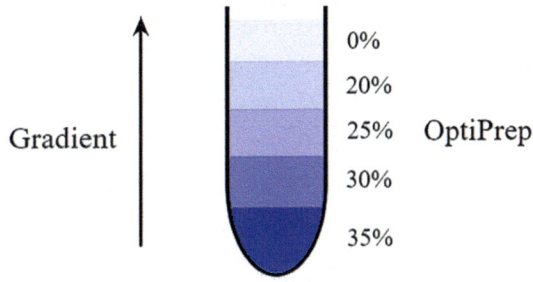

Gradient

0%
20%
25%   OptiPrep
30%
35%

Fig. 2 OptiPrep density gradient medium

5. Place 2 mL of gradient layer 1 at the bottom of the precooled ultracentrifuge tube.

6. Place each gradient layer over the other sequentially (*see* **Note 9**).

7. Balance the ultracentrifuge tubes.

8. Centrifuge at $200,000 \times g$ for 4 h at 4 °C.

9. Take the tubes out and place them on ice.

10. Mark 12 microcentrifuge tubes from 1–12. Tube 1 will be used for the lowest % of the gradient (the top of the ultracentrifuge tube).

11. Collect 1 mL fractions from top to bottom of the ultracentrifuge tube and transfer each fraction to a marked microcentrifuge tube.

12. Keep the fractions on ice (*see* **Note 10**).

13. Pre-label twelve 1.5 mL microcentrifuge tubes.

14. Aspirate 750 µL fractions from the top of the tube and transfer to the microcentrifuge tubes.

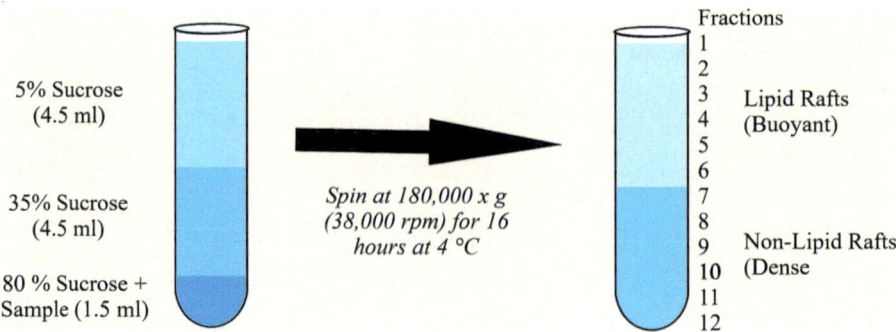

Fig. 3 After sequentially layering the sucrose solutions and centrifuging the tubes for 16 h, the buoyant lipid raft layer appears in the 5–35% range of sucrose gradient solutions

3.3 Isolation of Lipid Rafts Using the Detergent-Free Method

1. To the cell pellet, add 1.5 mL 500 mM sodium carbonate and vortex.

2. Homogenize the cell suspension by sonication using five 20-s bursts on ice.

3. Add 1.5 mL of 80% sucrose and mix by vortex and sonication (three 20-s bursts) on ice (*see* **Note 11**).

4. Sucrose gradient ultracentrifugation. Prepare 5%, 35%, and 80% sucrose solutions in Mes-buffered solution (MBS). The use of MBS with a pH close to 7.0 may be advantageous for most proteins.

5. Place 3 mL of cell homogenates into the bottom of precooled 12-mL ultracentrifuge tubes.

6. Overlay sequentially 4.5 mL of 35% sucrose and 4.5 mL of 5% sucrose to each tube.

7. With the tubes securely balanced in the swinging bucket rotor (Fig. 3), spin at 180,000 × g for 16 h at 4 °C (*see* **Note 12**).

8. Carefully aspirate twelve 1-mL fractions from the top of the tube and transfer into pre-labeled 1.5 microcentrifuge tubes.

9. Prepare 0.5 mL of each fraction by adding 0.1 mL 6X Laemmli sample buffer, vortex, and boil for 5 min before use for immunoblotting (*see* **Note 13**).

3.4 Localization of GPCR in Lipid Rafts Using Laser Scanning Confocal Microscopy

1. Culture cells on 12-mm coverslips placed in a 24-well tissue culture plate to ~50% confluence, as discussed in Subheading 3.1.

2. Serum-starve the cells and treat with the desired cholesterol depletor or replete with cholesterol as discussed in Subheading 3.1.

3. Wash cells with ice-cold PBS and place the 24-well plates on ice to suspend receptor endocytosis and endosomal trafficking.

4. Use a lipid raft marker (e.g., CTxB) tagged with Alexa Fluor® 488, Alexa Fluor® 555, or Alexa Fluor® 647 (total volume: 0.3 mL) at 4 °C for 10 min. Other protein markers can also be utilized as discussed above (e.g., caveolin-1, caveolin-2, flotillin-1). Several dyes with different wavelengths can also be used.

5. Draw off the solution and wash cells with ice-cold PBS.

6. Add 0.3 mL of the primary antibody against the GPCR of interest dissolved in 10% BSA (1:100–200 dilution) for 30–60 min.

7. Wash cells 3X with ice-cold PBS.

8. Add 0.3 mL of the secondary antibody (against the host of the primary antibody used in Subheading 3.1) in 10% BSA (*see* **Notes 14** and **15**).

9. Fix cells with 0.3 mL of 4% paraformaldehyde at room temperature for 15 min.

10. Wash cells with ice-cold PBS. Subsequent steps can be performed at room temperature.

11. Permeabilize the cells with 0.3 mL of 0.5% Triton X-100 in deionized water for 10 min (*see* **Note 16**).

12. Wash cells with PBS and once with deionized water. The use of deionized water washes away the residual NaCl crystals from PBS.

13. Mount coverslips using a mounting medium on a glass slide. Gently remove excess mounting medium by aspiration. Allow the mounting medium to harden completely.

14. Image the cells using a laser scanning confocal microscope. The appropriate filters should be used depending on the Alexa Fluor® dye that was used and whether DAPI was used as a nuclear stain.

15. Perform the colocalization of GPCR with lipid raft proteins (*see* **Note 17**).

16. Several other techniques are available for the detection and localization of GPCRs in lipid raft microdomains in cells (see Fig. 4). The most commonly employed approach utilizes cell fractionation procedures that break the cells apart and destroy cell morphology before GPCR analysis using biochemical or immunological assays. A complementary biophysical approach involves the visualization of GPCRs in intact cell membranes (*see* **Notes 18–20**). For immunohistochemistry, please see **Notes 21–24**. For live cell imaging, please *see* **Note 25–28**.

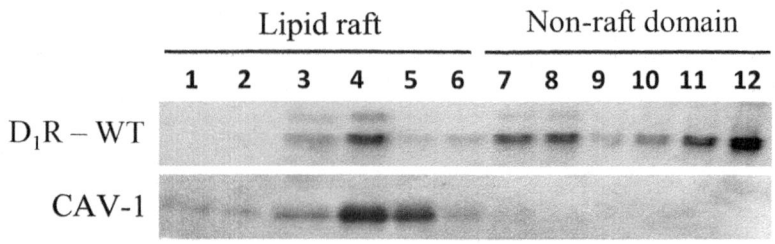

Fig. 4 Lipid raft distribution of caveolin-1 and D1R. Lipid raft and non-lipid raft fractions from human renal proximal tubule cells were prepared by detergent-free method and sucrose gradient ultracentrifugation. The distribution of caveolin-1, a lipid raft marker, and the dopamine D1 receptor (D1R), a GPCR, is shown in the immunoblots

4   Notes

1. Methyl-α-cyclodextrin (α-MCD) is recommended as a negative control.

2. To prepare this solution proceed as follows:

   (a) Dissolve cholesterol (20 mg/mL) in ethanol by sonication.

   (b) Dissolve βCD (2%) in DMEM/F12 SFM.

   (c) Prepare cholesterol-βCD solution by adding 20 μL of cholesterol solution to 10 mL cyclodextrin solution, mix by vortexing, and incubating the cholesterol-βCD solution at 40 °C for 30 min.

3. Separate dishes should be prepared for cholesterol depletion and cholesterol repletion groups.

4. For cell culture and pellet collection, make sure all buffers are kept cold prior to use.

5. The following protocols represent general steps to prepare the cells for the isolation of lipid rafts.

6. The pelleted nuclei have to be confirmed with protein markers (e.g., histone H4).

7. The pelleted mitochondria have to be confirmed with protein markers (e.g., citrate synthase).

8. Be careful when transferring the tubes to and from the bucket rotor to prevent spillage of contents especially that the tubes are almost full during this part.

9. When transferring the gradient, place the tip of the micropipette at the inner side of the tube to prevent splashes of the bottom gradient.

10. Samples can be stored at −20 °C for up to 1 month.

11. Protein concentration may be determined at this time using, for example, a BCA kit.

12. At this point, a light-scattering band that is enriched with caveolae/lipid rafts can be observed between the 5% and 35% sucrose gradients and corresponds to the fourth fraction.

13. These samples can be stored at $-20\ ^\circ$C, while the rest of the fractions without the 6X sample buffer can be stored at $-80\ ^\circ$C.

14. The secondary antibody should be tagged with a fluorophore other than the one used to label the CTxB. As counterstain, 300 nM DAPI may also be added to this working solution.

15. To confirm the appropriate distribution of proteins, use non-lipid raft markers for comparison (e.g., transferrin receptors, CD71, geranylated proteins).

16. Permeabilization provides access to intracellular antigens. Triton X-100 can effectively dissolve cellular membranes without disturbing protein–protein interactions. Other detergents such as saponin, Tween-20, or sodium dodecyl sulfate (SDS) may also be used.

17. For immunoblotting, use lipid raft markers to localize the lipid raft proteins which are non-uniformly distributed in the more buoyant fractions (top 5–6 fractions). These lipid raft markers include caveolin-1 (Abcam, Cambridge, MA, USA) [26, 69], flotillin-1, CD55, alkaline phosphatase, and pore-forming toxins, such as cholera toxin subunit B (CTxB), equinatoxin II, and perfringolysin [70–72].

18. Flotillin-1 has been used as a lipid raft marker protein in cells that do not contain caveolae, i.e., blood cells [70], neural cells [73], and rat renal proximal tubule cells [74, 75] and human embryonic kidney (HEK)-293 cells [9]. There is species specificity because human renal proximal tubule cells express caveolin-1 [27], while HEK293 cells express caveolin-2 [9] (Fig. 4).

19. If lipid rafts are disrupted through cholesterol depletion (e.g., β-MCD), the lipid raft markers would redistribute to the less buoyant fractions (bottom 7–12 fractions).

20. Cholesterol repletion should reconstruct the lipid rafts, thus shifting the lipid raft markers back to the buoyant fractions.

21. For immunohistochemistry, use Cholera Toxin B (CTxB) tagged with Alexa Fluor®488, Alexa Fluor® 555, or Alexa Fluor® 647 for the localization of lipid rafts. CTxB binds to the pentasaccharide chain of ganglioside GM1, which selectively partitions into lipid rafts.

22. Use an anti-CTxB antibody to visualize the protein distribution.

23. Multiple direct labeling of fluorophores on raft components through proximity enzymatic glyco-remodeling (PEGR). It is based on the transformation of one lipid recognition event to multiple lipid raft specific remodeling operations through raft resident glycans, which are closely linked proteins and lipids within the lipid raft. It relies on the lipid raft-recognition capability of CTxB and Gal/GalNAc-catalytic oxidation capability of GO on AuNPs [76].

24. The c-subunit of cytolethal distending toxin (cdt) may also be utilized for lipid raft colocalization experiments [77]. Other pore-forming toxins, besides CTxB, used to visualize lipid rafts include equinatoxin II which binds dispersed sphingomyelin, lysenin which binds clustered sphingomyelin, perfringolysin O which binds to cholesterol, and ostreolysin which binds to the combination of sphingomyelin and cholesterol [72, 78].

25. For live cell imaging, CTxB labeling may also be used to demonstrate lipid raft endocytosis upon agonist stimulation in live cells [79] and cultured explants [80]. There are limitations of CTxB in the visualization of lipid rafts which may be overcome by *Gaussia princeps* luciferase Protein-fragment Complementation Assay [81].

26. Single fluorophore tracking microscopy [82] and fluorescence recovery after photobleaching (FRAP) [83] may be used to monitor lateral diffusion of lipid raft-anchored GPCRs, while fluorescence lifetime imaging microscopy-fluorescence resonance energy transfer (FLIM-FRET) [84, 85] may be used to determine the proximity of GPCRs with other proteins of interest, or of lipid raft sizes depending on membrane composition [86].

27. Atomic force microscopy may be used to visualize the effects of detergent solubilization of membranes during lipid raft studies [87].

28. Lipid rafts can now be visualized using super-resolution imaging below the 200 nm limit of conventional microscopes, e.g., including structured illumination microscopy, stimulated emission depletion microscopy (STED), near field scanning optical microscopy (NSOM), photoactivated localization microscopy (PALM), and direct stochastic optical reconstruction microscopy (dSTORM) [53, 54, 88, 89].

References

1. Sonnino S, Prinetti A (2013) Membrane domains, and the "lipid raft" concept. Curr Med Chem 20:4–21

2. Karnovsky MJ, Kleinfeld AM, Hoover RL, Dawidowicz EA, McIntyre DE, Salzman EA et al (1982) Lipid domains in membranes. Ann N Y Acad Sci 401:61–75

3. Simons K, Ikonen E (1997) Functional rafts in cell membranes. Nature 387:569–572

4. Simons K, van Meer G (1998) Lipid sorting in epithelial cells. Biochemistry 27:6197–6202

5. Van Meer G, Simons K (1988) Lipid polarity and sorting in epithelial cells. J Cell Biochem 36:51–58

6. Levental I, Levental KR, Heberle FA (2020) Lipid rafts: controversies resolved, mysteries remain. Trends Cell Biol 30:341–353

7. Baumgart T, Hammond AT, Sengupta P et al (2007) Large-scale fluid/fluid phase separation of proteins and lipids in giant plasma membrane vesicles. Proc Natl Acad Sci U S A 104:3165–3170

8. Levental I, Wang HY (2020) Membrane domains beyond the reach of microscopy. J Lipid Res 61(5):592

9. Yu P, Yang Z, Jones JE, Wang Z, Owens SA et al (2004) D1 dopamine receptor signaling involves caveolin-2 in HEK-293 cells. Kidney Int 66:2167–2180

10. Yu P, Villar VA, Jose PA (2013) Methods for the study of dopamine receptors within lipid rafts of kidney cells. Methods Mol Biol 964:15–24

11. Quest AF, Leyton L, Párraga M (2004) Caveolins, caveolae, and lipid rafts in cellular transport, signaling, and disease. Biochem Cell Biol 82:129–144

12. Head BP, Patel HH, Insel PA (2014) Interaction of membrane/lipid rafts with the cytoskeleton: impact on signaling and function: membrane/lipid rafts, mediators of cytoskeletal arrangement and cell signaling. Biochim Biophys Acta 1838:532–545

13. Rajendran L, Le Lay S, Illges H (2007) Raft association and lipid droplet targeting of flotillins are independent of caveolin. Biol Chem 388:307–314

14. Hemler ME (2005) Tetraspanin functions and associated microdomains. Nat Rev Mol Cell Biol 6:801–811

15. Grove LM, Southern BD, Jin TH, White KE, Paruchuri S, Harel E, Wei Y, Rahaman SO, Gladson CL, Ding Q, Craik CS, Chapman HA, Olman MA (2014) Urokinase-type plasminogen activator receptor (uPAR) ligation induces a raft-localized integrin signaling switch that mediates the hypermotile phenotype of fibrotic fibroblasts. J Biol Chem 289:12791–12804

16. Palade GE (1953) The fine structure of blood capillaries. J Appl Phys 24:1424

17. Yamada E (1955) The fine structure of the gall bladder epithelium of the mouse. J Biophys Biochem Cytol 1:445–458

18. Parton RG, del Pozo MA (2013) Caveolae as plasma membrane sensors, protectors, and organizers. Nat Rev Mol Cell Biol 14:98–112

19. Barnett-Norris J, Lynch D, Reggio PH (2005) Lipids, lipid rafts, and caveolae: their importance for GPCR signaling and their centrality to the endocannabinoid system. Life Sci 77:625–639

20. Simons K, Toomre D (2000) Lipid rafts and signal transduction. Nat Rev Mol Cell Biol 1:31–39

21. Lefkowitz RJ (1998) G protein-coupled receptors III new roles for receptor kinases and beta-arrestins in receptor signaling and desensitization. J Biol Chem 273:18677–18680

22. Worby CA, Dixon JE (2002) Sorting out the cellular functions of sorting nexins. Nat Rev Mol Cell Biol 3:919–931

23. Von Zastrow M (2003) Mechanisms regulating trafficking of G protein-coupled receptors in the endocytic pathway. Life Sci 74:217–224

24. Ferguson SG (2001) Evolving concepts in G protein-coupled receptor endocytosis: the role in receptor desensitization and signaling. Pharmacol Rev 53:1–24

25. Barnett-Norris J, Lynch D, Reggio PH (2005) Lipids, lipid rafts, and caveolae: their importance for GPCR signaling and their centrality to the endocannabinoid system. Life Sci 77:625–639

26. Lingwood D, Simons K (2010) Lipid rafts as a membrane organizing principle. Science 327:46–50

27. Gildea JJ, Israel JA, Johnson AK, Zhang J, Jose PA, Felder RA (2009) Caveolin-1 and dopamine-mediated internalization of NaKATPase in human renal proximal tubule cells. Hypertension 54:1070–1076

28. Han W, Li H, Villar VA, Pascua AM, Dajani MI, Wang X et al (2008) Lipid rafts keep NADPH oxidase in the inactive state in human renal proximal tubule cells. Hypertension 51:481–487

29. Yu P, Sun M, Villar VA, Zhang Y, Weinman EJ, Felder RA, Jose PA (2014) Differential dopamine receptor subtype regulation of adenylyl cyclases in lipid rafts in human embryonic kidney and renal proximal tubule cells. Cell Signal 26:2521–2529

30. Webb Y, Hermida-Matsumoto L, Resh MD (2000) Inhibition of protein palmitoylation, raft localization, and T cell signaling by 2-bromopalmitate and polyunsaturated fatty acids. J Biol Chem 275:261–270

31. Simons M, Friedrichson T, Schulz JB, Pitto M, Masserini M, Kurzchalia TV (1999) Exogenous administration of gangliosides displaces GPI-anchored proteins from lipid microdomains in living cells. Mol Biol Cell 10:3187–3196

32. Ravacci GR, Brentani MM, Tortelli T Jr, Torrinhas RS, Saldanha T, Torres EA et al (2013) Lipid raft disruption by docosahexaenoic acid induces apoptosis in transformed human mammary luminal epithelial cells harboring HER-2 overexpression. J Nutr Biochem 24:505–515

33. Lipardi C, Nitsch L, Zurzolo C (2000) Detergent-insoluble GPI-anchored proteins are apically sorted in fischer rat thyroid cells, but interference with cholesterol or sphingolipids differentially affects detergent insolubility and apical sorting. Mol Biol Cell 11:531–542

34. Nakai Y, Kamiguchi H (2002) Migration of nerve growth cones requires detergent-resistant membranes in a spatially defined and substrate-dependent manner. J Cell Biol 159:1097–1108

35. Rentero C, Zech T, Quinn CM, Engelhardt K, Williamson D, Grewal T et al (2008) Functional implications of plasma membrane condensation for T cell activation. PLoS One 3: e2262

36. Schieffer D, Naware S, Bakun W, Bamezai AK (2014) Lipid raft-based membrane order is important for antigen-specific clonal expansion of CD4(+) T lymphocytes. BMC Immunol 15:58

37. Kyrikou I, Hadjikakou SK, Kovala-Demertzi D, Viras K, Mavromoustakos T (2004) Effects of non-steroid anti-inflammatory drugs in membrane bilayers. Chem Phys Lipids 132:157–169

38. Alsop RJ, Toppozini L, Marquardt D, Kučerka N, Harroun TA, Rheinstädter MC (2015) Aspirin inhibits formation of cholesterol rafts in fluid lipid membranes. Biochim Biophys Acta 1848:805–812

39. Gidwani A, Brown HA, Holowka D, Baird B (2003) Disruption of lipid order by short-chain ceramides correlates with inhibition of phospholipase D and downstream signaling by FcepsilonRI. J Cell Sci 116:3177–3187

40. Silvius JR (2003) Role of cholesterol in lipid raft formation: lessons from lipid model systems. Biochim Biophys Acta 1610:174–183

41. Brown DA, London E (2000) Structure and function of sphingolipid- and cholesterol-rich membrane rafts. J Biol Chem 275:17221–17224

42. Drake DR 3rd, Braciale TJ (2001) Cutting edge: lipid raft integrity affects the efficiency of MHC class I tetramer binding and cell surface TCR arrangement on CD8+ T cells. J Immunol 166:7009–7013

43. Oakley FD, Smith RL, Engelhardt JF (2009) Lipid rafts and caveolin-1 coordinate interleukin-1beta (IL-1beta)-dependent activation of NFkappaB by controlling endocytosis of Nox2 and IL-1beta receptor 1 from the plasma membrane. J Biol Chem 284:33255–33264

44. Wysoczynski M, Reca R, Ratajczak J, Kucia M, Shirvaikar N, Honczarenko M et al (2005) Incorporation of CXCR4 into membrane lipid rafts primes homing-related responses of hematopoietic stem/progenitor cells to an SDF-1 gradient. Blood 105:40–48

45. Schroeder RJ, Ahmed SN, Zhu Y, London E, Brown DA (1998) Cholesterol and sphingolipid enhance the triton X-100 insolubility of glycosylphosphatidylinositol-anchored proteins by promoting the formation of detergent-insoluble ordered membrane domains. J Biol Chem 273:1150–1157

46. Hering H, Lin CC, Sheng M (2003) Lipid rafts in the maintenance of synapses, dendritic spines, and surface AMPA receptor stability. J Neurosci 23:3262–3271

47. Oliferenko S, Paiha K, Harder T, Gerke V, Schwärzler C, Schwarz H et al (1999) Analysis of CD44-containing lipid rafts: recruitment of annexin II and stabilization by the actin cytoskeleton. J Cell Biol 146:843–854

48. Fernandez-Lizarbe S, Pascua LM, Gascon MS, Blanco A, Guerri C (2008) Lipid rafts regulate ethanol-induced activation of TLR4 signaling in murine macrophages. Mol Immunol 45:2007–2016

49. Vial C, Evans RJ (2005) Disruption of lipid rafts inhibits P2X1 receptor-mediated currents and arterial vasoconstriction. J Biol Chem 280:30705–30711

50. Meszaros P, Klappe K, Hummel I, Hoekstra D, Kok JW (2011) Function of MRP1/ABCC1 is not dependent on cholesterol or cholesterol-stabilized lipid rafts. Biochem J 437:483–491

51. Hoppe AD, Scott BL, Welliver TP, Straight SW, Swanson JA (2013) N-way FRET microscopy of multiple protein-protein interactions in live cells. PLoS One 8:e64760

52. Agarwal SR, Yang PC, Rice M, Singer CA, Nikolaev VO, Lohse MJ et al (2014) Role of membrane microdomains in compartmentation of cAMP signaling. PLoS One 9:e95835

53. Tobin SJ, Cacao EE, Hong DW, Terenius L, Vukojevic V, Jovanovic-Talisman T (2014) Nanoscale effects of ethanol and naltrexone on protein organization in the plasma membrane studied by photoactivated localization microscopy (PALM). PLoS One 9:e87225

54. Wu J, Gao J, Qi M, Wang J, Cai M, Liu S, Hao X, Jiang J, Wang H (2013) High-efficiency localization of Na(+)-K(+) ATPases on the cytoplasmic side by direct stochastic optical reconstruction microscopy. Nanoscale 5:11582–11586

55. Scarselli M, Annibale P, McCormick PJ, Kolachalam S, Aringhieri S, Radenovic A, Corsini GU, Maggio R (2015) Revealing GPCR oligomerization at the single-molecule level through a nanoscopic lens: methods, dynamics, and biological function. FEBS J 83:1197–1217. https://doi.org/10.1111/febs.13577

56. Dubach JM, Das S, Rosenzweig A, Clark HA (2009) Visualizing sodium dynamics in isolated cardiomyocytes using fluorescent nanosensors. Proc Natl Acad Sci U S A 106:16145–16150

57. Chang JC, Rosenthal SJ (2012) Visualization of lipid raft membrane compartmentalization in living RN46A neuronal cells using single quantum dot tracking. ACS Chem Neurosci 3:737–743

58. Miyagawa-Yamaguchi A, Kotani N, Honke K (2015) Each GPI-anchored protein species forms a specific lipid raft depending on its GPI attachment signal. Glycoconj J 32:531–540

59. Gaus K, Zech T, Harder T (2006) Visualizing membrane microdomains by Laurdan 2-photon microscopy. Mol Membr Biol 23:41–48

60. Kim HM, Choo HJ, Jung SY, Ko YG, Park WH, Jeon SJ, Kim CH et al (2007) A two-photon fluorescent probe for lipid raft imaging: C-laurdan. Chembiochem 8:553–559

61. Kim HM, Jeong BH, Hyon JY, An MJ, Seo MS, Hong JH et al (2008) Two-photon fluorescent turn-on probe for lipid rafts in live cell and tissue. J Am Chem Soc 130:4246–4247

62. Ohno-Iwashita Y, Shimada Y, Waheed AA, Hayashi M, Inomata M, Nakamur M et al (2004) Perfringolysin O, a cholesterol-binding cytolysin, as a probe for lipid rafts. Anaerobe 10:125–134

63. Schnitzer JE, McIntosh DP, Dvorak AM, Liu J, Oh P (1995) Separation of caveolae from associated microdomains of GPI-anchored proteins. Science 269:1435–1439. https://doi.org/10.1126/science.7660128

64. Pike LJ (2004) Lipid rafts: heterogeneity on the high seas. Biochem J 378:281–292

65. Song KS, Li S, Okamoto T, Quilliam LA, Sargiacomo M, Lisanti MP (1996) Co-purification and direct interaction of Ras with caveolin, an integral membrane protein of caveolae microdomains detergent-free purification of caveolae microdomains. J Biol Chem 271:9690–9697

66. Smart EJ, Ying YS, Mineo C, Anderson RG (1995) A detergent-free method for purifying caveolae membrane from tissue culture cells. Proc Natl Acad Sci U S A 92:10104–10108

67. Macdonald JL, Pike LJ (2005) A simplified method for the preparation of detergent-free lipid rafts. J Lipid Res 46:1061–1067

68. Murtazina R, Kovbasnjuk O, Donowitz M, Li X (2006) Na+/H+ exchanger NHE3 activity and trafficking are lipid raft-dependent. J Biol Chem 281:17845–17855

69. Insel PA, Head BP, Ostrom RS, Patel HH, Swaney JS, Tang CM et al (2005) Caveolae, and lipid rafts: G protein-coupled receptor signaling microdomains in cardiac myocytes. Ann N Y Acad Sci 1047:166–172

70. Salzer U, Prohaska R (2001) Stomatin, flotillin-1, and flotillin-2 are major integral proteins of erythrocyte lipid rafts. Blood 97:1141–1143

71. Foster LJ, De Hoog CL, Mann M (2003) Unbiased quantitative proteomics of lipid rafts reveals high specificity for signaling factors. Proc Natl Acad Sci U S A 100:5813–5818

72. Skočaj M, Bakrač B, Križaj I, Maček P, Anderluh G, Sepčić K (2013) The sensing of membrane microdomains based on pore-forming toxins. Curr Med Chem 20:91–501

73. Huang P, Xu W, Yoon SI, Chen C, Chong PL, Liu-Chen LY (2007) Cholesterol reduction by methyl-beta-cyclodextrin attenuates the delta opioid receptor-mediated signaling in neuronal cells but enhances it in non-neuronal cells. Biochem Pharmacol 73:534–549

74. Breton S, Lisanti MP, Tyszkowski R, McLaughlin M, Brown D (1998) Basolateral distribution of caveolin-1 in the kidney absence from H+-atpase-coated endocytic vesicles in intercalated cells. J Histochem Cytochem 46:205–214

75. Riquier AD, Lee DH, McDonough AA (2009) Renal NHE3 and NaPi2 partition into distinct membrane domains. Am J Physiol Cell Physiol 296:C900–C910

76. Tao J, Yu X, Guo Y, Wang G, Ju H, Ding L (2020) Proximity enzymatic Glyco-Remodeling enables direct and highly efficient lipid raft imaging on live cells. Anal Chem 92:7232–7239

77. Boesze-Battaglia K (2006) Isolation of membrane rafts and signaling complexes. Methods Mol Biol 332:169–179

78. Makino A, Abe M, Murate M, Inaba T, Yilmaz N, Hullin-Matsuda F, Kishimoto T, Schieber NL, Taguchi T, Arai H, Anderluh G, Parton RG, Kobayashi T (2015) Visualization of the heterogeneous membrane distribution of sphingomyelin associated with cytokinesis, cell polarity, and sphingolipidosis. FASEB J 29:477–493

79. Qi R, Mullen DG, Baker JR, Holl MM (2010) The mechanism of polyplex internalization into cells: testing the GM1/caveolin-1 lipid raft mediated endocytosis pathway. Mol Pharm 7:267–279

80. Hansen GH, Dalskov SM, Rasmussen CR, Immerda LL, Niels-Christiansen LL, Danielsen EM (2005) Cholera toxin entry into pig enterocytes occurs via a lipid raft- and clathrin-dependent mechanism. Biochemistry 44:873–882

81. Merezhko M, Pakarinen E, Uronen RL, Huttunen HJ (2020) Live-cell monitoring of protein localization to membrane rafts using protein-fragment complementation. Biosci Rep 40:BSR20191290. https://doi.org/10.1042/BSR20191290

82. Schütz GJ, Kada G, Pastushenko VP, Schindler H (2000) Properties of lipid microdomains in a muscle cell membrane visualized by single molecule microscopy. EMBO J 19:892–901

83. Kenworthy AK (2007) Fluorescence recovery after photobleaching studies of lipid rafts. Methods Mol Biol 398:179–192

84. Kenworthy AK, Petranova N, Edidin M (2000) High-resolution FRET microscopy of cholera toxin B-subunit and GPI-anchored proteins in cell plasma membranes. Mol Biol Cell 11:1645–1655

85. Thaa B, Herrmann A, Veit M (2010) Intrinsic cytoskeleton-dependent clustering of influenza virus M2 protein with hemagglutinin assessed by FLIM-FRET. J Virol 84:12445–12449

86. de Almeida RF, Loura LM, Fedorov A, Prieto M (2005) Lipid rafts have different sizes depending on membrane composition: a time-resolved fluorescence resonance energy transfer study. J Mol Biol 346:1109–1120

87. Garner AE, Smith DA, Hooper NM (2008) Visualization of detergent solubilization of membranes: implications for the isolation of rafts. Biophys J 94:1326–1340

88. Owen DM, Gaus K (2013) Imaging lipid domains in cell membranes: the advent of super-resolution fluorescence microscopy. Front Plant Sci 4:503

89. Nieves DJ, Owen DM (2020) Analysis methods for interrogating spatial organisation of single molecule localisation microscopy data. Int J Biochem Cell Biol 123:105749

90. Whitehead SN, Gangaraju S, Aylsworth A, Hou ST (2012) Membrane raft disruption results in neuritic retraction prior to neuronal death in cortical neurons. Biosci Trends 6:4183–4191

91. Szoke E, Börzsei R, Tóth DM et al (2010) Effect of lipid raft disruption on TRPV1 receptor activation of trigeminal sensory neurons and transfected cell line. Eur J Pharmacol 628:67–74. https://doi.org/10.1016/j.ejphar.2009.11.052

92. Somanath PR, Ciocea A, Byzova TV (2009) Integrin and growth factor receptor alliance in angiogenesis. Cell Biochem Biophys 53:53–64. https://doi.org/10.1007/s12013-008-9040-5

93. Smythies LE, White CR, Maheshwari A et al (2009) Apolipoprotein A-I mimetic 4F alters the function of human monocyte-derived macrophages. Am J Physiol Cell Physiol 298:C1538–C1548. https://doi.org/10.1152/ajpcell.00467.2009

94. Sánchez-Wandelmer J, Dávalos A, Herrera E et al (2009) Inhibition of cholesterol biosynthesis disrupts lipid raft/caveolae and affects insulin receptor activation in 3T3-L1 preadipocytes. Biochim Biophys Acta 1788:1731–1739. https://doi.org/10.1016/j.bbamem.2009.05.002

95. Burger HM, Abel S, Gelderblom WCA (2018) Modulation of key lipid raft constituents in primary rat hepatocytes by fumonisin B1 - implications for cancer promotion in the liver. Food Chem Toxicol 115:34–41. https://doi.org/10.1016/j.fct.2018.03.004

Chapter 2

Detection of GPCR mRNA Expression in Primary Cells Via qPCR, Microarrays, and RNA-Sequencing

Krishna Sriram, Cristina Salmerón, Anna Di Nardo, and Paul A. Insel

Abstract

A workflow is described for assaying the expression of G protein-coupled receptors (GPCRs) in cultured cells, using a combination of methods that assess GPCR mRNAs. Beginning from the isolation of cDNA and preparation of mRNA, we provide protocols for designing and testing qPCR primers, assaying mRNA expression using qPCR and high-throughput analysis of GPCR mRNA expression via TaqMan qPCR-based, GPCR-selective arrays. We also provide a workflow for analysis of expression from RNA-sequencing (RNA-seq) assays, which can be queried to yield expression of GPCRs and related genes in samples of interest, as well as to test changes in expression between groups, such as in cells treated with drugs or from healthy and diseased subjects. We place priority on optimized protocols that distinguish signal from noise, as GPCR mRNAs are typically present in low abundance, necessitating techniques that maximize sensitivity while minimizing noise. These methods may also be applicable for assessing the expression of members of families of other low abundance genes via high-throughput analyses of mRNAs, followed by independent confirmation and validation of results via qPCR.

Key words qPCR, RNA-seq, Arrays/microarrays, Screening, Omics

---

1 Introduction

G protein-coupled receptors (GPCRs) include approximately 800 proteins, forming the largest family of cell surface receptors in the human genome. Based on their membrane association, selectivity of expression, and broad range of cellular processes that they regulate, GPCRs are the largest family of proteins targeted by approved drugs [1]. Approximately 360 of the ~800 human GPCRs are endoGPCRs, i.e., expressed in tissues and activated by endogenously produced ligands [2]. This subset represents the majority of current drug targets and drug discovery efforts. The remaining GPCRs are involved in taste, vision, and olfaction, although certain of those receptors are also expressed in tissues other than the primary sensory organs [3]. Most studies of GPCRs focus on the endoGPCRs, although efforts are also

Sofia Aires M. Martins and Duarte Miguel F. Prazeres (eds.), *G Protein-Coupled Receptor Screening Assays: Methods and Protocols*, Methods in Molecular Biology, vol. 2268, https://doi.org/10.1007/978-1-0716-1221-7_2,

directed at chemosensing GPCRs. Of the ~360 endoGPCRs, ~100 are primary targets for approved drugs and ~ 100 are currently orphans (i.e., without known endogenous agonists [1, 2]. Hence, studies of GPCRs are an active area of research, with various approaches used to define their role in health and disease in cultured cells, as well as in cells and tissues from experimental animals and human subjects.

Due to their low abundance at the protein level, combined with the general paucity of well-validated antibodies [3, 4], identification and quantification of GPCR proteins is challenging. Detection of GPCRs thus typically involves the use of methods for quantification of mRNA, followed by validation of these GPCRomics data with quantitative polymerase chain reaction (qPCR) analyses and, importantly, signaling and functional assays to verify that the receptors detected in cells are physiological receptors. These techniques may be applied to primary cells isolated from humans or experimental animals or by using cell lines. We recently published a report in which we compared three omics methods for high-throughput detection of GPCR mRNA expression in primary cells: qPCR-based GPCR microarrays, Affymetrix arrays, and RNA-seq [4]. A key conclusion of our study was that both qPCR-based GPCR microarrays and RNA-seq provide useful, mutually consistent data for detection of GPCRs, whereas Affymetrix arrays do not. Hence, in the subsequent sections, we discuss protocols for the detection of GPCR mRNA abundance and briefly review approaches for analysis of data generated by use of qPCR, qPCR-based GPCR microarrays, and RNA-seq.

GPCR expression data can provide valuable information, especially for drug discovery efforts, by identifying novel roles for GPCRs in disease [5]. In addition, such data are available in the public domain via servers, such as NCBI GEO (Gene Expression Omnibus), or large database studies such as TCGA (The Cancer Genome Atlas) [3]. These findings are mineable by the broader scientific community, thereby facilitating research efforts beyond the individual laboratories that generated the data [3, 4]. Such data have applications in exploring and elucidating roles for GPCRs, including in novel settings such as in cancers [3] and infectious disease, for example, in the COVID-19 pandemic [6].

Figure 1 shows a general schematic of the steps involved in analysis of GPCR mRNA expression. Topics in solid boxes are discussed below while topics in boxes with dashed lines are outside the scope of this chapter. In the protocols described here, a primary goal is to minimize the amount of noise in the data, which can be introduced by artifacts such as sample contamination, improper reagents, and poorly optimized protocols, thereby compromising one's ability to detect GPCR mRNA. This is of particular importance for GPCRs, due to their relatively low magnitude of expression.

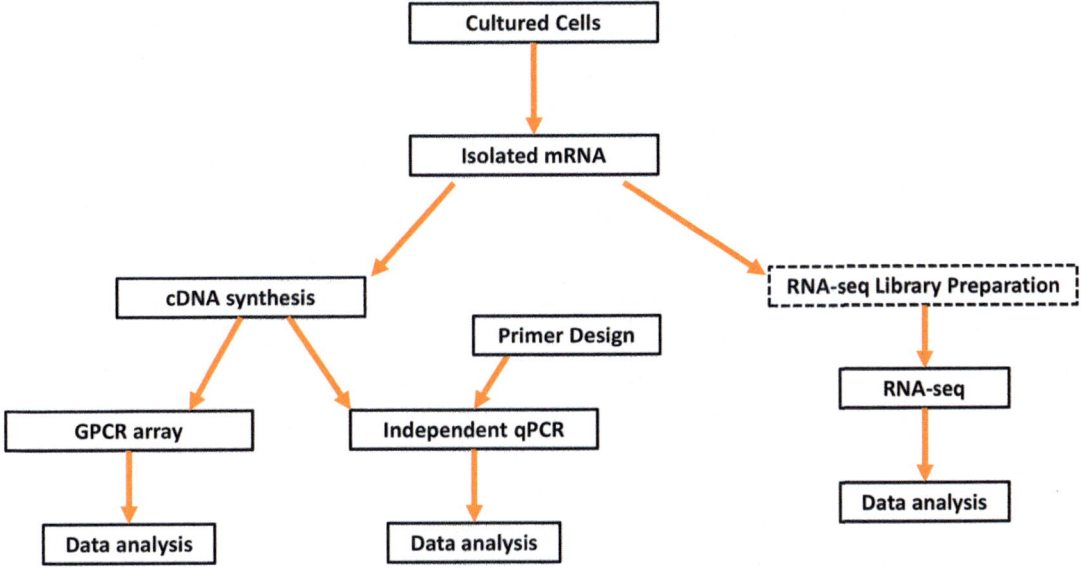

Fig. 1 The sequence of steps to identify expression of GPCR mRNA. Steps in solid boxes are discussed in this text

2 Materials

2.1 Cell Lines

1. Cells in culture (e.g., pancreatic cancer-associated fibroblasts [4] and see notes for cell numbers and plate layouts).

2.2 Equipment

1. Thermocycler for reading 96-well qPCR plates, compatible with SYBR green chemistry.
2. Thermocycler for reading 384-well qPCR plates, compatible with TaqMan chemistry.
3. Thermocycler for standard end point PCR.
4. Tabletop microcentrifuge able to maintain up to $10,000 \times g$.
5. Nanodrop.
6. UVP imager for imaging of gels.
7. A desktop computer running current versions of any standard operating system, for data analysis.
8. 1000, 200, and 20 μL pipettes.

2.3 Reagents and Kits

1. Qiagen RNeasy Mini Kit.
2. Molecular biology grade ethanol, 200 proof.
3. β-Mercaptoethanol (β-ME) or 2 M dithiothreitol (DTT).
4. RNase-free DNase-1 kit.
5. Ultrapure molecular biology-grade water.

6. 70% Ethanol (200 proof molecular biology-grade ethanol, diluted to 70% with ultrapure nuclease-free water).

7. iScript cDNA synthesis kit.

8. SYBR Green qPCR master mix.

9. qPCR primers (*see* **Note 1**).

10. TaqMan Universal PCR Master Mix.

11. TaqMan GPCR array.

12. SYBR Safe DNA Gel Stain.

13. RNase away.

2.4 Consumables

1. Nuclease-free 1.7 and 0.6 mL microcentrifuge tubes.

2. Filtered pipette tips, RNase/DNase free for 1000, 200, and 20 μL pipettes.

3. PCR strip tubes with caps.

4. 96-Well plates for qPCR with adhesive seal.

5. Suitable wipes for cleaning of surfaces.

2.5 Software Tools

1. Manufacturer-supplied software for qPCR thermocyclers.

2. FASTQC (https://www.bioinformatics.babraham.ac.uk/projects/fastqc/, Babraham bioinformatics).

3. BBDUK [U.S. Department of Energy (DOE) Joint Genome Institute (JGI)].

4. Kallisto [Lior Pachter Lab, https://pachterlab.github.io/kallisto/ [7]].

5. Reference transcriptome for relevant species, obtained from Ensembl (ensembl.org/info/data/ftp/index.html).

6. R, preferably with R Studio (https://www.r-project.org/; https://rstudio.com/).

7. R package tximport [https://bioconductor.org/packages/release/bioc/html/tximport.html, [8]].

8. R package edgeR [https://bioconductor.org/packages/release/bioc/html/edgeR.html, [9]].

9. R package BioMart [https://bioconductor.org/packages/release/bioc/html/biomaRt.html, [10]].

3 Methods

3.1 mRNA Isolation from Cultured Cells

1. Harvest cells, typically using 350 μL of RLT Lysis buffer (included with Qiagen kit) for a semi-confluent cell culture 10 cm dish, or smaller (*see* **Note 2**). For a highly confluent dish, or in larger dishes or flasks, increase the volume of lysis buffer to 700 μL (*see* **Notes 3** and **4**).

2. Use a cell scraper to ensure removal of all cells in lysis buffer and collect lysate in an RNase-free 1.7 mL microcentrifuge tube.

3. Vortex the contents vigorously for ~20 s and allow to rest at room temperature ~ 5 min, to allow for complete lysis.

4. Prepare RNase-free DNase-1. Dissolve the bottle of lyophilized enzyme in 550 μL of nuclease-free water. Store 10 μL aliquots at −20 °C; avoid repeated freeze-thawing. Aliquots are usable for ~3 months.

5. Add 1 volume of 70% ethanol to the lysate from **step 2** and mix well by pipetting. Do not centrifuge (*see* **Note 5**).

6. Transfer up to 700 μL of the sample, including any precipitate, to a RNeasy Mini spin column (pink color) placed in a 2 mL collection tube (supplied with Qiagen RNeasy mini kit). Close the lid and centrifuge for 20 s at ≥8000 × g (*see* **Note 6**). Discard the flow-through.

7. Add 350 μL Buffer RW1 (included with Qiagen RNeasy mini kit) to the RNeasy spin column. Close the lid, and centrifuge for 20 s at ≥8000 × g. Discard the flow-through.

8. Add 10 μL DNase stock solution from **step 4** to 70 μL Buffer RDD (included with DNase-1 kit).

9. Mix by gently inverting the tube, and centrifuge briefly to collect residual liquid from sides of tube. Add to RNeasy spin column membrane, and incubate for 15 min at 20–30 °C.

10. Add 350 μL Buffer RW1 to the RNeasy column and centrifuge for 20 s at ≥8000 × g. Discard flow-through.

11. Add 500 μL Buffer RPE (included with Qiagen RNeasy mini kit; see **Note 7**) to the RNeasy column. Centrifuge for 20 s at ≥8000 × g and discard flow-through.

12. Add 500 μL Buffer RPE to RNeasy column and centrifuge for 2 min at ≥8000 × g.

13. Place the RNeasy column in new 2 mL tube and centrifuge at full speed for 1 min.

14. Place RNeasy column in new 1.7 mL tube. Add 34 μL RNase-free water, close lid and wait ~1 min to allow water to saturate filter membrane. Centrifuge for 1 min at ≥8000 × g.

15. Test RNA quantity and quality via a nanodrop (*see* **Note 8**). RNA concentrations in ng/μL are necessary for the cDNA synthesis steps, so as to estimate the amount of RNA sample to load per cDNA synthesis reaction.

3.2   cDNA Synthesis

1. In PCR strip tubes, prepare 20 μL reactions for cDNA synthesis, by adding 4 μL of 5X iScript Reaction mix, 1 μL iScript Reverse Transcriptase, Input RNA template (20–1000 ng),

molecular biology-grade, nuclease-free water to make up volume to 20 μL (*see* **Note 9**).

2. Perform PCR in thermocycler, using the following reaction parameters: priming 5 min at 25 °C reverse; transcription (RT) 20 min at 46 °C; RT inactivation 1 min at 95 °C; hold at 4 °C.

3.3 qPCR Primer Design (Bioinformatics Protocol)

Due to their low magnitude of expression, signal-to-noise ratios in qPCR detection of GPCRs can be poor, i.e., these assays are heavily compromised by primer-dimer formation. The steps below for primer design are intended to minimize this phenomenon.

1. Query the gene of interest at NCBI (https://www.ncbi.nlm.nih.gov/) using the option "gene" in the drop-down menu.

2. In search results, select the gene for the species of interest.

3. Inspect the figure showing the various refseq-annotated transcripts; transcripts with the prefix "NM" correspond to well-validated transcripts, for which evidence exists of their expression in cells as mRNA.

4. Unless the aim is to assess individual splice variants, when examining GPCR mRNA expression as part of a screening exercise, it is preferable to focus on regions of the gene common to all well-validated transcripts. This is typically straightforward for GPCRs, as they usually possess few, if any, introns.

5. In the drop-down menu lower down the page titled "NCBI Reference Sequences (RefSeq)," click on one of the validated refseq transcripts. This opens a new page that has specific details about the transcript. In the top of the page, click on "Graphics."

6. This opens a graphical view of the transcript highlighting regions of interest, including exons, protein-coding regions, etc. The coordinates at the top of the graphical view are critical for the next step. Based on regions that are present in all validated transcripts from the previous view, identify a region of the transcript over which primers should be identified. For example, if a transcript is 5000 bp long, you may wish to design primers for a region in the middle of the transcript, perhaps from 1000 to 3500, a region conserved in all variants/transcripts of the gene.

7. Under the menu "Analyze this sequence," select "Pick Primers." This opens the Primer-blast primer design tool.

8. In "Range," for forward primer, enter the first/start coordinate (1000 in the example above), which is the beginning of the region of interest from the graphical view discussed above. Leave the second box in Forward Primer blank.

9. Similarly, for the reverse primer, leave the first box blank, and enter the coordinate corresponding to the end of the region of interest (3500 in the example above) in the second box.

10. Under PCR product size, change "Max" to 200. If difficulty is encountered in identifying suitable primers, this can be increased to 250–300.

11. Do not select the primer to be either exon-junction spanning or to include an intron. In general, this places excessive constraints on primer design which force a compromise in identifying primer pairs with minimal primer-dimer formation, which is critical when studying GPCRs. In addition, the inclusion of DNase-1 digestion during the RNA isolation removes the need for intron-including primer pairs.

12. Check the box "Allow Splice Variants."

13. In "Advanced Parameters," in "Secondary Structure Alignment Methods," select "Use Thermodynamic Oligo Alignment."

14. Run the tool with all other parameters as default.

15. The tool may produce a prompt that the region identified corresponds to multiple transcripts. Select "All" and proceed.

16. After a few moments, the tool will produce a graphical view that illustrates regions where the primer pairs bind to the transcript of interest, as well as various properties of the primers. An example of this graphical output is shown in Fig. 2.

17. Prioritize primer pairs with GC content closest to 50% and with lowest self-complementarity and Self 3′ complementarity. Typically, Primer-Blast produces primer pairs with both complementarity scores of 0, or < 2.

18. Using the oligo-analyzer tool from IDT (https://www.idtdna.com/calc/analyzer/), analyze the primer pairs for their self-dimer formation and their heterodimer characteristics.

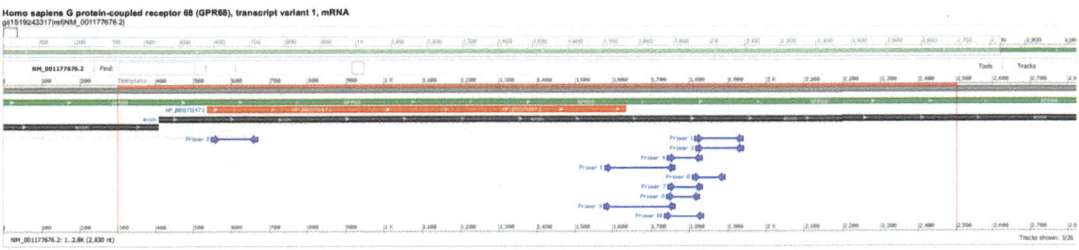

Fig. 2 Graphical output from Primer BLAST showing candidate primer pairs (arrows indicate forward and reverse). Red genomic region: protein coding portion of mRNA; Green: the gene of interest; Black: exons for this gene; vertical red lines: demarcating region of interest for primer design

19. For self-dimers, we recommend primers with Gibbs free energy >-9.0 kcal/mol (i.e., values closer to 0).

20. For heterodimers, we suggest primer pairs with Gibbs free energy >-6.0 kcal/mol, ideally >4 kcal/mol.

21. Once a primer pair is selected, test each primer using NCBI nucleotide BLAST(https://blast.ncbi.nlm.nih.gov/) (Fig. 2). This is to ensure (a) the resulting primer has a 100% match to the gene of interest and (b) within the species of interest, the extent of a match of the primer sequences to other genes. Preferably, to minimize nonspecific amplification, primers should have a $< 85\%$ match to any other regions on coding regions of other genes.

22. Repeat this process varying different parameters above (e.g., amplicon size, region of interest) until suitable primer pairs are obtained, with low Gibbs free energies for dimer formation.

3.4 Primer Validation

For examples of melting curves and amplification curves, see Figs. 3–5. Figure 4 shows an example of a suitable amplification curve, along with examples of amplification curves for a nonoptimal assay.

1. Prepare stock (100 µM) and working (1–10 µM) solutions of the oligonucleotides in nuclease-free water with aliquots stored at $-20\,^{\circ}$C (*see* **Note 10**).

2. Prepare cDNA as above in Subheading 3.2, from 1 µg of species-appropriate universal RNA (*see* **Note 11**).

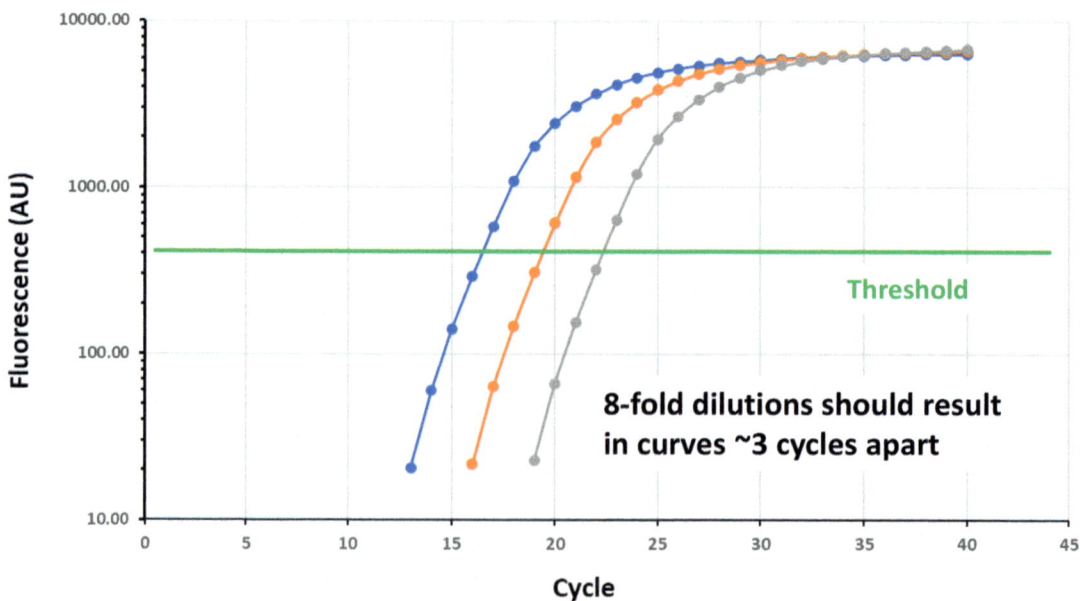

Fig. 3 Amplification curves, corresponding to eightfold dilutions to produce a standard curve for evaluating qPCR primer efficiency. An example of a suitable amplification threshold is also shown

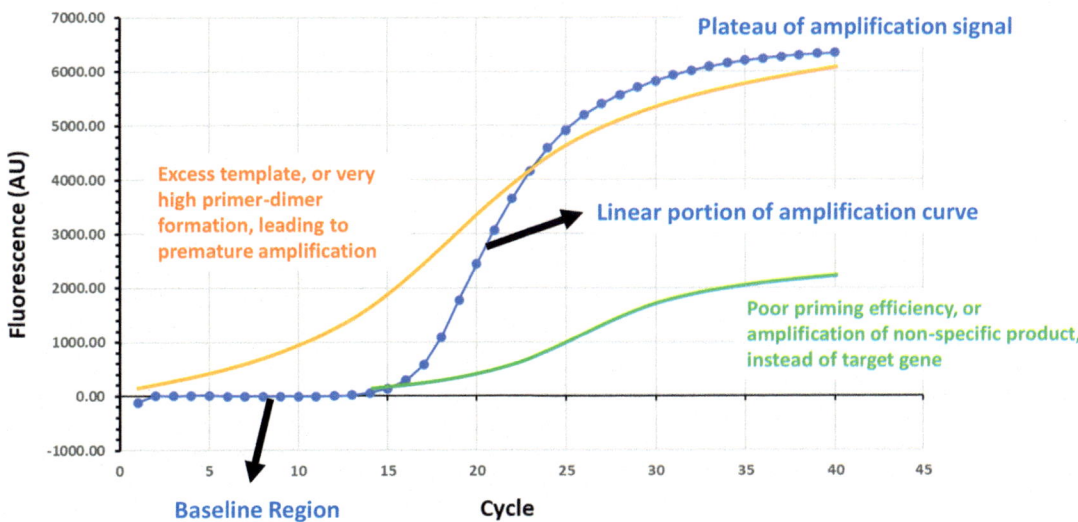

Fig. 4 An example of a typical qPCR amplification curve for an assay performing as designed (Blue), compared to examples (orange, green) of amplification curves associated with commonly encountered technical difficulties

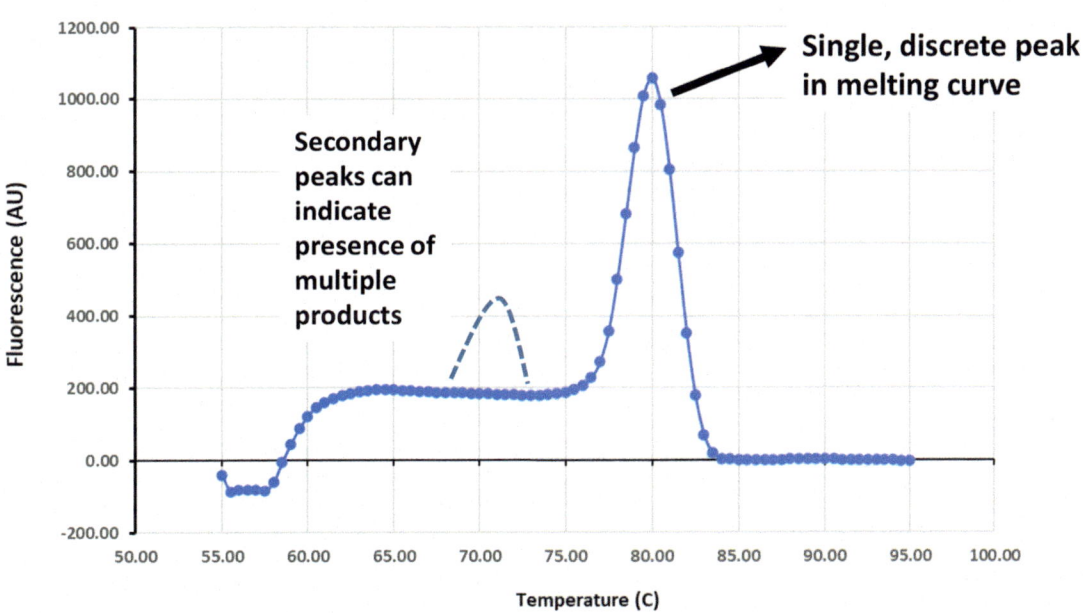

Fig. 5 An example of a melting curve of qPCR products. A single discrete peak indicates presence of a single product, whereas multiple peaks may indicate presence of primer dimers or other artifacts

3. Using a 6-point serial dilution, with two-fold dilutions, prepare qPCR reactions beginning with initial input of 100 ng universal cDNA template, with the final cDNA input of 3.125 ng (*see* **Note 12**). *See* Subheading 3.5 for preparation of qPCR reactions.

4. Perform qPCR, using the steps described in Subheading 3.5.

5. Determine the efficiency of the qPCR assay by plotting log of cDNA concentration against the cycle quantification (Cq) values for each point in the standard curve (*see* **Note 13**). An efficiency of 100% and a gradient (slope) of −3.323 indicates perfect doubling of product in each cycle; primer efficiency scores between 90% and 110% are typically accepted. The efficiency is represented by the following equation: Efficiency (%) $= (10^{-1/\text{slope}} - 1) \times 100$. An example dilution curve for serial dilutions is shown in Fig. 3; for clarity, curves are shown for eight-fold dilutions, which will be separated by ~3 cycles, for a primer pair with ~100% efficiency (*see* **Note 14**).

6. Load the qPCR products from the wells containing 100 ng of template and the PCR control with DNA loading buffer diluted to 1X, in a 2% agarose gel (with SYBR Safe DNA Gel Stain, at 10,000 X stock concentration, diluted in agarose gel to 1X), along with a DNA ladder in a separate lane.

7. Perform gel electrophoresis at 100 V for 45 min. Expected results (visualized in an appropriate imager for DNA gels) are a single band of the predicted size in the cDNA template samples and no background in the PCR control samples.

3.5 Independent qPCR

1. Prepare 10–50 μL of qPCR reactions in 96-well plates. For every 10 μL of reaction mix, the reaction contains 5 μL of 2X SYBR Green Mastermix, 1 μL of diluted mixture of Forward and Reverse Primers, and 4 μL of diluted cDNA template.

2. Perform the two-step qPCR reaction protocol in the thermocycler according to the following steps: Initial denaturing at 95 °C for 3 min, denaturation at 95 °C for 10 s, and annealing and extension at 60 °C for 30 s.

3. Repeat the former steps for 40 cycles (*see* **Notes 14–20**).

4. Perform the melting curve analysis over 55–95 °C. Increase temperature in 0.5 °C increments (*see* **Note 21**). Specific instructions for holding/reading at each temperature increment will vary among machines.

3.6 qPCR-Based TaqMan Array

Figure 6 shows the primary result of interest from such an array; the ΔCt values vs. housekeeping gene, for the highest expressed GPCRs in a sample, thus quantifying their magnitudes of expression.

1. Prepare a reaction mixture by combining 500 μL TaqMan mastermix, 1 μg of cDNA template, and nuclease-free water, to a final volume of 1 mL.

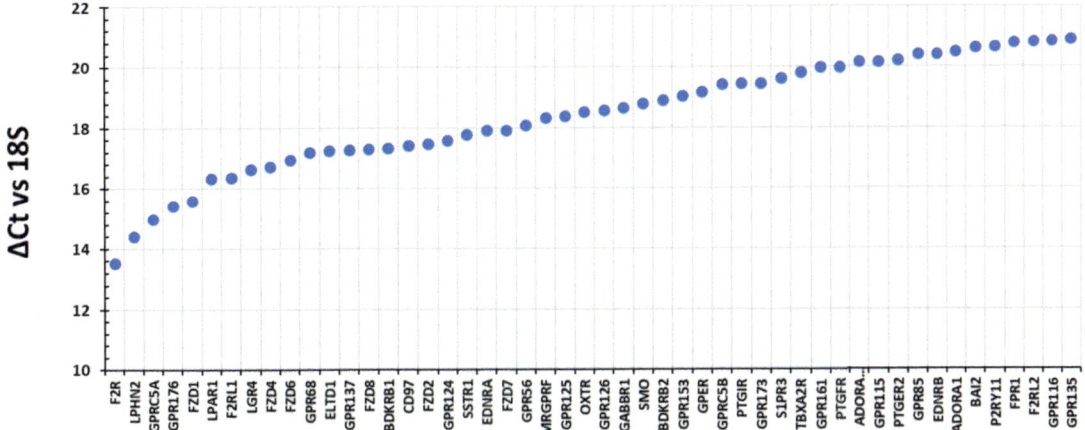

Fig. 6 Sample output data from a TaqMan GPCR array. The key results are the different in cycle threshold (ΔCt) values compared to 18S rRNA, for all detected GPCRs. The data shown are for the 50 most highly expressed GPCRs in a pancreatic cancer-associated fibroblast sample, from data presented in [4]. For identification of GPCRs expressed in a sample, this is the primary end point of a TaqMan GPCR array experiment

2. Using ports for pipetting reaction mixture into microchannels on the TaqMan array, evenly pipette the reaction mixture into each of the five microchannels.

3. Use the following qPCR reaction protocol: incubation at 50 °C for 2 min; incubation at 95 °C for 10 min; denaturation at 95 °C for 15 s; annealing and extension at 60 °C for 60 s.

4. Repeat the protocol steps for 40 cycles (*see* **Note 22**).

3.7 RNA-Sequencing Data Analysis for GPCR Expression (Bioinformatics Protocol)

Following preparation of RNA samples and validation via qPCR, preparation of libraries for RNA-sequencing and the subsequent sequencing of these libraries is typically undertaken externally, ideally by core facilities available at most research universities or by commercial vendors. Use of such resources is typically done based on the specialized nature of these protocols and the equipment used, in particular the high costs of sequencers. Specific details of protocols for these steps for RNA sequencing are beyond the scope of this discussion (*see* **Note 23**). Where applicable, we provide links to additional resources. Many of these tools and alternatives can also be accessed via Galaxy, at https://usegalaxy.org/.

Once sequencing is completed, users will typically be provided with a download link, to obtain the raw data files from a server. These files will be in FASTQ/FASTQ.GZ format. Figure 7 provides a schematic describing the steps needed to proceed from these raw RNA-seq data, to analyzed data that provide information regarding magnitudes of GPCR expression and changes in expression between groups, where applicable. We describe these steps in brief; users may find more details in the corresponding notes and in

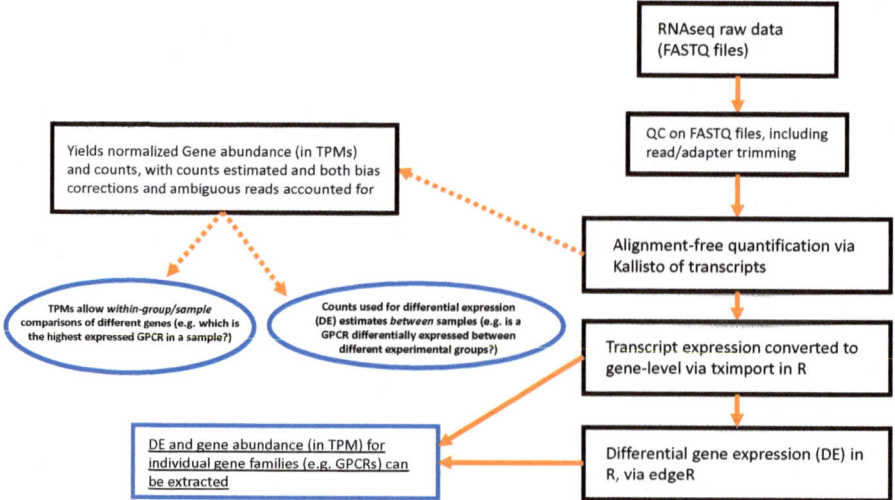

Fig. 7 A workflow for RNA-seq data analysis

links and resources provided below. The steps below will work in any operating system; all tools are open source.

1. Inspect the files for quality using the FASTQC tool (Babraham Bioinformatics). Specifically, one should verify that the quality of sequencing is adequate (i.e., that one has high confidence that each base in a sequenced read was identified correctly). Quality scores >30 are typically considered adequate (see Fig. 8). If the quality scores are poor, trimming for low quality bases may be needed in subsequent steps. In general, with modern sequencing technology, quality scores are usually satisfactory; we have rarely encountered the need for quality trimming. In some cases, detection of bases at the 5′ or 3′ end can be of poor quality. If this is the case, trimming of the final few or first few bases may be performed. In addition, FASTQC also provides useful information about whether certain sequences are overrepresented in the data, which may indicate contamination with sequencing adapters or from other sources, such as ribosomal RNA. For additional information on FASTQC output and "good" vs. "bad" FASTQC results, consult https://www.bioinformatics.babraham.ac.uk/projects/fastqc/.

2. Perform trimming of reads for quality, or adapter trimming as necessary, using BBDUK. The tool may be downloaded at https://sourceforge.net/projects/bbmap/, containing executable files and a database of standard sequencing adapters, which can be used for adapter trimming. The tool can be run from the command line, on raw FASTQ files, using commands

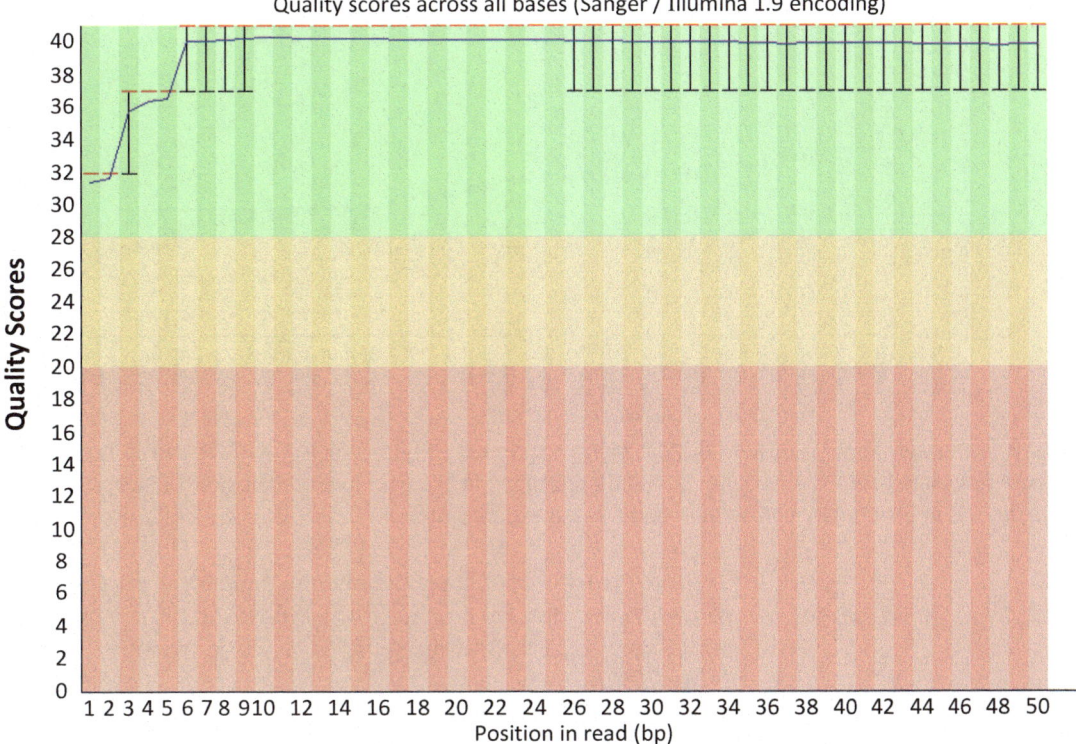

Fig. 8 Example Output, graphing quality scores across the length of reads in the experiment, from the FASTQC software, for RNA-seq raw FASTQ data files. Blue line indicates averages, with ranges highlighted in black

described at https://jgi.doe.gov/data-and-tools/bbtools/bb-tools-user-guide/bbduk-guide/ (*see* **Note 24**). This will yield new, trimmed FASTQ files for downstream use. Adapter trimming is especially relevant for paired-end sequencing runs.

3. Using the trimmed FASTQ files, analyze the files via Kallisto [7]. The steps are as follows:

   (a) Download the reference transcriptome for the relevant species from ensembl, via ensembl.org/info/data/ftp/index.html. This will produce a "Fasta" file (see **Note 25**).

   (b) Download Kallisto from https://pachterlab.github.io/kallisto/download to desired folder.

   (c) Run Kallisto (calling the Kallisto executable file) on the reference transcriptome Fasta file downloaded above. This generates an index file for the next step (see **Note 26**).

   (d) Run Kallisto to quantify transcript expression, using the constructed index and trimmed fastq files as input. This will yield an abundance.tsv file, with quantification at the transcript-level of expression in TPM (transcripts per million) and estimated counts.

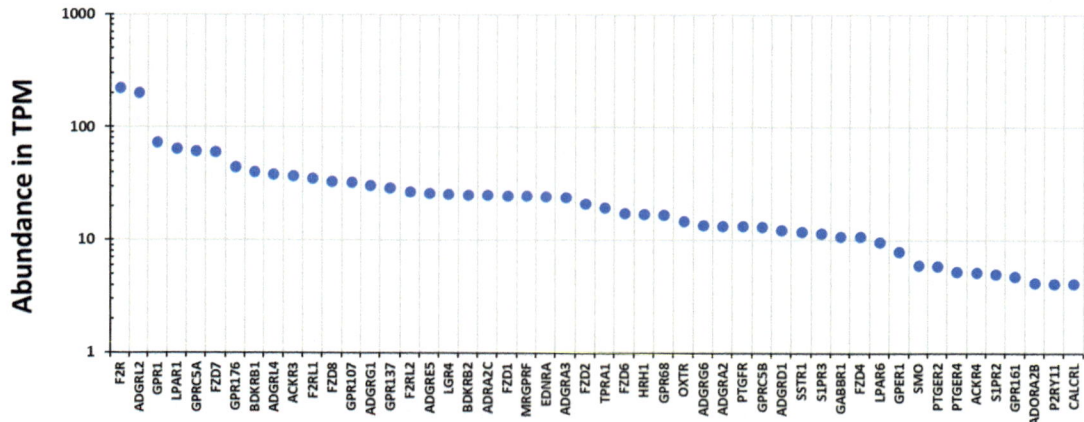

Fig. 9 Expression in TPM for the 50 most highly expressed GPCRs detected via RNA-seq in a pancreatic cancer-associated fibroblast sample, from data presented in [4]. For identification of GPCRs expressed in a sample, this is the primary end point of an RNA-seq experiment

4. To convert transcript-level data to the gene level, run tximport [8], with the abundance file from Kallisto and a mapping file containing corresponding gene IDs for transcript IDs (see **Note 27**).

5. The resulting file from tximport is ready to query for GPCR expression. A list of gene IDs corresponding to human GPCRs can be found at https://insellab.github.io/data. Expression in TPM for these genes can be queried from the file generated by tximport using a variety of tools as convenient, including R, excel, etc. An example of GPCR expression in TPM from the file produced by tximport is shown in Fig. 9, for the 50 highest expressed GPCRs in a cancer-associated fibroblast sample, analyzed using this approach. Sample commands and syntax for running tximport (including how to generate a mapping file of transcript IDs to gene IDs) on output from Kallisto can be found in **Note 28**.

6. The gene-level data from tximport also contain estimated counts for each gene. A matrix of these counts for all samples in an experiment can be supplied to the edgeR package [9] within R, to perform differential expression (DE) analysis. Do not use standard statistical methods such as T-tests for RNA-seq data. The result from edgeR is the fold-change between test conditions and the corresponding statistical significance (expressed as a false discovery rate, FDR) for all genes. FDR < 0.05 is typically used to infer statistical significance. Results for GPCRs can be queried from the DE analysis. The edgeR Bioconductor page can be accessed by users for detailed tutorials, https://bioconductor.org/packages/release/bioc/html/edgeR.html; the details of how to perform such analysis

are beyond the scope of this text. For plotting GPCR expression between different samples, one can use edgeR to generate gene expression in CPM (Counts per Million).

4   Notes

1. Custom oligos were purchased from IDT. Subheading 3.3 describes the design of oligo sequences.

2. We recommend use of the Qiagen RNeasy Min Kit. This method, using spin columns with on-column DNase digestion, is widely used for producing RNA as input for omics methods. Alternative kits/isolation methods can also be used, for instance methods using trizol (or equivalents), in particular, in cell types with high levels of RNases, such as immune cell types [11]. As an example, we tested the "Directzol" kit from Zymo Research and found similar results to samples prepared with the Qiagen kits, as assessed by qPCR. This method can also be adapted for use with RNA-later stabilized tissue. Please refer to notes below for additional details.

3. For most adherent mammalian cell types, a single well of a 6-well plate, grown to confluency is typically adequate for the assays described here. Independent qPCR analysis of GPCR mRNA expression requires ~200 ng of total RNA, TaqMan arrays and other qPCR-based arrays typically require ~1000 ng of total RNA; RNA-seq can typically be performed using 200 ng of total RNA.

4. Tissue samples stabilized in RNAlater can be homogenized using tools such as a tissue tearor homogenizer, as per the manufacturer's instructions. Lysates should be resuspended in RLT buffer, with either 10 µL β-mercaptoethanol or 20 µL dithiothreitol per mL of RLT lysis buffer. While keeping samples on ice between steps, the lysates should then be passed through a QIAshredder per the manufacturer's instructions, to obtain a homogenate from which RNA can be isolated. Lysates resuspended in RLT buffer, stored at −80 °C are stable for many months or even years in samples with low RNase content.

5. It is critical to minimize RNA degradation due to presence of RNases in samples, especially in some RNase-rich cell types. As a first step, clean down all surfaces and pipettes with RNase-away, as well as doing the same for the tabletop centrifuge, gloves, etc. As far as possible, work in a PCR hood or a dedicated bench intended as a clean area for working with RNA.

6. The protocol for RNA isolation via RNeasy spin columns should be performed at room temperature, i.e., ~20–25 °C. Priority is on working quickly, to minimize opportunities for sample degradation and/or contamination. Following completion of the isolation procedure, RNA samples should be stored at −80 °C. RNase-free samples stored at these low temperatures are stable for several years.

7. Upon first use, the RPE buffer needs to be reconstituted with an appropriate volume of 200-proof molecular-biology grade ethanol. Add the indicated volume of ethanol (typically 4 volumes of ethanol to 1 volume of stock RPE buffer concentrate) to the bottle of RPE buffer, mix well and check the box on the lid that ethanol was added.

8. The nanodrop should yield a clear peak at 260 nm wavelength. Ideal 260/280 ratios (indicating protein contamination, a particular hazard for RNA degradation) are 2.0–2.1, for samples generated via the Qiagen kit. 260/230 ratios, indicating presence of salt content and/or alcohol, which can inhibit downstream PCR reactions, should typically be >1.8. However, for low RNA concentrations (especially <50 ng/uL), the 260/230 ratio is rarely reliable and will often be <1.0.

9. To produce larger amounts of cDNA for experiments needing larger amounts of cDNA template downstream, this reaction can be scaled-up to 50 µL.

10. With primers designed by the steps in Subheading 3.3, concentrations of primers between ~0.1 and 1 µM are suitable to produce amplification while avoiding excess primer-dimer formation. Lower primer concentrations can reduce the likelihood of primer-dimer formation. A stock of primers, containing a 1:1 ratio of forward and reverse primers, can be prepared at ~10 µM each, and diluted as needed, to yield the final concentrations noted above. Primer stocks and working solutions stored at −20 °C are usable for many years, as long as contamination is avoided.

11. For qPCR reactions downstream, the cDNA product should usually be diluted at least tenfold, as some contents of the reaction (from the iScript cDNA synthesis kit) can inhibit qPCR reactions. cDNA concentration should not be measured using a Nanodrop; these values will not be accurate due to reaction contaminants within the reaction mix. The amount of cDNA template should be estimated on the basis of the amount of input RNA in the cDNA synthesis reaction. cDNA samples can be stored at −20 °C and usable for qPCR for many years. As an alternative to the iScript kit, we have obtained comparable results with the qScript kit and the SuperScript 3 kit.

12. For common species (e.g., human, mouse or rat), universal RNA is commercially available. If it is not available for the species of interest, use lysates from a mixture of different tissues, to obtain a representative pool of RNA expressed in that species.

13. Controls in a qPCR experiment, in particular negative controls, ensure that spurious signals are minimized. This is of particular importance when studying expression of low-abundance targets such as GPCRs. No-template Controls (NTCs) are qPCR reactions with all reagents added, except for the cDNA template. This allows estimation of primer-dimer formation. This control should be used especially when initially validating primers. Once a lack of primer-dimer formation has been verified for a given mixture of forward and reverse primers, this control need not be included in every subsequent qPCR assay. In addition, a Reverse Transcription minus control [RT(−)] contains all components of the cDNA synthesis step, except for the presence of the reverse transcriptase. This RT(−) template is then assayed via qPCR as normal; amplification indicates the presence of genomic DNA contamination, or potentially unforeseen primer-dimer formation between primers for cDNA synthesis and qPCR. This RT(−) control should be included with every qPCR assay.

14. For primers with ~100% efficiency, every two-fold dilution should yield a delay of amplification to a set threshold, of 1 cycle. Correspondingly, for every eight-fold dilution, curves should be separated by 3 cycles. An example is shown in Fig. 3. The cycle threshold should be set approximately in the middle of the linear portion of the amplification curve, where all amplification curves correspond to phases of the qPCR reaction occurring with high-efficiency amplification.

15. The designed primers may be prepared as a mix of forward and reverse primers, at 1–10 μM each, diluted in nuclease-free molecular biology grade water. This will be further diluted at a 1:10 ratio, giving a final primer concentration of 100 nM–1 μM in the qPCR reaction. Lower concentrations are generally preferable, in order to minimize primer-dimer formation.

16. cDNA template from the earlier cDNA synthesis step (Subheading 3.2) should be diluted, in order to obtain a typical amount (1–10 ng) of input cDNA in each qPCR reaction. Input cDNA in these quantities typically yields amplification of both housekeeping genes and GPCRs within acceptable numbers of qPCR cycles.

17. The 2-step qPCR protocol works best with modern, high efficiency thermocyclers. In older machines, it is often preferable to add an extra annealing step (after the incubation at

~60 °C), often at ~68–72 °C, for 30-60 sec. Similarly, in situations involving large amplicons, e.g., those >200 bp in length, this extra annealing step, yielding a 3-step qPCR protocol, may be preferable.

18. Figure 4 shows a typical amplification curve for a qPCR assay (Blue) and examples of amplification curves (in orange and green) associated with common technical difficulties. An ideal amplification curve should have three distinct phases: an S-curve formed by a baseline region with little signal, a rapidly increasing linear portion and a plateau region as amplification ends and the signal stabilizes.

19. In order to normalize qPCR data, one uses a housekeeping gene. Common choices include ACTB (β-actin), GAPDH (Glyceraldehyde 3-phosphate dehydrogenase) and 18S ribosomal RNA. These housekeeping genes are present on qPCR-based GPCR arrays, including TaqMan arrays. We have found that 18S rRNA is effective for normalizing qPCR data, in particular, if comparing estimates of GPCR expression from qPCR and RNA-seq [4]. Expression of a GPCR relative to 18S rRNA may be computed as their difference in cycle threshold (i.e., ΔCt). The relative expression of two GPCRs A and B within the same sample is then: Expression of GPCR A/Expression of GPCR $B = 2^{\wedge}(\Delta Ct$ for $B)/2^{\wedge}(\Delta Ct$ for $A) = 2^{\wedge}(\Delta Ct$ for $B - \Delta Ct$ for $A)$. Similarly, differences in expression of a GPCR between two samples $A1$ and $A2$ are calculated as: Expression in $A1$/Expression in $A2 = 2^{\wedge}(\Delta Ct$ for $A2 - \Delta Ct$ for $A1)$, i.e., the "$\Delta\Delta Ct$" method. This method may be used to analyze data from independent qPCR and qPCR-based arrays. Besides this semiquantitative approach using housekeeping genes, spiked-in control genes can be used to generate quantitative expression data.

20. In general, we consider GPCRs to be detected if they show linear amplification curves within 25 cycles of that for 18S [4]. Below this limit, one encounters amplification close to the 40-cycle limit. Data at such low mRNA abundance can be highly inconsistent among technical replicates, yielding non-reproducible, non-rigorous data at these extremely delayed amplification levels.

21. An example of a melting curve is shown in Fig. 5, showing a typical melting curve for a primer behaving as designed, yielding a single peak, implying the presence of a single PCR product. The presence of multiple peaks indicates multiple products, either as nonspecific binding or formation of primer-dimers. Such melting curves can also be observed in certain rare cases where multiple variants are amplified, including with inclusion of small introns or exon skipping of small

exons. If an anomalous melting curve is observed, the PCR product should be assessed by gel electrophoresis. Formation of primer-dimers should yield bands of lower molecular weight products than the amplicon of interest. A further application of melting curves is in identification of genomic DNA contamination, via the presence of a peak in RT($-$) control qPCR reactions.

22. Similar qPCR-based microarrays are available from other vendors, using SYBR green chemistry. Examples include arrays for "PrimePCR Pathways" from Bio-Rad laboratories and "RT^2 Profiler PCR Arrays" from Qiagen.

23. The general parameters of a sequencing run to generate RNA-seq data sufficient for obtaining data on GPCR expression are as follows:

    (a) Library type: typically, stranded mRNA. These libraries are most frequently prepared via Illumina Truseq stranded mRNA library kits and protocols.

    (b) Sequencing depth: 25–30 million reads, with 50–75 base single reads is generally sufficient, to identify which GPCRs are expressed in a given sample.

24. For detailed instructions, refer to the guide at https://jgi.doe. gov/data-and-tools/bbtools/bb-tools-user-guide/bbduk-guide/. For adapter trimming, a list of adapters in the file "*adapters.fa*" is downloadable with the BBDUK package and contains many standard adapter sequences. Enter the command—.

    (a) *bbduk.sh in=raw_data.fastq.gz out=clean_data.fastq. gz ref=adapters.fa ktrim=r k=23 mink=11 hdist=1 tpe tbo.*

    (b) Rename files in the example above, as necessary. Once complete, rerun FASTQC to ensure that adapter-trimming was successful. Refer the BBDUK guide above for more settings and options for the command above.

    (c) Similar commands can also be used for quality trimming or quality filtering of reads, the details are beyond the scope of this text and can be found in the BBDUK user guide.

25. Reference transcriptomes for Kallisto should be obtained from ensembl. Ensembl annotations are preferable to refseq, as they tend to be more complete; these differences are most noticeable for nonhuman species. Reference transcriptomes can be downloaded from http://uswest.ensembl.org/info/data/ftp/index.html, by downloading the corresponding FASTA file of cDNAs for the species of interest. Alternately, reference

transcriptomes and pre-built indexes for Kallisto for several species can be downloaded from https://github.com/pachterlab/kallisto-transcriptome-indices/releases

26. Once the reference transcriptome has been obtained as a FASTA file, one runs Kallisto by entering in the command line—.

    (a) *kallisto index -i name_of_index_file.idx name_of_fasta_file.gz,*

    (b) This will create an index file, with name as specified, with extension ".idx".

    (c) For additional options for running this command, refer to the Kallisto manual, at https://pachterlab.github.io/kallisto/manual.

27. Once an index file has been generated, this can be used along with the trimmed FASTQ files as input for quantification of transcript expression.

    (a) For paired-end data, enter into the command line—.

    (b) *kallisto quant -i name_of_index_file.idx   -o output_folder_name pairA_1.fastq pairA_2.fastq,*

    (c) For single read data, enter:
    kallisto quant -i name_of_index_file.idx -o output_folder_name --single -l 200 -s 20 file 1.fastq.gz file 2.fastq.gz file 3.fastq.gz.

    where file 1, file 2, and file 3 are the FASTQ files corresponding to a single sample. Depending on how the sequencing was performed, each sample may have only a single file, or be divided among multiple FASTQ files. Since Kallisto processes one sample at a time, one does not supply it with FASTQ files corresponding to multiple samples in a single command. The "l" and "s" parameters correspond to the average and standard deviation of fragment lengths in the sequencing run. The 200 and 20 values indicated above for each correspond to standard values for Illumina libraries; values for a specific experiment can be obtained from bioanalyzer data. Communication with a sequencing center will likely be needed to obtain this information.

    (d) For additional options for running these commands, refer to the Kallisto manual, at https://pachterlab.github.io/kallisto/manual. In particular, the *--bias* and *--pseudobam* options may be relevant to specific projects.

    (e) The output from Kallisto is stored in the output folder as named by the user, in three files. The *abundance.tsv* file is most relevant to the present exercise; it contains abundance quantifications in TPM for all annotated transcripts, effective length and estimated counts. The other two files are described in the Kallisto manual and not discussed here.

28. tximport is a package in R, which uses as input, quantification data from Kallisto (the abundance.tsv file) and a file containing a map of transcript IDs to gene IDs. To install tximport in R, enter in the R or R Studio command line—,

(a) *if (!requireNamespace("BiocManager", quietly = TRUE)),*

(b) *install.packages("BiocManager"),*

(c) *BiocManager::install("tximport").*

(d) To obtain a map of transcript IDs to gene IDs, via BioMart [10], first install BioMart—.

(e) *if (!requireNamespace("BiocManager", quietly = TRUE))*

(f) *install.packages("BiocManager").*

(g) *BiocManager::install("biomaRt").*

(h) And then, to generate the mapping file for transcript IDs to gene IDs—

(i) *library(biomaRt).*

(j) *ensembl = useEnsembl(biomart="ensembl", dataset="hsa-piens_gene_ensembl", version=96).*

(k) *t2g <- biomaRt::getBM(attributes = c("ensembl_transcript_id", "ensembl_gene_id", "external_gene_name"), mart = ensembl).*

(l) *t2g <-- dplyr::rename(t2g, target_id = ensembl_transcript_id, ens_gene = ensembl_gene_id, ext_gene = external_gene_name).*

(m) *save(t2g, file = "t2g.RData").*

This provides a file named "t2g" with mapping of transcript IDs to both ensembl gene IDs and gene symbols, such as those used by HUGO. The species name and ensembl version in the commands above can be modified as needed. If so, one uses the external gene symbols for gene level data. Hence, one removes the ensembl IDs and then runs tximport. The commands in R are as follows:

(a) *library(tximport).*

(b) *t2g$ens_gene <- NULL.*

(c) *txi = tximport("abundance.tsv", type = c("kallisto"), countsFromAbundance = c("scaledTPM"), tx2gene =t2g).*

(d) *write.table(txi, file = "gene_abundance ", sep = "\t", col. names= NA).*

This creates a file, named "gene abundance," containing the data from abundance.tsv for transcript level to gene level, i.e., for each gene symbol the corresponding magnitude of expression in TPM and estimated counts. These can be used for downstream gene-level analyses, including differential expression (DE) analysis.

Acknowledgements

Support was provided by Academic Senate of the University of California, San Diego and NIH grant RO1A1093957.

References

1. Sriram K, Insel PA (2018) G protein-coupled receptors as targets for approved drugs: how many targets and how many drugs? Mol Pharmacol 93:251–258

2. Alexander SP, Christopoulos A, Davenport AP, Kelly E, Mathie A, Peters JA, Veale EL, Armstrong JF, Faccenda E, Harding SD, Pawson AJ (2019) The concise guide to PHARMACOLOGY 2019/20: G protein-coupled receptors. Br J Pharmacol 176:S21–S141

3. Sriram K, Moyung K, Corriden R, Carter H, Insel PA (2019) GPCRs show widespread differential mRNA expression and frequent mutation and copy number variation in solid tumor. PLoS Biol 17:e3000434

4. Sriram K, Wiley SZ, Moyung K, Gorr MW, Salmerón C, Marucut J, French RP, Lowy AM, Insel PA (2019) Detection and quantification of GPCR mRNA: an assessment and implications of data from high-content methods. ACS Omega 4:17048–17059

5. Insel PA, Sriram K, Gorr MW, Wiley SZ, Michkov A, Salmerón C, Chinn AM (2019) GPCRomics: an approach to discover GPCR drug targets. Trends Pharmacol Sci 40:378–387

6. Sriram K, Insel PA (2020) A hypothesis for pathobiology and treatment of COVID-19: the centrality of ACE1/ACE2 imbalance. Br J Pharmacol. https://doi.org/10.1111/bph.15082

7. Bray NL, Pimentel H, Melsted P, Pachter L (2016) Near-optimal probabilistic RNA-seq quantification. Nat Biotechnol 34:525–527

8. Soneson C, Love MI, Robinson MD (2015) Differential analyses for RNA-seq: transcript-level estimates improve gene-level inferences. F1000Res 4:1521

9. Robinson MD, McCarthy DJ, Smyth GK (2010) edgeR: a bioconductor package for differential expression analysis of digital gene expression data. Bioinformatics 26:139–140

10. Durinck S, Spellman P, Birney E, Huber W (2009) Mapping identifiers for the integration of genomic datasets with the R/bioconductor package biomaRt. Nat Protoc 4:1184–1191

11. Gupta SK, Haigh BJ, Griffin FJ, Wheeler TT (2013) The mammalian secreted RNases: mechanisms of action in host defence. Innate Immun 19:86–97

Chapter 3

Construction of Recombinant Cell Lines for GPCR Expression

Philip J. Reeves

Abstract

Large-scale recombinant expression of G protein-coupled receptors (GPCRs) is required for structure and function studies where there is a need for milligram amounts of protein in pure form. Here we describe a procedure for the construction of human embryonic kidney 293S (HEK293S) stable cell lines for inducible expression of the gene encoding bovine rhodopsin. The HEK293S cell line is particularly suitable for this application because of several favorable properties as a recombinant host including: its ease of transfection, its capacity for handling large amounts of protein cargo, and its ability to perform the necessary co- and post-translational modifications required for correct folding and processing of complex membrane proteins such as GPCRs. The procedures described here will focus on the HEK293S GnTI⁻ cell line, an HEK293S derivative that is widely used for the production of glycoproteins modified homogeneously with truncated N-glycans.

Key words Glycosylation, Tetracycline, Sodium butyrate, Calcium phosphate

1   Introduction

Advanced recombinant gene expression systems have enabled the application of powerful but often insensitive biophysical methods for structure and function studies of GPCRs and other membrane proteins. Recombinant protein expression technologies also enable the construction of site-directed mutants and the addition of fusion tags for purification or facilitation of crystallization [1–4]. Additionally, recombinant expression enables the incorporation of stable-isotope-labeled or unnatural amino acids into proteins [5, 6]. While many soluble proteins can be produced conveniently using recombinant expression hosts such as *Escherichia coli*, mammalian membrane proteins including GPCRs, which have more complex folding requirements, usually require a eukaryotic host for successful expression [4]. Yeast, baculovirus infected insect cells, and mammalian cell lines are some of the systems available

Sofia Aires M. Martins and Duarte Miguel F. Prazeres (eds.), *G Protein-Coupled Receptor Screening Assays: Methods and Protocols*, Methods in Molecular Biology, vol. 2268, https://doi.org/10.1007/978-1-0716-1221-7_3,
© Springer Science+Business Media, LLC, part of Springer Nature 2021

[4, 7]. These are especially attractive when attempts using *E. coli* have failed to produce correctly folded proteins [8]. For high-level continuous gene expression and the capacity for arbitrary scale-up by suspension growth, stable mammalian cell lines are advantageous compared to transient gene expression platforms [9]. The large-scale constitutive production of secreted proteins is relatively straightforward and is often a preferred choice for the production of pharmaceutical proteins such as antibodies [10]. However, we have found this approach unsatisfactory for producing larger amounts of rhodopsin (>2 mg/L) or soluble cytosolic proteins such as rhodopsin kinase, both of which are likely to impart cytotoxicity when produced at very high levels [9]. We overcame this limitation by optimization of an inducible gene expression system assembled from a strong viral promoter and control elements from the *E. coli* transposon *Tn10* tetracycline operon [11, 12]. This system still exploits the powerful cytomegalovirus promoter but it is regulated by internal tandem tetO repressor DNA sequences (Fig. 1). In stable cell lines constructed using the pACMVtetO vector, the expression of the target gene is repressed by TetR repressor protein in stable cell lines previously transfected with pcDNA6/TR (*see* **Note 1**). The addition of tetracycline in combination with sodium butyrate to cultures upon reaching high cell density leads to robust expression of the target gene [11, 13].

Glycoproteins expressed recombinantly using standard HEK293 cells are decorated with complex N-glycans [11, 14]. The method described here is for the construction of cell lines expressing GPCRs containing N-glycans of type $GlcNAc_2Man_5$ [14] (*see* **Note 2**). This is achieved by using HEK293S-GnTI$^-$-TetR cells, a derivative that has no GlcNAc transferase I (GnTI) activity. The advantage of making GPCRs as well as other membrane or secreted mammalian proteins with simple N-glycans is that they run as discrete bands on SDS PAGE and are potentially better candidates for crystallization experiments due to their reduced heterogeneity. Recombinant GPCRs made using this system are highly useful for NMR and FTIR spectroscopy as well as crystallization experiments [15–17].

2   Materials

Prepare all buffers and solutions using ultrapure water and sterilize by filtration (0.2 μM) unless stated otherwise. Store all media and buffers at 4 °C unless indicated. Warm all cell culture media and buffers to 37 °C immediately prior to use.

2.1 Growth and Storage of HEK293S-GnTI$^-$-TetR Cells

1. Two humidified incubators adjusted to 37 °C, one set at 5% CO_2, 95% air; the other set to 1% CO_2, 99% air.

2. Cell culture plasticware.

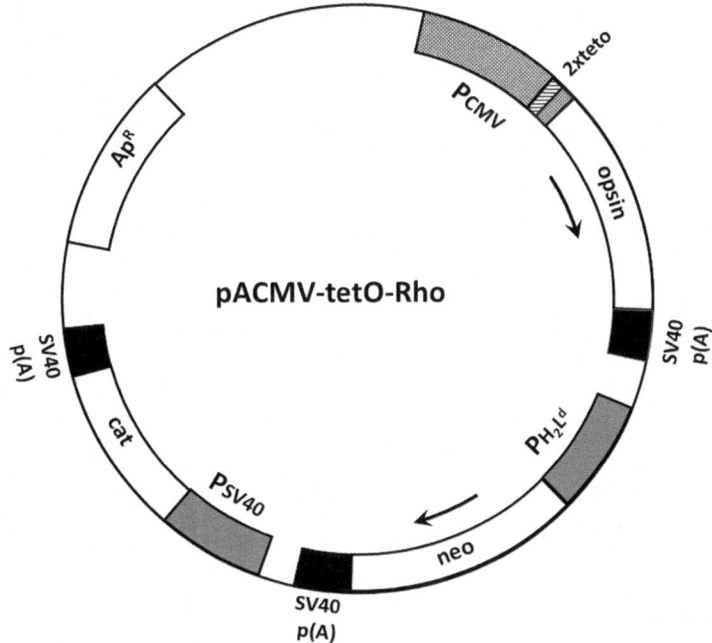

Fig. 1 Plasmid pACMV-tetO-Rho for inducible expression of the bovine rhodopsin gene. This plasmid is delivered to mammalian cells by transfection. Stable genomic integration events are selected using G418 resistance conferred by the gene encoding neomycin (neo). This neo gene is expressed from the weak H^2L_d promoter ensuring that stably transfected cells will only survive G418 selection when the plasmid integrates at genomic positions that are transcriptionally active [19]. The target gene, in this case rhodopsin (opsin), is expressed under the control of a modified human cytomegalovirus promoter that contains two tandem tetO operator sequences. Upon integration of this plasmid into the genome, expression from this CMV-tetO promoter is prevented in cell lines expressing the TetR repressor protein but can be induced by the addition together of tetracycline and sodium butyrate to the growth medium

3. HEK293S-GnTI⁻-TetR (*see* **Note 3**).

4. DMEM is Dulbecco's Modified Eagle's Medium (high glucose).

5. DMEM/F12: DMEM/Ham's Nutrient Mixture (1:1).

6. Complete DMEM and DMEM/F12: DMEM and DMEM/F12 supplemented with 10% heat-treated Fetal Bovine Serum (FBS), 100 I.U./mL penicillin, 100 μg/mL streptomycin, 2 mM L-glutamine.

7. Blasticidin: 500 μg/mL in water.

8. Phosphate-buffered saline (PBS): 137 mM NaCl, 2.7 mM KCl, 1.5 mM KH_2PO_4, 8 mM Na_2HPO_4, pH 7.2.

9. Trypsin: 2.5 mg/mL in PBS.

10. Freezing media: 90% complete medium, 10% cell culture grade DMSO.

11. Cryovials.

12. Styrofoam box for freezing and storing cells.

2.2 Construction of HEK293S GnTI⁻-TetR Cells Inducibly Expressing GPCRs

1. HEK293S-GnTI⁻-TetR cells.

2. Mammalian expression plasmid such as pACMVtetO (Fig. 1) containing the GPCR gene and a selectable marker (other than blasticidin resistance). High-purity plasmid DNA is prepared using a commercial kit. Stored at $-20\ ^\circ$C.

3. BES buffer 2X: 50 mM BES, 250 mM NaCl, 1.5 mM Na_2HPO_4, pH 7.02 adjusted with NaOH. Store at $-20\ ^\circ$C.

4. 2.5 M $CaCl_2$. Store at $-20\ ^\circ$C.

5. Conditioned complete DMEM/F12 medium.

6. Geneticin (G418): 2 mg/mL in water. Store at $-20\ ^\circ$C.

7. Tetracycline: 2 mg/mL (w/v) in water. Store at $-20\ ^\circ$C, foil wrapped.

8. 1 M Sodium butyrate. Store at $-20\ ^\circ$C.

9. Brass ferrules (¼ in. front) or cloning rings sterilized in a container also containing a slug of silicone grease.

10. Bent tip forceps immersed tip down in a 200 mL beaker containing 50 mL ethanol.

2.3 Identification of Cell Lines Expressing Rhodopsin

1. Dodecylmaltoside (DDM): 10% w/v in water, unfiltered. Stored at $-20\ ^\circ$C.

2. Dot blot apparatus.

3. 96-Well ELISA plate and 12-channel multichannel pipette (5–200 µL volume).

4. Nitrocellulose membrane (cut to size of dot blot membrane support).

5. Rho-1D4 anti-rhodopsin antibody solution: 100 mM K-PO4 buffer pH 7.4, 0.5% (w/v) gelatine, 0.25% (w/v) BSA, 0.05% (w/v) sodium azide, 10 µg/ml of Rho-1D4 antibody.

6. Blocking solution: PBS, 10% (w/v) semi-skimmed powdered milk, 0.1% (v/v) Triton X100, 0.01% (w/v) sodium azide.

7. Membrane washing solution: PBS, 0.1% (v/v) Triton-X100.

8. Secondary antibody solution: membrane washing solution, 20 µg/mL hydrogen peroxidase conjugated goat anti-mouse antibody.

9. ECL reagents or kit.

2.4 Determination of Rhodopsin Expression Levels in HEK293S GnTI⁻-TetR-Cell Lines Stably Transfected with the Bovine Rhodopsin Gene

1. 11-*cis* retinal, 10 mM in ethanol. Stored at −70 °C.

2. Split beam UV–vis spectrometer (Perkin Elmer λ 35) equipped with a water-jacketed cuvette holder.

3. Pair of matched Suprasil™ Quartz semi-microcuvettes, 1 cm path length, with self-masking walls.

4. 250 W light-source with fiber optic light guide.

5. Long-pass filter (>495-nm) to fit 250 W light source.

3 Methods

Carry out all cell culture methods in a designated cell culture room equipped with laminar flow cabinets. For routine growth of HEK293S-GnTI⁻-TetR cells use a humidified incubator set at 5% CO_2, 95% air at 37 °C.

3.1 Growth and Long-Term Storage of HEK293S-GnTI⁻-tetR Cells

1. Propagate HEK293S GnTI⁻-tetR cells as adherent monolayers on 10-cm TC dishes in complete DMEM/F12 containing blasticidin (5 µg/mL) (*see* **Note 4**).

2. Once coverage of the dish with cells reaches 90–100% confluence, remove the spent culture medium by aspiration and wash the cell monolayer gently with 10 mL of PBS.

3. Remove the PBS by aspiration and add 1 mL of trypsin solution to the monolayer.

4. Place the dish in the incubator set at 37 °C until sheets of cells begin to lift off from the surface of the dish (1–2 min). Rap the edge of the dish against a firm surface several times to dislodge all the cell sheets.

5. Add 9 mL of complete DMEM/F12 to the dish and disaggregate the cells by pipetting up and down gently using a 10-mL pipette controlled with an electronic pipette aid.

6. Dilute the cell suspension (typically 1:6–1:8) in 10 mL of fresh medium containing blasticidin solution and return TC dishes to the incubator. Incubate until the surface of the dish is almost covered (about 90–95% confluence) and repeat this entire process.

7. Bring up fresh stocks of cells from liquid nitrogen storage every 2 months (*see* **Note 5**).

8. To prepare freeze-down cells, dislodge cells from the 10-cm TC dish surface and resuspend them in 10 mL of fresh growth medium (without blasticidin) exactly as described above for standard propagation.

9. Centrifuge the 15-mL tubes for 2 min at 161 × g in a clinical centrifuge.

10. Carefully resuspend the cells in 10 mL of freezing medium.

11. Transfer 1 mL aliquots to labeled and dated cryovials and store in a Styrofoam box at -80 °C for a minimum of 24 h (*see* **Note 6**). Transfer the cryovials to a liquid nitrogen container for long-term storage.

12. Remove cryovials containing cells from liquid nitrogen (using adequate safety precautions) and keep on dry ice until needed.

13. Place cryovials on a float in a water bath set at 37 °C until the cell suspension is completely defrosted.

14. Surface-sterilize the cryovials using ethanol or equivalent and transfer them to the laminar flow cabinet. Use a 1-mL serological pipet to transfer the cells to a 15-mL tube and add 9-mL of complete DMEM-F12 slowly over 1 min with mixing.

15. Pellet the cells by centrifugation in a clinical centrifuge and remove the medium by aspiration.

16. Resuspend the cell pellet gently in 10 mL of complete DMEM/F12 using a 10-mL pipette and transfer the cell suspension to a 10-cm TC dish.

17. Incubate the cells as described above for 24 h. Once cells are attached, feed them with medium containing selective antibiotic.

3.2 Transfection of HEK293S-GnTI⁻-TetR Cells Using Calcium Phosphate/N,N[Bis(2-Hydroxyethyl)-2-Aminoethanesulphonic Acid] Precipitation

The procedures described here can be readily adapted for transient transfection of HEK293S cells (*see* **Note 7**). The expression plasmid is prepared beforehand at medium scale (500–1000 μg/mL) using a commercial kit. Transfections are performed when cells are growing exponentially in medium containing the selective antibiotic needed to maintain the TetR gene (blasticidin). This is typically a week after being brought up from liquid nitrogen storage. For consistent transfections the following routine is followed.

1. The Friday before the week of transfection, split a near confluent 10-cm dish of cells 1:4 in complete DMEM/F12 containing blasticidin (5 μg/mL).

2. Incubate dishes over the weekend so they reach about 90% confluence.

3. The day before transfection (Monday), remove the HEK293S-GnTI⁻-TetR cells using 1 mL of trypsin as described previously for propagation.

4. Once cell sheets have detached fully add 9 mL of complete DMEM (not complete DMEM/F12) without blasticidin.

5. Disperse the cells by pipetting up and down ten times using an electronic pipette-aid equipped with a 10-mL pipette. Transfer the cell suspension to a 50-ml tube containing 26 ml of complete DMEM (*see* **Note 8**).

6. Mix the cell suspension and distribute 9 mL to four 10-cm TC dishes. Incubate the cells for 14–16 h, after which time 90% of the cells should have adhered to the surface of the dish and covering about 30–40% of the dish surface.

7. On transfection day 1 (Tuesday), prepare the plasmid DNA/-CaCl$_2$/N,N[bis(2-hydroxyethyl)-2-aminoethanesulphonic acid] (BES) transfection cocktail immediately prior to transfection by transferring a volume of plasmid DNA equivalent to 30 μg to a 15-mL tube and add sterile water to give a final volume of 450 μL.

8. Mix this solution for 30 s by vortex. Add 50 μL of calcium chloride to the diluted plasmid DNA and mix for a further 30 s by vortex.

9. Add 500 μL of 2X BES buffer dropwise to the plasmid DNA/-calcium chloride solution over a time period of 60 s, with continuous mixing by vortexing throughout.

10. Once addition is complete, vortex this transfection cocktail for a further 60 s.

11. Immediately add this transfection cocktail dropwise to the medium covering the cells and gently rock the dish to ensure its even distribution. Transfer the dish to the humidified incubator set at 37 °C and 1% (v/v) CO$_2$/air atmosphere (*see* **Note 9**).

12. About 16 h after addition of the transfection cocktail (transfection day 2), remove the spent medium by aspiration.

13. Rinse the cells thoroughly but carefully with 10 mL of un-supplemented DMEM/F12 (wash medium).

14. Remove this wash medium by aspiration and feed the cells with 10 mL of complete DMEM/F12 before transferring the dish to the humidified incubator set at 5% CO$_2$ and 37 °C for a further 24 h.

15. On transfection day 3 (Thursday), remove the complete DMEM/F12 medium and split the cells using trypsin as described previously in a volume of 10 mL of complete DMEM/F12.

16. Prepare dilutions of the cell suspension in complete DMEM/F12 containing 20% (v/v) conditioned medium as follows: 1:5, 1:10, 1:50, 1:100, and 1:1000 each in a final volume of 20 mL (*see* **Note 10**).

17. Transfer 10 mL of each dilution to two 10-cm dishes and incubate the cells for 24 h with no antibiotic selection.

18. On transfection day 4 (Friday), remove the spent medium and feed the cells with complete DMEM/F12-containing 20%

(v/v) conditioned medium and Geneticin (200 μg/mL) until discrete colonies 2–4 mm in diameter form, typically 2–3 weeks from the start of selection.

19. Replace the selective medium every 5–6 days (*see* **Notes 11** and **12**).

3.3 Isolation and Expansion of G418-Resistant Colonies

1. Identify 8–10 large isolated colonies per TC dish by visual inspection. Mark their position on the underneath of the TC dish by circling them with a marker pen, then transfer the dishes to the laminar flow cabinet (*see* **Note 13**).

2. Remove the spent medium and rinse the TC dish very gently with 10 mL of PBS so as not to disturb the colonies.

3. Use bent tipped forceps, sterilized by immersion in ethanol and then air-dried, to dip the base of a cloning ring into the silicone grease.

4. Place this cloning ring over one of the marker-pen-circled colonies and press the angle of the forceps down on the cloning ring to make a watertight seal with the base of the TC dish.

5. Repeat this procedure for the other colonies using the marker pen circles on the underside of the dish as a guide.

6. Take up 200 μL of trypsin using a P200 pipette and dispense two drops of trypsin to each sealed ring.

7. Place the TC dish in the 37 °C incubator for 1–2 min and then add four drops of complete DMEM/F12 to each cloning ring in the same order as the trypsin was added.

8. Pipette up and down 5 times using a P200 pipette to dislodge cells contained within the cloning ring. Take care to not move the cloning ring.

9. Transfer the suspension to a single chamber of a 24-well TC dish containing 1 mL of complete DMEM/F12 containing G418.

10. Repeat this process for the remaining colonies until the first row of chambers (A1–A6) in the 24-well dish is used. Do not use the remaining chambers, B, C, D (1–6) as they will be used for expansion of the cell lines in row A chambers.

11. Feed the cells every 2–3 days with medium containing G418 until they reach near confluence (*see* **Note 14**).

12. Once near confluent, remove the spent medium, wash with 1-mL of PBS, and add 100 μL of trypsin.

13. Incubate at 37 ° C for 1–2 min or until the cells start to detach.

14. Rap the dishes gently against a firm object and return to the laminar flow cabinet.

15. Add 200 µL of complete DMEM/F12 containing G418, disperse the cells gently using a P1000, and transfer ~100 µL of this cell suspension to each of the three chambers in the same column (e.g., for cell suspension volume ~ 300 µL in chamber A1, transfer to chamber B1, C1, and D1 each containing 500 µL of growth medium with G418).

16. Return the dish to the 37 °C incubator and feed the cells every 2–3 days until they reach near confluence.

17. At this stage make freeze-downs for cells growing in chambers B1 to B6. To do this, remove the spent medium, wash with 1 mL of PBS, and add 100 µL of trypsin.

18. Allow trypsinization to proceed as described earlier but once the cell sheets have started to dislodge stop the reaction with 900 µL of freezing medium (*see* **Note 15**).

19. Disperse the cells using a P1000 and transfer each of the chamber contents to a carefully labeled cryovial.

20. Transfer the cryovials to a Styrofoam container and store at −80 °C for 24 h before transferring to liquid nitrogen.

21. Cells in chambers C and D are used to investigate the inducible expression profiles of the cell lines. To do this add 1 mL of complete DMEM (no G418) to the cells in chamber C.

22. To the cells in chamber D add 1 mL of complete DMEM (no G418) containing sodium butyrate (5 mM) and tetracycline (2 µg/mL).

23. Incubate the cells for 48 h at 37 °C.

24. Harvest the cells in the spent growth media by pipetting up and down using a P1000 pipette.

25. Transfer the cell suspension to a 1.5 mL tube and pellet by centrifugation (14,000 × g, 3 min in a Microfuge).

26. Remove the supernatant and store the cell pellets in the same tube at −20 °C prior to analysis (Subheading 3.3).

3.4 Identification of Cell Lines Exhibiting High-Level Inducible Expression of Rhodopsin by Dot Blot and Immunodetection

1. The dot blot immunodetection procedure described below allows a large number of cell lines to be processed simultaneously in order to identify the best candidates for further analysis. Cell lines expressing rhodopsin are identified by immune detection using the Rho-1D4 monoclonal antibody against the C-terminus of rhodopsin (*see* **Note 16**).

2. Defrost the frozen cell pellets (rows C and D) from the inducible expression experiment on ice and resuspend in 100 µL of PBS containing dodecylmaltoside (DDM; 1% w/v) (*see* **Note 17**).

3. Place samples on a mixing platform (nutator) and allow solubilization to proceed for 30 min at 4 °C.

4. Centrifuge the tubes at 20,000 × g for 30 min at 4 °C and keep the supernatants.

5. Prepare a 96-well dish for a serial dilution procedure by pipetting 75 μL of PBS containing 0.1% (w/v) DDM to rows B, C, and D using a multichannel pipette.

6. Transfer 50 μL of the clarified supernatants to the chambers (1–12) in row A of this 96-well dish.

7. Perform a serial dilution of each lysate by transferring, using a multichannel pipette, 25 μL of cell lysate solution from row A chambers into row B chambers (each containing 75 μL of PBS, 0.1% (w/v) DDM).

8. Mix the contents of row B chambers and transfer 25 μL to the chambers immediately below in row C. Repeat this procedure for row D.

9. Soak a sheet of nitrocellulose filter in PBS containing 0.1% Triton X100 and assemble this securely into a 96-well dot blot apparatus without application of a vacuum.

10. Transfer 25 μL of each serial dilution (high dilution in row D to low dilution in row A) using the multichannel pipette.

11. Apply vacuum to the dot blot apparatus until samples are fully transferred to the nitrocellulose filter and then remove the vacuum.

12. Apply 150 μL of PBS containing 0.1% (v/v) Triton X100 to each chamber of the dot blot apparatus and then reapply the vacuum and allow the wash buffer to drain.

13. Repeat this washing step three times taking care to ensure the filter does not dry out. Disassemble the dot blot apparatus and place the NC filter in 50 mL of blocking solution and incubate with gentle mixing using a rocking platform for 16 h at 4 °C.

14. Rinse the NC filter in PBS briefly and then transfer the membrane to 50 mL Rho-1D4 primary antibody solution.

15. Incubate for 5–6 h at room temperature on a rocking platform.

16. Collect the primary antibody solution and store at 4 °C for future use.

17. Add 50 mL of washing buffer to the container containing the NC filter and place back on the rocking platform for 5 min at room temperature.

18. Repeat this washing step four more times discarding the wash buffer each time.

19. Add 30 mL of secondary antibody solution to the NC filter and place the container back on the rocking platform for 1 h at room temperature.

20. Discard the secondary antibody solution and wash the NC filter five times with wash buffer as described previously.

21. Process the NC filter by enhanced chemiluminescence (ECL). A result from such an experiment is presented in Fig. 2.

3.5 Determination of Rhodopsin Expression Levels in HEK293S GnTI⁻-TetR-Rho Cell Lines

In order to determine quantitatively the expression levels of GPCRs it is often necessary to measure the number of surface receptor sites in the cell line population. For most GPCRs this is most accurately measured with ligand-binding experiments, typically using a radiolabeled antagonist or agonist. For cell lines expressing correctly folded rhodopsin, the rhodopsin pigment is formed upon treating cell line suspensions directly with the rhodopsin inverse-agonist, 11-*cis* retinal ligand. The amount of rhodopsin in the sample can then be accurately determined by using UV–visible absorption spectroscopy scans before and after photobleaching the sample.

1. From the dot blot result identify those cell lines exhibiting high-level expression under inducible conditions but with minimal background expression in the absence of induction.

2. Bring up these cell lines and propagate them initially in 3 mL of complete DMEM/F12 containing G418 (250 μg/ml) using 6-well TC dishes.

3. Expand these cell lines under G418 selection until three confluent 10 cm dishes are obtained.

4. Prepare 2–3 freeze-downs with one of these dishes as described earlier.

5. Feed the other two dishes with 10 ml of complete DMEM (no G418), one with and one without tetracycline and sodium butyrate.

6. Incubate these dishes for 48 h (*see* **Note 18**) and then harvest the cells in their spent medium by pipetting up and down with a pipette-aid equipped with a 10-mL pipette until all the cells are dislodged.

7. Pellet the cells by centrifugation (4000 × g, 5 min), remove the supernatant, and resuspend the pellet in 10 mL of PBS.

8. Recentrifuge as before and resuspend the pellet in 500 μL of PBS and transfer the cell suspension to 1.5 mL tubes. The samples can be frozen and stored at −20 °C at this step and analyzed later.

9. Transfer the cell suspensions to a darkroom illuminated with dim red safelights.

10. Treat each cell suspension with 1 μL of 10 mM 11 *cis*-retinal and incubate on a rocking platform at 4 °C for 3 h.

11. Add 50 μL of DDM (10% w/v in water) to each tube and incubate on a rocking platform at 4 °C for 1 h.

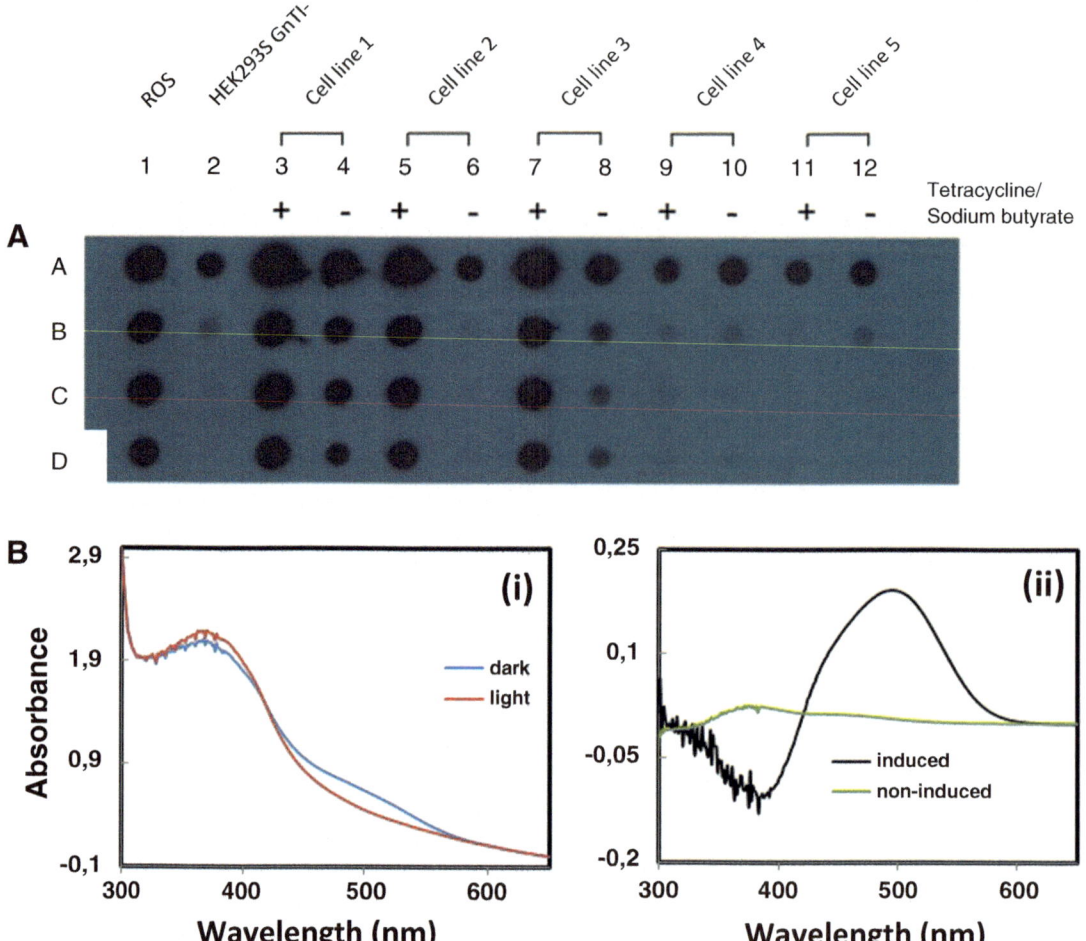

Fig. 2 Dot blot analysis and UV–vis absorption spectroscopy for identification of stable cell lines. Panel (**a**) shows the results from a dot blot experiment to screen HEK293S-GnT⁻-TetR-Rho cell lines for inducible gene expression of rhodopsin. Solubilized cell extracts prepared as described in Subheading 3.3 were applied to the dot blot apparatus for processing. Samples in row A (1–12) were serially diluted 1:4 (A through to D) prior to loading onto the dot blot apparatus. Column 1 is a positive control (~1 μg bovine rhodopsin in well A1, then serially diluted as indicated). Column 2 is a negative control (lysate from non-transfected HEK293S-GnTI⁻-TetR cells). The remaining columns contained cell extracts from candidate cell lines grown in the presence or absence of both tetracycline and sodium butyrate as indicated. Cell line #1 shows inducible expression of rhodopsin (column 3) but measurable rhodopsin is also produced in the absence of inducers (column 4). Cell lines # 2 and #3 displaying inducible expression with reduced background expression when compared to cell line #1. Cell lines #4 and #5 have no measurable inducible expression of rhodopsin. Panel **B(i)** presents UV–visible absorbance spectra from a candidate cell line identified from a dot blot experiment. This cell line was expanded to two 10-cm dishes of cells. Upon reaching confluence one dish was fed with complete DMEM containing tetracycline and sodium butyrate, the other dish with complete DMEM alone as described in Subheading 3.5. UV–vis absorbance spectra of prepared cell lysates were recorded before (blue) and after (red) photobleaching. Shown is the induced sample. The dark–light difference spectrum for this (induced) sample is shown by the black trace in panel **b(ii)**. The peak difference in absorbance at 500 mm (black trace) is used to calculate the rhodopsin concentration. The green trace in panel **b(ii)** presents the difference spectrum of a cell lysate prepared from the same cell line candidate but grown in the absence of tetracycline and sodium butyrate (uninduced)

12. Once solubilized, centrifuge the samples for 30 min at 20,000 × g and 4 °C.

13. Set the spectrometer to record absorbance using a scan speed of 480 nm/min, 2 nm bandwidth, and 2 s response time between 650 and 250 nm.

14. Blank the spectrometer and make a baseline recording using both cuvettes holding PBS containing DDM (1%, w/v) at 20 °C.

15. Record the UV–vis absorption spectrum of a sample and then illuminate the sample, while it still is in the cuvette, for 1 min with light delivered through a fiber optic light guide from a light-source fitted with a > 495-nm long-pass filter. Immediately record the UV–vis absorbance spectrum of the photobleached sample.

16. A typical result for such an experiment is shown in Fig. 2. This cell line produces about 100 μg of rhodopsin per 10-cm dish (~10e^7 cells total). This would correspond to a cell line expressing about 250 micrograms from a 15 cm dish or 10 mg per liter when grown in a 1 l suspension culture using a bioreactor (*see* **Note 19**). For purification of rhodopsin from detergent-solubilized recombinant cell lines extracts the rho-1D4 antibody can be utilized (*see* **Note 20**).

4   Notes

1. HEK293S-GnTI$^-$-TetR and HEK293S-TetR cell lines can be constructed by stable transfection using plasmid pcDNA6/TR followed by selection with the antibiotic blasticidin (5 μg/mL) using the same method described herein. Cell lines pools stably expressing the tetR repressor protein can be collected after stable transfection and used for transfection with pACMVtetO expression plasmids. However, it is preferable to isolate individual blasticidin-resistant clones and screen them using an anti-TetR antibody to identify those expressing high levels of the TetR protein. The dot blot procedure described in Subheading 3.4 is applicable.

2. The methods described here are for construction of tetracycline inducible cell lines using HEK293S-GnTI$^-$-TetR cells. These methods are also applicable for making tetracycline inducible cell lines using HEK293S-TetR cells for situations where the presence of complex N-glycans on the target GPCR is not a concern. However, it is advisable to make the following changes to the protocol. First, HEK293S-TetR do not seem to be transfected as efficiently as the GnTI$^-$ cell line derivative. For this reason, it is necessary to perform lower dilutions

(1:5–1:100) when splitting the cells after transfection. Secondly, HEK293S-TetR have higher levels of resistance to certain antibiotics such as G418. For this reason, it is recommended to use higher doses of antibiotic during selection. For example, for G418 we typically use concentrations of 2 mg/mL during selection. Once cell lines have been established, we reduce the concentration of G418 from 2 mg/mL to 500 μg/mL for their routine propagation.

3. The HEK293S GnTI⁻ cell line is a HEK293S ricin-resistant derivative that has no detectable N-acetylglucosaminyltransferase I activity. These cell lines have been further modified to express the TetR repressor protein by stable transfection with pcDNA6/TR (Invitrogen).

4. HEK293S and its derivative cells will grow as adherent monolayers on TC dishes and flasks. However, this cell line does not stick strongly to cell culture dishes. It is important to wash the cells carefully, for example, when washing with PBS during cell splitting procedures. The HEK293S-GnTI⁻ cell line appears to be more fragile than the HEK293S parental line, so it is important to pipette cell suspensions slowly. All HEK293S cell lines will also grow as suspension cultures under suitable conditions. This property enables their growth in spinner flasks or in bioreactor vessels designed for mammalian cell culture. To encourage growth in suspension culture, it is necessary to alter the composition of the growth medium and to slowly and gently agitate the cells by stirring to keep them in suspension. The important changes to the medium composition are to reduce calcium concentration (to discourage clumping of cells) and to include Pluronic acid F-68 to improve the robustness of cells in suspension culture [9]. Companies that manufacture cell culture media will sometimes prepare customized growth media by request.

5. HEK293S-GnTI⁻-TetR cells can be stored in cryovials at −80 °C for 2–3 months or ideally in liquid nitrogen to ensure long-term viability. To prepare freeze-downs, grow cell monolayers in 10-cm TC dishes to about 70–80% coverage.

6. When making freeze-downs of mammalian cells in freezing medium it is important that cells are frozen slowly until they reach −80 °C. This can be achieved using specialized equipment or by placing the cryovials in an insulated foam container overnight in a −80 °C freezer. When defrosting the cells, it is important to warm them up fast. One way to do this is to secure the cryovials in a foam floatation device and place this carefully in a 37 °C water bath ensuring the cryovials do not fully submerge. It is important to carefully sterilize the surface of the cryovials with 70% ethanol or a suitable disinfectant before transferring them to the laminar flow cabinet.

7. This transfection method [18] can also be used for high-level transient gene expression experiments using HEK293S cells. It is advisable to determine expression levels of the target gene in a transient system before making stable cell lines in order to gauge how well the gene is expressed. Genes that are expressed at high levels after transient transfection are likely to be expressed effectively in stable cell lines. For transient expression it is preferable to use an expression plasmid containing an SV40 origin of replication as well as a strong promoter, for example, pCDNA3 or similar. For high-level transient expression, it is necessary to co-transfect the expression plasmid together with a plasmid expressing the SV40 large T-antigen, such as pRSVTAg (expression plasmid:pRSVTAg ratio 10:1 w/w). Alternatively, use HEK293T cell lines as a host which produce the T-antigen protein.

8. It is important to use complete DMEM instead of complete DMEM/F12 during the treatment with the DNA transfection cocktail because DMEM/F12 contains HEPES buffer, which might interfere with the pH of the growth medium, which could in turn affect the DNA/calcium phosphate precipitate formation.

9. The low CO_2 (1%) concentration during the transfection step promotes the slow formation of a fine sand-like DNA/calcium phosphate precipitate. This fine precipitate should be visible between the cells when viewed by phase contrast microscopy about 30 min after addition of the transfection cocktail. If there is little or no precipitate, this is usually because the cell density at the start of transfection was too high. A heavy fast forming clumpy, often brown, precipitate alongside crystal formation usually indicates that the initial cell density was too low and the culture medium had not sufficiently acidified at the start of transfection. The remedy for this is to optimize the seeding density on the day before transfection. For example, cells could be split 1:3, 1:4, and 1:5. Another reason for inadequate precipitate formation could be the pH of the BES buffer which should be checked carefully and prepared again if necessary. It is advisable to have spare dishes of near confluent cells on the day of transfection. These can be split and used the following day for repeat transfection experiments if there are concerns about the observed precipitate formation.

10. The HEK292S GnTI⁻ cell line has a higher level of transfection efficiency, and for this reason it is necessary to dilute transfected cells to very high dilutions (up to 1:1000) in order to obtain single colonies. At such high dilutions, the complete medium must be supplemented with conditioned medium (20% (v/v)) in order to support growth of colonies from single cells during antibiotic selection and to continue

using it until colonies form. It is a good practice to use conditioned medium for all dilutions in the experiment. Conditioned medium (typically prepare 200 ml) is made in advance of transfection experiments as follows. HEK293S or HEK292S GnTI⁻ cells are grown to 90% confluence in 15 cm TC dishes in 25–30 mL of complete DMEM/F12. The spent medium is replaced with fresh complete medium followed by incubation for 24 h. This conditioned medium is then collected and filtered (0.2 μM) before storage at 20 °C for up to a year. These cells can be further refed and incubated with fresh medium and this second batch conditioned medium can be collected and processed 24 h later.

11. An unexpected property of the HEK293S GnTI⁻ cell line (and its TetR derivative) was super-sensitivity to G418; for this reason, this antibiotic is used at a concentration of 200–250 μg/mL during selection of G418-resistant colonies. If other plasmid vectors containing different selectable markers are used, it is recommended that the sensitivity of the HEK293S GnTI⁻ cell line to the antibiotic is determined.

12. Mark the side of each cell culture dish with a marker pen to ensure the same position can be used for aspiration and addition of fresh media during the several weeks it takes for colonies to form.

13. It is important to have all equipment necessary for this procedure ready and carefully organized within the laminar flow cabinet. Once the dishes containing the colonies have been rinsed with PBS it is necessary to work fast and methodically in order to prevent the cells from drying out. With practice it is feasible to isolate 8–10 colonies from a single dish, but very difficult to harvest any more. This is also why it is necessary to have multiple dishes containing colonies.

14. The cells that survive this process will grow as clumps and should be fed every 2–3 days until reaching near confluence. The dishes need to be observed on a daily basis because the cell lines will need to be fed at different times and will reach confluence at different times. The color of the growth medium is a good indicator to guide the feeding process. Once the medium has turned from red to orange it is a good point to feeds the cells.

15. This modified freeze-down procedure omits the washing step and a centrifugation step, thus enabling large numbers of samples to be processed.

16. To screen for cell lines exhibiting inducible high-level expression of the rhodopsin receptor it is convenient to use a dot blot procedure followed by immuno-detection with the anti-

rhodopsin antibody (Rho-1D4). This procedure can be used for any GPCR engineered to contain the Rho-1D4 epitope (TETSQVAPA) at the C-terminus of the protein.

17. Each 96-well plate can be used for the analysis of the solubilized pellets from 12 cell lines grown in the absence or presence of both sodium butyrate and tetracycline.

18. It is useful to observe the response of individual cell lines to the presence of growth medium containing tetracycline and sodium butyrate and to record careful notes. We find that cell lines exhibiting very high levels of inducible gene expression lose viability and lift off from the surface of the TC dish. This information is useful when determining the time period of induction for individual cell lines during production runs.

19. For large-scale production of rhodopsin and other GPCRs, cell lines prepared in this manner can be scaled up to multiple 15-cm dishes. Alternatively, these cell lines can be grown in suspension by using multiple 1 L spinner-flasks (500-mL growth medium) or a bioreactor using fed batch approaches. The conditions for growth in suspension have been described elsewhere [9].

20. For purification of rhodopsin it is convenient to perform single-step immunoaffinity chromatography using the Rho-1D4 antibody coupled to Sepharose beads and the cognate 9-mer peptide (TETSQVAPA) for elution [9].

References

1. He Y, Wang K, Yan N (2014) The recombinant expression systems for structure determination of eukaryotic membrane proteins. Protein Cell 9:658–672. https://doi.org/10.1007/s13238-014-0086-4.

2. Lyons JA, Shahsavar A, Paulsen PA, Pedersen BP, Nissen P (2016) Expression strategies for structural studies of eukaryotic membrane proteins. Curr Opin Struct Biol 38:137–144. https://doi.org/10.1016/j.sbi.2016.06.011.

3. Milic D, Veprintsev DB (2015) Large-scale production and protein engineering of G protein-coupled receptors for structural studies. Front Pharmacol 6:66. https://doi.org/10.3389/fphar.2015.00066

4. Stevens RC (2000) Design of high-throughput methods of protein production for structural biology. Structure 9:R177–RR85. https://doi.org/10.1016/s0969-2126(00)00193-3

5. Kimata N, Pope A, Sanchez-Reyes OB, Eilers M, Opefi CA, Ziliox M et al (2016) Free backbone carbonyls mediate rhodopsin activation. Nat Struct Mol Biol 8:738–743. https://doi.org/10.1038/nsmb.3257

6. Ye S, Kohrer C, Huber T, Kazmi M, Sachdev P, Yan EC et al (2008) Site-specific incorporation of keto amino acids into functional G protein-coupled receptors using unnatural amino acid mutagenesis. J Biol Chem 3:1525–1533. https://doi.org/10.1074/jbc.M707355200.

7. Mollaaghababa R, Davidson FF, Kaiser C, Khorana HG (1996) Structure and function in rhodopsin: expression of functional mammalian opsin in Saccharomyces cerevisiae. Proc Natl Acad Sci U S A 21:11482–11486. https://doi.org/10.1073/pnas.93.21.11482.

8. Sarramegna V, Talmont F, Demange P, Milon A (2003) Heterologous expression of G-protein-coupled receptors: comparison of expression systems from the standpoint of large-scale production and purification. Cell Mol Life Sci 8:1529–1546. https://doi.org/10.1007/s00018-003-3168-7.

9. Reeves PJ, Thurmond RL, Khorana HG (1996) Structure and function in rhodopsin: high level expression of a synthetic bovine opsin gene and its mutants in stable mammalian cell lines. Proc Natl Acad Sci U S A 21:11487–11492. https://doi.org/10.1073/pnas.93.21.11487.

10. Voronina EV, Seregin YA, Litvinova NA, Shvets VI, Shukurov RR (2016) Design of a stable cell line producing a recombinant monoclonal anti-TNFalpha antibody based on a CHO cell line. Springerplus 1:1584. https://doi.org/10.1186/s40064-016-3213-2.

11. Reeves PJ, Kim JM, Khorana HG (2002) Structure and function in rhodopsin: a tetracycline-inducible system in stable mammalian cell lines for high-level expression of opsin mutants. Proc Natl Acad Sci U S A 21:13413–13418. https://doi.org/10.1073/pnas.212519199.

12. Yao F, Svensjo T, Winkler T, Lu M, Eriksson C, Eriksson E (1998) Tetracycline repressor, tetR, rather than the tetR-mammalian cell transcription factor fusion derivatives, regulates inducible gene expression in mammalian cells. Hum Gene Ther 13:1939–1950. https://doi.org/10.1089/hum.1998.9.13-1939.

13. Chelikani P, Reeves PJ, Rajbhandary UL, Khorana HG (2006) The synthesis and high-level expression of a beta2-adrenergic receptor gene in a tetracycline-inducible stable mammalian cell line. Protein Sci 6:1433–1440. https://doi.org/10.1110/ps.062080006

14. Reeves PJ, Callewaert N, Contreras R, Khorana HG (2002) Structure and function in rhodopsin: high-level expression of rhodopsin with restricted and homogeneous N-glycosylation by a tetracycline-inducible N-acetylglucosaminyltransferase I-negative HEK293S stable mammalian cell line. Proc Natl Acad Sci U S A 21:13419–13424. https://doi.org/10.1073/pnas.212519299.

15. Pope AL, Sanchez-Reyes OB, South K, Zaitseva E, Ziliox M, Vogel R et al (2020) A conserved proline hinge mediates helix dynamics and activation of rhodopsin structure. https://doi.org/10.1016/j.str.2020.05.004

16. Standfuss J, Edwards PC, D'Antona A, Fransen M, Xie G, Oprian DD et al (2011) The structural basis of agonist-induced activation in constitutively active rhodopsin. Nature 7340:656–660. https://doi.org/10.1038/nature09795.

17. Varma N, Mutt E, Muhle J, Panneels V, Terakita A, Deupi X et al (2019) Crystal structure of jumping spider rhodopsin-1 as a light sensitive GPCR. Proc Natl Acad Sci U S A 29:14547–14556. https://doi.org/10.1073/pnas.1902192116.

18. Chen C, Okayama H (1987) High-efficiency transformation of mammalian cells by plasmid DNA. Mol Cell Biol 8:2745–2752. https://doi.org/10.1128/mcb.7.8.2745

19. Velan B, Kronman C, Ordentlich A, Flashner Y, Leitner M, Cohen S et al (1993) N-glycosylation of human acetylcholinesterase: effects on activity, stability and biosynthesis. Biochem J 296(Pt 3):649–656. https://doi.org/10.1042/bj2960649.

Chapter 4

Recombinant Expression and Purification of Cannabinoid Receptor CB$_2$, a G Protein-Coupled Receptor

Alexei A. Yeliseev

Abstract

G protein-coupled receptors (GPCR) are integral membrane proteins that regulate multiple cellular processes. To obtain insights into structural properties of GPCR and mechanism of activity, these proteins should be isolated in significant (milligram) quantities, in a pure, homogenous, and stable form. Here we describe the expression and purification of type II human cannabinoid receptor CB$_2$, a class A GPCR, in two different types of expression hosts: in *Escherichia coli* and in mammalian suspension cell culture Expi293. Our method allows preparation of milligram quantities of the purified receptors suitable for a wide array of downstream applications including high-resolution structural studies and functional assays.

Key words Membrane protein, CB$_2$ receptor, Recombinant expression, Thermostability

1 Introduction

G protein-coupled receptors (GPCR) belong to a large class of integral membrane proteins that regulate a wide array of physiological processes and became of the main targets of pharmaceutical drug development. Important prerequisite for rational design of specific pharmaceuticals targeting GPCR is availability of these proteins in pure, homogenous, and stable form in milligram quantities for high-resolution structural characterization and functional analysis. Here, we describe protocols for expression of the functional human cannabinoid receptor CB$_2$, a class A GPCR, in two different expression hosts: bacterial cell culture and mammalian suspension cell line. The efficient purification using two small affinity tags flanking the receptor result in preparation of highly pure functional protein with milligram yield.

In this chapter we present protocols for expression of peripheral cannabinoid receptor CB$_2$ in *E. coli* and in suspension HEK cell line Expi293F™, extraction from cell membranes, solubilization and stabilization of the recombinant receptor in detergent micelles,

Sofia Aires M. Martins and Duarte Miguel F. Prazeres (eds.), *G Protein-Coupled Receptor Screening Assays: Methods and Protocols*, Methods in Molecular Biology, vol. 2268, https://doi.org/10.1007/978-1-0716-1221-7_4,
© Springer Science+Business Media, LLC, part of Springer Nature 2021

and affinity purification using His-tag followed by purification via twin-Streptag. To achieve efficient purification, a C-terminal His_{10} as well as two identical Streptag sequences bridged by a short linker is attached to the N-terminus of CB_2 to increase the binding affinity to their respective resins [1]. Purification is performed under mild conditions to maintain the functional fold of the receptor. The purity of the resulting CB_2 can be determined by SDS-PAGE followed by Coomassie blue staining or Western blot, its biophysical properties in micelles by a variety of techniques including circular dichroism spectroscopy and NMR, and its functionality by binding of radioligands and its ability to activate cognate G proteins.

The purification is performed in the presence of the nonionic detergent dodecyl maltoside (DDM), the zwitterionic detergent CHAPS as well a derivative of cholesterol, cholesteryl hemisuccinate, required for stabilization of CB_2 in micelles. The necessity to perform purification in the presence of detergents constitutes a big challenge since they may affect the affinity and specificity of interaction between the tag and the resin, by covering the tag and decreasing its accessibility [2].

The first step of chromatographic purification of GPCR is based on a Ni-NTA technology [3]. The binding step is performed at a pH of 7–8, to ensure that all histidine residues are deprotonated, and the elution takes place in the presence of imidazole, which competes with histidine for the metal ions. Both the affinity and specificity of binding can vary dramatically depending on the target protein, location of the tag, composition of the binding buffer, and accessibility of the tag. To improve the efficiency of interaction with the resin, the tag is placed at either N- or C-terminus of the target protein, where it is also less likely to interfere with the biological activity of the recombinant protein.

The second chromatographic step of the protocol involves purification on a StrepTactin XT resin using the repeat sequence of twin-Streptag (Fig. 1). Streptag is an 8-amino acid polypeptide consisting of Trp-Ser-His-Pro-Gln-Phe-Glu-Lys. This octapeptide, and especially its double repeat separated by a small linker, which is termed twin-Streptag, exhibits high affinity for an engineered streptavidin StrepTactin XT [4–6]. Because of its small size, the twin-Streptag usually does not interfere with the biological activity of its fusion partner. Furthermore, the purification is performed under mild conditions that preserve the bioactivity of most proteins. The elution of the purified recombinant protein from the resin by biotin is specific, since only Strep-tagged proteins are displaced from the interaction with the StrepTactin XT resin. The resin can be reused after regeneration with the NaOH and re-equilibration with chromatography buffer.

A

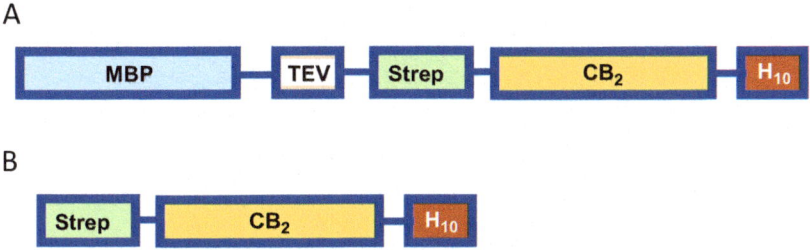

B

Fig. 1 Fusion protein CB$_2$–130 (**a**) and twin-Streptag-CB$_2$-His$_{10}$ (**b**)

The protein is eluted in the presence of Façade-TEG, a non-ionic detergent that is characterized by a very small aggregation number [7]. CB$_2$ solubilized in Facade-TEG/CHS micelles forms small, homogenous particles and retains functional activity at experimental conditions [8]. The protein in Facade detergent can be concentrated on spin-concentrators without concomitant co-concentration of detergents. This allows preparation of multi-milligram GPCR samples for structural studies that require high protein concentration such as NMR, crystallography, and cryo-electron microscopy.

2   Materials

Prepare all solutions using ultrapure water prepared by purifying deionized water, with conductivity 17–18 MΩ-cm at 25 °C and analytical grade reagents. Media for cell culture should be sterilized on the day of preparation and stored at room temperature. All buffers for protein purification should be prepared prior to the experiment and stored at 4 °C. Solutions of antibiotic and ligands should be stored at −20 or −80 °C.

2.1   GPCR Expression in E. coli Cells

1. Plasmid for expression of CB$_2$ fusion protein (Fig. 1).

2. *Escherichia coli* BL21(DE3) competent cells.

3. Shaker-incubator with a temperature range capability between 20 °C and 37 °C.

4. Ampicillin stock solution, 50 mg/mL. Dissolve ampicillin in water and filter the solution though a 0.22 μm filter.

5. Isopropyl-thiogalactoside (IPTG), 0.5 mM stock solution. Dissolve IPTG in water and filter through a 0.22 μm filter.

6. Glucose stock solution, 20% (w/v). To 250 mL water slowly add 100 g of glucose while stirring, complete to 500 mL with water and sterilize by autoclaving.

7. Luria-Bertani broth (LB) medium: 10 g/L casein digest, 5 g/L yeast extract, 5 g/L NaCl. Dissolve powder in water and sterilize by autoclaving.

8. Baffled, 125-mL shake flasks. Sterilize by autoclaving.

9. Double-strength YT-medium (2 × YT): 16 g/L casein digest, 10 g/L yeast extract, 5 g/L NaCl. Autoclave 500 mL medium in 2-L baffled shake flasks.

10. Phosphate-buffered saline (PBS): 10 mM sodium phosphate, 150 mM NaCl, pH 7.4.

11. Stock solution of 26 mM CP-55,940 ligand, in methanol.

2.2 GPCR Expression in Expi293F™ Cells

1. Expi293™ Expression medium.

2. ExpiFectamine 293 Transfection kit consisting of Expifectamine 293 reagent, ExpiFectamine 293 Transfection enhancer 1, and ExpiFectamine 293 Transfection enhancer 2.

3. Opti-Plex complexation buffer.

4. Plasmid DNA preparation sterile, free from phenol and sodium chloride, and containing mostly supercoiled DNA. For expression of CB_2 protein we use plasmid based on pCDNA 3.4 vector harboring twin-Streptag-CB_2-His_{10} construct under control of full-length CMV promoter (Fig. 1b).

5. Expi293F™ cells (1×10^7 cells/mL).

6. Incubator with ability to control temperature at 37 °C, 80% or higher relative humidity, and CO_2 concentration of 6.5%.

7. Orbital shaker capable of accommodating shake flasks of 125 mL to 3 L volume.

8. Cell counter.

9. Polycarbonate or PETG, non-baffled, disposable, sterile, vented shaker flasks for culturing suspension cells, such as Nalgene Single use PETG Erlenmeyer Flasks with plain bottom.

2.3 Solubilization of GPCR in Detergent Micelles

1. Tris-buffered saline (TBS): 25 mM Tris–HCl, 130 mM NaCl, 2.7 mM KCl, pH 7.5.

2. EDTA-free protease inhibitor cocktail tablets.

3. DNAse I stock solution, 5 mg/mL in water, approximately 1000 U/mL.

4. $MgCl_2$ stock solution, 1 M.

5. Anti-foam A.

6. 2× Solubilization buffer:100 mM Tris–HCl, 400 mM NaCl, 60% glycerol, pH 7.5.

7. 3-[(cholamidopropyl) dimethylammonio]-1-propanesulfonate (CHAPS)/cholesteryl hemisuccinate (CHS) stock solution (12x), 6% CHAPS, 1.2% CHS, store at 4 °C (*see* **Note 1**).

8. *n*-Dodecyl-β-D-maltoside (DDM), 10% stock solution. Store at 4 °C.

9. CP-55,940 stock solution, 26 mM stock solution in MeOH. Store at −20 °C.

10. "Complete" protease inhibitor cocktail tablets.

11. Sucrose solution 20% (v/v) in PBS.

12. Avestin Emulsiflex C5 cell homogenizer.

13. Handheld glass homogenizer.

2.4 Purification

1. Ni-NTA agarose.

2. StrepTactin XT agarose.

3. Facade-TEG.

4. Buffer A: 50 mM Tris–HCl, 200 mM NaCl, 0.5% [w/v] CHAPS, 0.1% [w/v] CHS, 1% [w/v] DDM, and 30% glycerol, pH 7.5.

5. Buffer B: 50 mM Tris–HCl, 200 mM NaCl, 0.5% [w/v] CHAPS, 0.1% [w/v] CHS, 1% [w/v] DDM, and 30% glycerol, 250 mM imidazole, pH 7.5.

6. Buffer C: 50 mM Tris–HCl, 100 mM NaCl, 0.25 mM Façade-TEG/0.025 mM CHS, pH 7.5.

7. Buffer D: 50 mM Tris–HCl, 100 mM NaCl, 0.25 mM Façade-TEG/0.025 mM CHS, pH 7.5, 50 mM biotin (*see* **Note 2**).

8. TEV protease. Either commercially available or in-house-prepared recombinant tobacco etch virus (TEV) protease can be used. We express and purify the recombinant poly-His-poly-Arg-tagged protease according to established protocols [9–11]. Expression in BL21(DE3) harboring the expression plasmid and purification via Ni-NTA affinity chromatography followed by cation exchange chromatography are described in [12].

9. 5 M Imidazole stock solution in H_2O.

10. Avidin.

11. Dowex 1x4 chromatography resin.

12. 0.45 μm pore size PES filters.

13. Ultracentrifuge.

14. AKTA Purifier chromatography system or similar automated chromatography system.

15. Protein concentration measurement Kit.

3 Methods

3.1 Expression of GPCR in E. coli

The protocols described here were developed for the peripheral cannabinoid receptor CB_2, expressed as a fusion with the maltose-binding protein and two small affinity tags flanking receptor sequence [1, 13] (see **Note 3**). The corresponding DNA construct, CB_2–130 is represented in Fig. 1a.

1. Inoculate 25 mL of LB medium supplemented with 25 µL of 50 mg/mL ampicillin in 125-mL baffled flask with *E. coli* BL21 (DE3) cells harboring the expression plasmid (see **Note 4**).

2. Incubate overnight at 37 °C and 230 rpm agitation.

3. Prepare expression medium by adding 5 mL glucose solution and 500 µL ampicillin solution to the 500 mL of double-strength YT-medium flask immediately before inoculation. Use an appropriate number of flasks assuming yield of ~500 µg of purified CB_2 protein from each flask.

4. Add 1–2 mL of the overnight culture to each 2 L flask and incubate at 37 °C under agitation for 3–4 h until the optical density of the culture at 600 nm reached 0.4–0.5 units.

5. Lower the temperature of incubation to 20 °C (see **Note 5**).

6. Add 500 µL of IPTG solution to induce synthesis of the recombinant protein.

7. Add 62.5 µL of a solution of CP-55,940 ligand (26 mM in Methanol, see **Note 6**).

8. Incubate for additional 38–40 h (see **Note 5**).

9. Collect cells by centrifugation at $4000 \times g$ for 30 min at 4 °C.

10. Resuspend and wash cells once in ice-cold PBS.

11. Collect cells by centrifugation and store pellet at −80 °C.

3.2 Expression of GPCR in Mammalian Suspension Cell Culture

The following protocol is for expression of CB_2 in 480 mL of cell culture. It can be scaled up or down, depending on the need of the experiment.

1. Thaw and establish Expi293F™ cell line according to manufacturer's protocol (ThermoFisher Scientific). Use the viable cell density to calculate the volume of cell suspension required to seed a new shake flask according to the recommended seeding densities by the manufacturer.

2. Transfer the calculated volume of cells to fresh, prewarmed Expi293 expression medium in a shake flask.

3. Incubate flasks in a 37 °C incubator with 80% relative humidity and 6.5% CO_2 on an orbital shaker platform until cultures reach a density of 4.5–5.4×10^6 viable cells/mL (see **Note 7**).

4. Dilute the cells in 2 L non-baffled flask to a final density of 3×10^6 viable cells/mL with fresh, prewarmed Expi293 expression medium, so that the final volume is 400 mL (*see* **Note 8**).

5. Add 75 mL of 26 mM ligand (CP-55,940) to the flask.

6. Gently dispense 46.5 mL of Opti-Plex to 125 mL sterile flask. Add 1.3 mL of ExpiFectamine transfection reagent. Gently swirl and wait 5 min at room temperature.

7. Add 400 μL of plasmid DNA (concentration 1 mg/mL), gently swirl and wait 5 min at room temperature (*see* **Note 9**).

8. Slowly add complexation mix to the 2 L flask with cell culture while swirling. Place the flask in the incubator on a shaker at 120 rpm.

9. 18–22 h post-transfection, add ExpiFectamine 293 transfection enhancer 1 and ExpiFectamine 293 transfection enhancer 2 to the transfection flask, gently swirling the flask. Immediately return the flask to the 37 °C incubator with 80% relative humidity and 6.5% CO_2 on an orbital shaker platform.

10. Collect cells 48 h post-transfection by centrifugation at $100 \times g$ for 5–10 min. Small aliquots (30 mL) of cell suspension can be collected separately for subsequent preparation of membranes.

11. Wash cell pellets once with cold PBS by gently pipetting the cell suspension up and down. Centrifuge again.

12. Freeze cell pellets at −80 °C.

3.3 Preparation of Membranes

We recommend testing the expression levels of the target protein in cell-membrane preparations obtained from small-scale cultures before proceeding with the large-scale fermentation and purification. The following protocol is based on a previously published method for CB_2 expression [13]. Carry out all steps of membrane preparation on ice or at 4 °C.

1. Perform expression as described in Subheadings 3.1 or 3.2 and collect cells from 25 to 30 mL of culture in a conical tube by centrifugation.

2. To 20 mL of cold PBS add 1 tablet of "complete" protease inhibitor cocktail.

3. Resuspended cells in a small volume (3–5 mL) of the cold PBS prepared in **step 2**.

4. Disrupt cells by passing the cell paste twice through a French Press.

5. Remove unbroken cells and cell debris by centrifugation at $20,000 \times g$ for 30 min at 4 °C.

6. Subject supernatant to a high-speed centrifugation (26,500 × g, 1 h, 4 °C).

7. Prepare a 10 mL solution of 20% sucrose in PBS and add 1 tablet of "complete" protease inhibitor cocktail.

8. Wash the resulting membrane pellet with cold PBS and resuspended in a small volume of PBS solution prepared in **step 7** using a handheld glass homogenizer (*see* **Note 10**).

9. Flash-freeze small aliquots of the membrane suspension in liquid nitrogen and store at −80 °C until use.

10. Proceed with evaluation of expression levels and functional activity of the target protein (*see* **Note 11**).

3.4 Solubilization of Recombinant GPCR

The following protocol describes the processing of 100 g of wet biomass. Depending on the size of cell pellet adjust the quantities of reagents accordingly. Perform all the following procedures on ice or at 4 °C.

1. Defrost frozen cell biomass on ice and resuspend in 150 mL of cold TBS buffer (*see* **Note 12**).

2. Supplement the cell suspension with 3.25 mL of 1 M $MgCl_2$ (*see* **Note 13**).

3. Add 2–3 tablets of EDTA-free protease inhibitor cocktail solubilized in 10 mL water.

4. Homogenize cell suspension using 200 mL PKontes brand Potter-Elvehjem Pyrex brand tissue grinder with Teflon pestle (*see* **Note 14**).

5. Transfer the suspension on ice and start stirring.

6. Add 50 µL of DNAse I solution.

7. Add Anti-foam A until the foam disappears (20–40 µL).

8. Pass the suspension two times through EmulsiFlex-C3 cell homogenizer at 15,000 psi pressure setting (*see* **Note 15**).

9. Place the homogenized suspension in a 1 L beaker on ice and start stirring.

10. Under continuous stirring, add 122 µL of 26.6 mM stock solution of CP-55,940 in methanol, 325 mL of double-concentrated solubilization buffer and 65 mL of 10X solution of detergents CHAPS, CHS, and DDM (concentration 5%, 1%, and 10%, respectively) (see **Note 16**).

11. Continue stirring for 40–60 min. The final composition of the buffer in which CB_2 is solubilized is Buffer A, supplemented with DNAse I (0.5 µg/mL) and EDTA-free protease-inhibitor cocktail.

12. Remove cell debris by centrifugation (20,700 × g, 1 h, 4 °C).

13. Prepare suspension of 70–80 g of Dowex 1x 4 in chromatography buffer by the following protocol: resuspend resin in 400–500 mL of water and remove water by filtration on 0.45 mm PES filter. Repeat twice. Finally, wash the resin with 100 mL of buffer A supplemented with stabilizing ligand and resuspend in a minimal volume of the same buffer.

14. Add suspension of Dowex 1x4 to the cell supernatant from **step 12** and incubate on a shaker at 4 °C for 30 min.

15. Pass the suspension through a 0.45-μm filter and proceed with purification.

3.5 Ni-NTA Chromatography

We recommend that all buffers used for purification should be supplemented with a suitable ligand that stabilizes GPCR in detergent micelles (for purification of CB_2 receptor we supplement all detergent buffers with the high-affinity agonist CP-55,940 to a final concentration of 10 μM). All procedures should be performed on ice or at 4 °C. Small aliquots can be taken at various purification steps to follow the purification process by SDS-PAGE and Western blot (*see* **Note 11**).

1. Pack 5 mL of Ni-NTA resin to a suitable column compatible with an available chromatography workstation (we use AKTA Purifier 100) allowing gradient-based chromatography and automated fraction collection.

2. Equilibrate the resin using 5 column volumes (CV) of buffer A + 10 μM CP-55,940.

3. Load the filtered CB_2 solution at a flow rate of 0.5–0.7 mL/min. Maintain this flow rate for the entire purification.

4. Wash the resin with 20 CV of buffer A + 10 μM CP 55,940 supplemented with 40 mM imidazole.

5. Elute protein with 10 CV of buffer B+ 10 μM CP-55,940 and collect 4 mL fractions (*see* **Notes 17** and **18**).

6. Select the CB_2-containing fractions by electrophoresis followed by Coomassie blue staining or by dot blot using anti-CB_2 mAb or anti-His-tag Ab (*see* **Note 19**).

3.6 Removal of Fusion Partners by Tobacco Etch Virus (TEV) Protease

This section refers to the treatment of MBP-CB_2 fusion protein expressed in *E. coli*. If there is no need in removing the fusion partner from CB_2, proceed directly to Subheading 3.7.

1. Combine CB_2-containing fractions and load onto 5 mL Strep-Tactin XT Superflow column (using sample pump in AKTA Purifier 100 system). The suggested flow rate is 0.2–0.3 mL/min.

2. Wash the column with 4 CV of buffer C, at a flow rate of 0.3 mL/min.

3. Dissolve 1–2 mg of TEV protease in 7 mL of buffer C. Load this solution slowly (0.3 mL/min) over the column using sample pump of the chromatography system. Measure the volume of the buffer displaced from the column and stop the flow once it reached 1 CV (5 mL).

4. Allow digestion to proceed at 4 °C for 4 h or overnight.

5. Apply 20 CV of buffer C at a flow rate of 0.5 mL/min to remove impurities.

6. Apply 3–4 CV of buffer D (elution buffer with biotin) at a flow rate of 0.1 mL/min. Apply additional 6 CV of buffer D at a flow rate of 0.3 mL/min. (see **Notes 20 and 21**).

7. Combine elution fractions and concentrate protein on a 15 mL spin concentrator with 30 kDa MWCO to 150–200 µL.

8. Remove biotin by diluting the protein to 3–4 mL with buffer C and concentrating again to 15–200 µL. Repeat two more times (see **Note 22**).

9. Add glycerol to the final concentration of 15%.

10. Measure protein concentration (BioRad DC kit or similar).

11. Aliquot protein solution in Eppendorf tubes and freeze in liquid nitrogen.

12. Store at −80 °C.

3.7 Purification Via StrepTactin Affinity Tag

The protocol described in this section can be applied to partially purified CB_2 protein from Subheading 3.5 if no removal of expression tag by action of TEV protease is required.

1. Connect 5 mL StrepTactin XT Superflow cartridge to the AKTA Purifier chromatography system.

2. Equilibrate the resin with 5 CV of buffer A+ 10 µM CP-55,940.

3. Load the combined protein fractions from Subheading 3.5 onto the StrepTactin XT column by sample pump at a flow rate of 0.1–0.2 mL/min.

4. Wash with 20 CV of buffer C+ 10 µM CP-55,940 at a flow rate of 0.5 mL/min.

5. Elute CB_2 with 8–10 CV of buffer D + 10 µM CP-55,940 at a flow rate of 0.1 mL/min (*see* **Note 19**).

6. Combine elution fractions and concentrate in centrifugal spin concentrators with a 30 kDa molecular mass cutoff.

7. Wash the concentrated protein solution 3–4 times with buffer C to remove biotin.

8. Add glycerol to the final concentration of 15%.

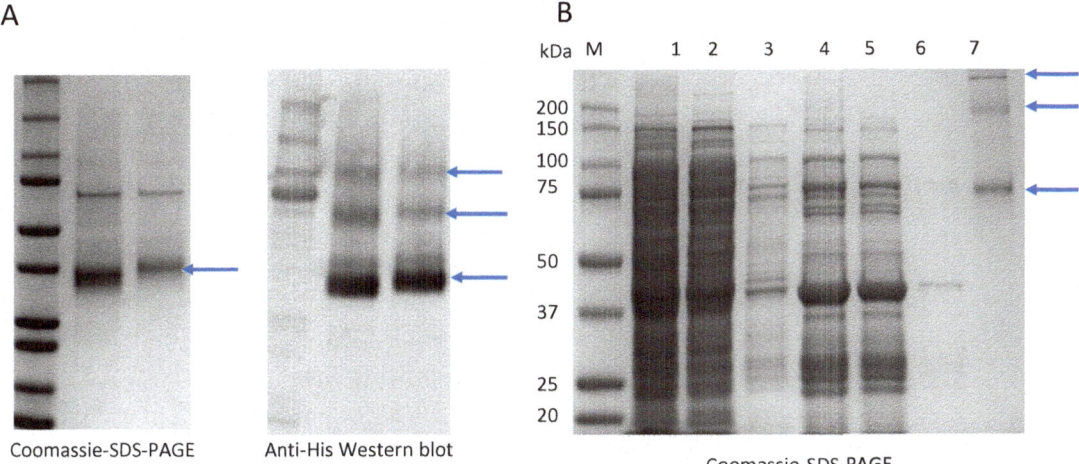

Fig. 2 Purification of CB₂. (**a**) Purification of CB2 protein from Expi293F™ cells. Coomassie blue staining and a Western blot probed with anti-His-tag antibody. Arrows indicate position of monomer CB₂ protein as well as dimer and higher oligomers. Please note that CB2 in detergent micelles stains very poorly by Coomassie dye; therefore oligomeric forms of CB₂ are not well visible in Coomassie stain [19]. (**b**) Purification of MBP-CB₂ fusion protein from *E. coli* BL21 (DE3) cells. M, molecular weight marker; 1—load of Ni-NTA column; 2—flowthrough from Ni-NTA column; 3—wash of Ni-NTA column; 4—combined elution fraction from Ni-NTA column loaded onto StrepTactin XT column; 5—flowthrough from StrepTactin XT column; 6—wash of StrepTactin XT column; 7—eluted purified MBP-CB2 protein. Arrows indicate positions of monomer and oligomers of purified protein

9. Determine protein concentration with the Bio-Rad DC kit (*see* **Note 18**).

10. Aliquot the concentrated protein into Eppendorf tubes.

11. Freeze aliquots in liquid nitrogen and store at −80 °C until use.

12. Evaluate purity of the preparation (*see* **Note 22** and Fig. 2).

13. Characterize functional state of the purified protein (*see* **Note 23**).

14. The purified protein can be analyzed by various biophysical techniques (*see* **Note 24**).

4   Notes

1. CHS will not solubilize when resuspended in water directly, and it has to be dissolved in a concentrated detergent solution. For preparation of 400 mL of the 12× stock solution of 6% CHAPS and 1.2% CHS we recommend the following protocol:

   (a) Take 4.8 g CHS and resuspend in 200 mL of water in a 250-mL beaker under vigorous stirring. Continue stirring through the entire solubilization procedure.

(b) Dissolve 24 g CHAPS in 150 mL of water with stirring.

(c) Add CHS suspension drop by drop to the CHAPS solution while stirring, until the solution is clear.

(d) Adjust volume to 400 mL with water. Solution can be stored at 4 °C for at least 2–3 months.

2. Weigh the required amount of powder biotin (IBA Lifesciences or other supplier) in a small Eppendorf tube and carefully dissolve in 100–200 μL of 0.1 N NaOH. Mix with a desired volume of buffer C (usually 15–20 mL) to achieve the final concentration of biotin of 50 mM.

3. CB$_2$–130 includes the following fusion partners:

(a) N-terminal Maltose-binding protein (MBP): facilitates proper folding of CB$_2$ as well as the insertion of the hydrophobic polypeptide into the cytoplasmic membrane of *E. coli*.

(b) Tobacco etch virus protease recognition site (TEV): is a sequence recognized by TEV protease and is required for cleavage of the fusion protein.

We use the recognition sequence E-N-L-Y-F-Q-S, which is cleaved in between the Q and S residues, leaving the following non-native sequences attached to CB$_2$: S (the remainder of the TEV site) followed by twin-Strep-tag: WSHPQFEK, GSGGAS (linker) followed by WSHPQFEK (second Streptag) followed by GGGS (linker) upstream of CB$_2$, and a decahistidine tag downstream of CB$_2$.

(a) StrepTag (Strep): required for affinity chromatography on a StrepTactin-agarose (final chromatographic step in our protocol).

(b) Polyhistidine tag (His$_{10}$): required for affinity purification of the recombinant protein on a Ni-NTA resin (first chromatographic step in our protocol).

4. Work under sterile conditions and wear gloves on all occasions where cultivation of cells or handling of potentially harmful chemicals is involved.

5. Expression conditions were optimized previously [13]. While cells can be grown at 37 °C until they reach OD$_{600}$ of 0.5, expression of the recombinant protein is performed at 20 °C to facilitate production of functional receptor. Incubation proceeds for 40 h after induction at which time the accumulation of cellular biomass reaches a maximum.

6. The addition of CP-55,940 or another high-affinity ligand to the growth medium enhances levels of expression and stabilizes functional CB$_2$ receptor [14].

7. Do not let cells grow above 5×10^6 viable cells/ mL during routine culture. Cells that are cultured at densities outside this early log-phase growth stage can show longer doubling times and lower protein titers over time. Modify the initial seeding density to attain the target cell density of $3–5 \times 10^6$ viable cells/mL at the time of subculturing.

8. Discard the remaining cells. Do not reuse high-density cells for routine subculturing.

9. Total plasmid DNA concentration of 1 μg/mL of culture volume to be transfected is appropriate.

10. Use a volume of 400–600 μL to resuspend the pellet. Transfer the partially resuspended pellet to a handheld glass homogenizer with an appropriate transfer pipette. Homogenize and transfer to an Eppendorf tube. Repeat the process with half of the original volume. Combine both fractions. Avoid formation of air bubbles to prevent losing membranes.

11. We recommend testing the effects of tags on expression levels and functional state of the target protein before proceeding with a large-scale expression and purification. For CB_2, the expression levels of various constructs and functionality of the resulting recombinant fusion CB_2 are typically evaluated by Western blot analysis, ligand-binding assays, and G-protein activation assay performed on membrane preparations [13]. Incorporation of the N-terminal twin-Streptag and C-terminal His_{10} tag does not significantly affect the expression levels of the CB_2 as shown by Western blot analysis using anti-CB_2, anti-His, and anti-Streptag antibodies [1, 13].

    The ability to detect small affinity tags on fusion CB_2 in membrane preparations subjected to SDS-PAGE and Western blot by assessing interaction with their respective antibodies does not guarantee that these same tags will be accessible for interaction with the chromatographic resin. However, we believe that the Western blot analysis is an accurate predictor of the accessibility of a tag (including poly-histidine and twin-Streptag) for interaction with the respective resin, for subsequent chromatographic purification of CB_2 (unpublished observations).

12. Some prewarming of the cell pellet (at room temperature for 20–30 min) may be required to facilitate transfer to the blender.

13. The addition of DNA and magnesium will help to digest the released DNA and decrease the viscosity of the extract. The protease inhibitor cocktail is needed to prevent the proteolytic degradation of the expressed protein.

14. Homogenization of the cell paste prior to cell disruption is necessary to speed up the preparation of cell-free extract.

15. Alternatively, if a cell homogenizer is not available, the cell suspension can be subjected to sonication for 15 min (Branson Sonifier 250, 1/2-inch flat tip, output 6, duty cycle 50%) [13]. Make sure that the solution is not overheated, place it on ice and stir during sonication.

16. In order to stabilize the CB_2 and minimize loses of active protein during purification, several parameters are critical [14]. Temperature has to be maintained at 4 °C, and the entire purification procedure should take no longer than 2 days to achieve the yield of functional protein $\geq$90%. Solubilization of the fusion protein is performed in a mixture of CHAPS (0.5%), DDM (1%), and CHS (0.1%) in the buffer as described previously [13]. The presence of 0.1% CHS during solubilization and purification is critical to maintain the functional structure of CB_2 in detergent micelles. In addition, the buffer contains 30% glycerol. Supplementation with the high-affinity ligand such as CP-55,940 during purification was also shown to significantly increase the recovery of active protein. We recommend 10 μM CP-55,940 in all purification buffers.

17. Take samples from load, flowthrough, washes, elution fractions (separated or combined) for each chromatographic step as well as before and after TEV cleavage. In order to prepare CB_2 for SDS-PAGE it is important not to subject the sample to boiling or even high-temperature treatment since this will result in aggregation of the protein. We recommend mixing the protein sample with equal volume of the 2x Laemmli sample buffer and incubating at 37 °C for 30 min.

18. For regeneration of the Ni-NTA resin, wash extensively with imidazole buffer and remove imidazole by extensive washing with binding buffer before use, according to the manufacturer's instructions.

19. This step can be omitted once the pattern of elution of the target protein is determined. We typically observe elution of CB_2 in fractions 3–7.

20. The elution of the CB_2 protein from StrepTactin XT with 50 mM biotin is a slow process at 4 °C and is most efficient when performed at a slow flow rate. We recommend the initial flow rate of 0.1 mL/min for several CV. The flow rate then can be increased to 0.3 mL/min without any detrimental effects [15].

21. To regenerate the StrepTactin resin, wash it with NaOH and buffers according to manufacturer's instructions (IBA Lifesciences).

22. Purity of the final preparation can be evaluated by Coomassie blue staining. Be aware of the fact that the efficiency of staining

of highly hydrophobic membrane proteins (like CB_2) may be much lower than that of soluble proteins, resulting in a weaker than expected signal in Coomassie-stained gels.

23. The functional state of the purified CB_2 protein can be assessed either upon its reconstitution into liposomes [14, 16] or upon dilution of the concentrated protein 100–200-fold into a buffer containing 0.3 mM Facade-TEG/0.03 mM CHS, supplemented with an agonist of CB_2, CP-55,940 [8]. Then, a G-protein activation assay can be performed on these preparations as described earlier [8, 13, 17]. Reaction conditions were optimized for 3–10 ng of CB_2 per reaction to ensure that less than 30% of the available [^{35}S] GTPγS is consumed.

24. Purified CB_2 in detergent micelles is suitable for characterization by high-resolution NMR (ligand-binding studies [13] and diffusion experiments [18]) or other biophysical techniques [17].

Acknowledgments

This work was supported by the Intramural Research Program of the National Institute on Alcoholism and Alcohol Abuse, National Institutes of Health. Authors thank Ms. L Zoubak and Mr. K. Hines who contributed to the project and Dr. K. Gawrisch for continuous support.

References

1. Yeliseev A, Zoubak L, Gawrisch K (2007) Use of dual affinity tags for expression and purification of functional peripheral cannabinoid receptor. Protein Expr Purif 53:153–163. https://doi.org/10.1016/j.pep.2006.12.003

2. Block H, Maertens B, Spriestersbach A, Brinker N, Kubicek J, Fabis R, Labahn J, Schafer F (2009) Immobilized-metal affinity chromatography (IMAC): a review. Methods Enzymol 463:439–473. https://doi.org/10.1016/j.pep.2011.08.021

3. Locatelli-Hoops SC, Yeliseev AA (2014) Use of tandem affinity chromatography for purification of cannabinoid receptor CB(2). Methods Mol Biol 1177:107–120. https://doi.org/10.1007/978-1-4939-1034-2_9

4. Vani D, Geetanjali S, Punja GM, Bharathi M (2015) A case of invasive papillary breast carcinoma: fierce facade with favorable prognosis. J Cancer Res Ther 11:1029. https://doi.org/10.4103/0973-1482.154086

5. Voss S, Skerra A (1997) Mutagenesis of a flexible loop in streptavidin leads to higher affinity for the strep-tag II peptide and improved performance in recombinant protein purification. Protein Eng 10:975–982. https://doi.org/10.1093/protein/10.8.975

6. Schmidt TG, Batz L, Bonet L, Carl U, Holzapfel G, Kiem K, Matulewicz K, Niermeier D, Schuchardt I, Stanar K (2013) Development of the twin-strep-tag(R) and its application for purification of recombinant proteins from cell culture supernatants. Protein Expr Purif 92:54–61. https://doi.org/10.1016/j.pep.2013.08.021

7. Mineev KS, Nadezhdin KD, Goncharuk SA, Arseniev AS (2016) Characterization of small isotropic Bicelles with various compositions. Langmuir 32:6624–6637. https://doi.org/10.1021/acs.langmuir.6b00867

8. Beckner RL, Zoubak L, Hines KG, Gawrisch K, Yeliseev AA (2020) Probing thermostability of detergent-solubilized CB2 receptor by parallel G protein-activation and ligand-binding assays. J Biol Chem

295:181–190. https://doi.org/10.1074/jbc. RA119.010696

9. Kapust RB, Waugh DS (1999) Escherichia coli maltose-binding protein is uncommonly effective at promoting the solubility of polypeptides to which it is fused. Protein Sci 8:1668–1674. https://doi.org/10.1110/ps.8.8.1668

10. Kapust RB, Tozser J, Fox JD, Anderson DE, Cherry S, Copeland TD, Waugh DS (2001) Tobacco etch virus protease: mechanism of autolysis and rational design of stable mutants with wild-type catalytic proficiency. Protein Eng 14:993–1000. https://doi.org/10. 1093/protein/14.12.993

11. Kapust RB, Routzahn KM, Waugh DS (2002) Processive degradation of nascent polypeptides, triggered by tandem AGA codons, limits the accumulation of recombinant tobacco etch virus protease in Escherichia coli BL21(DE3). Protein Expr Purif 24:61–70. https://doi.org/ 10.1006/prep.2001.1545

12. Raran-Kurussi S, Cherry S, Zhang D, Waugh DS (2017) Removal of affinity tags with TEV protease. Heterologous gene expression in E coli. Methods Protoc 1586:221–230. https:// doi.org/10.1007/978-1-4939-6887-9_14

13. Yeliseev AA, Wong KK, Soubias O, Gawrisch K (2005) Expression of human peripheral cannabinoid receptor for structural studies. Protein Sci 14:2638–2653. https://doi.org/10.1110/ ps.051550305

14. Vukoti K, Kimura T, Macke L, Gawrisch K, Yeliseev A (2012) Stabilization of functional recombinant cannabinoid receptor CB(2) in

detergent micelles and lipid bilayers. PLoS One 7:e46290. https://doi.org/10.1371/ journal.pone.0046290

15. Yeliseev A, Zoubak L, Schmidt TGM (2017) Application of strep-Tactin XT for affinity purification of twin-strep-tagged CB2, a G protein-coupled cannabinoid receptor. Protein Expr Purif 131:109–118. https://doi.org/10. 1016/j.pep.2016.11.006

16. Kimura T, Yeliseev AA, Vukoti K, Rhodes SD, Cheng K, Rice KC, Gawrisch K (2012) Recombinant cannabinoid type 2 receptor in liposome model activates g protein in response to anionic lipid constituents. J Biol Chem 287:4076–4087. https://doi.org/10.1074/ jbc.M111.268425

17. Yeliseev A, Gawrisch K (2017) Expression and NMR structural studies of isotopically labeled cannabinoid receptor type II. Methods Enzymol 593:387–403. https://doi.org/10.1016/ bs.mie.2017.06.020

18. Kimura T, Vukoti K, Lynch DL, Hurst DP, Grossfield A, Pitman MC, Reggio PH, Yeliseev AA, Gawrisch K (2014) Global fold of human cannabinoid type 2 receptor probed by solid-state 13C-, 15N-MAS NMR and molecular dynamics simulations. Proteins 82:452–465. https://doi.org/10.1002/prot.24411

19. Yeliseev AA (2013) Methods for recombinant expression and functional characterization of human cannabinoid receptor CB2. Comput Struct Biotechnol J 6:e201303011. https:// doi.org/10.5936/csbj.201303011

Chapter 5

Screening for Serotonin Receptor 4 Agonists Using a GPCR-Based Sensor in Yeast

Emily A. Yasi and Pamela Peralta-Yahya

Abstract

More than 30% of all pharmaceuticals target G-protein-coupled receptors (GPCRs). Here, we present a GPCR-based screen in yeast to identify ligands for human serotonin receptor 4 (5-HTR$_4$). Serotonin receptor 4 agonists are used for the treatment of irritable bowel syndrome with constipation. Specifically, the HTR$_4$-based screen couples activation of 5-HTR$_4$ on the yeast cell surface to luciferase reporter expression. The HTR$_4$-based screen has a throughput of one compound per second allowing the screening of more than a thousand compounds per day.

Key words Serotonin receptor, Yeast, GPCR, High-throughput assay

1 Introduction

There are over 800 G-protein-coupled receptors (GPCRs) expressed in humans [1]. These receptors have a plethora of functions from sensing odorants [2] and flavors [3] to binding neurotransmitters [4] to controlling blood pressure and heart rate [5]. Currently, approximately 34% of FDA approved drugs target GPCRs [6]. Notably, levodopa targets dopamine receptors in the brain to treat Parkinson's disease [7], sumatriptan targets serotonin receptors 1B and 1D to treat migraines [8], and propranolol targets beta-adrenergic receptors to treat hypertension [9]. Some over-the-counter supplements also target GPCRs, such as melatonin, a sleep aid [10]. With the vast consequences that GPCRs have on human health and physiology, high-throughput assays are needed to rapidly identify ligands, synthetic and endogenous, to further understand their downstream effects.

Mammalian-based GPCR assays have long been used to detect GPCR activation. Indeed, several commercial kits are available to read out endogenous GPCR activation, including cAMP accumu-

Sofia Aires M. Martins and Duarte Miguel F. Prazeres (eds.), *G Protein-Coupled Receptor Screening Assays: Methods and Protocols*, Methods in Molecular Biology, vol. 2268, https://doi.org/10.1007/978-1-0716-1221-7_5,

lation, which can be linked to cell fluorescence (e.g., Catch-PointTM cAMP, Molecular Devices), calcium flux (e.g., FLIPR® Calcium Assay Kits, Molecular Devices), or luminescence [11–14]. A general challenge of using mammalian cell-based assays for the identification of GPCR ligands is that mammalian cells endogenously express tens of GPCRs (e.g., 75 different GPCRs in HEK293 [15]). Thus, an assay phenotype can be trigged by a ligand that activates a different GPCR that transduces via the same G_α. Additionally, mammalian-based assays can be easily contaminated and are inherently slow due to the long doubling time of mammalian cells and the need for passaging.

The yeast *Saccharomyces cerevisiae* is an ideal host for the development of GPCR-based assays. Haploid yeasts have only two GPCRs: Ste2 (Matα) or Ste3 (Mata) responds to pheromones resulting in yeast mating, and Gpr1 responds to glucose [16]. Detection of GPCR activation can be achieved by expressing a heterologous GPCR on the yeast cell surface, coupling it to the yeast mating pathway, and expressing a reporter gene (e.g., green fluorescent protein, luciferase) under control of a mating pathway responsive promoter. To increase the heterologous GPCR signal three deletions are needed. Deletion of *ste2/3* [17] increases signal transduction, and deletions of *far1* and *sst2* prevent cell cycle arrest and desensitization of the Gα subunit, respectively [18]. The receptor Gpr1 couples to a distinct adenylate cyclase cascade [19]. Traditionally, the reporter gene has been an auxotrophic marker that enables growth selection [20, 21], requiring up to 3 days to see results. More recently, reporters that allow same day readouts, including green fluorescent protein (4 h) [22, 23] and luciferase (1–2.5 h) [24, 25], have been developed.

Recently, we presented a GPCR-based assay that links serotonin receptor 4b (5-HTR$_{4b}$) to luciferase (NanoLuc) [26] expression [24]. 5-HTR$_{4b}$ is expressed in the brain but is more prominently expressed throughout the gastrointestinal tract [27], where over 95% of serotonin is found [28]. 5-HTR$_4$ is a drug target for irritable bowel syndrome with constipation; currently, only a few FDA-approved treatments target 5-HTR$_4$, such as prucalopride (Motegrity™) [29] and metoclopramide (Reglan®) [30]. The luciferase-based 5-HTR$_4$ assay is fast, going from chemical incubation to signal readout in 2.5 h. The ability to use a luminescence plate reader to detect reporter gene expression enables a screening rate of 1 well per second [24]. This method enables the high-throughput screening of vast chemical libraries for 5-HTR$_{4b}$ activation.

2 Materials

2.1 Plasmids and Strains

1. GPCR plasmid: PPY1192: pESC-His3-P$_{TEF1}$-HTR$_{4b}$-T$_{Cyc1}$.

2. No GPCR plasmid: PPY111: pESC-His3- P$_{TEF1}$- T$_{Cyc1}$.

3. GPCR microscopy plasmid: PPY1133: pESC-His3- P$_{TEF1}$-HTR$_{4b}$-GFP- T$_{Cyc1}$.

4. Reporter plasmid: PPY1740: pRS415-Leu2-P$_{FIG1}$-NanoLuc.

5. Yeast sensor strain: PPY140: *S. cerevisiae* W303–1a MATa *ade2–1, ura3–1, his3–11, trp1–1, leu2–3, leu2–112, can1–100, Δfar1, Δsst2, Δste2.*

6. *5-HTR$_{4b}$*-sensing strain: PPY140 carrying plasmids PPY1192 + PPY1740

7. Control sensing strain: PPY140 carrying plasmids PPY111 + PPY1740.

8. *5-HTR$_{4b}$-GFP* fusion strain: PPY140 carrying plasmid PPY1133.

2.2 Yeast Media and Transformation Supplies

1. Synthetic Dropout media lacking histidine and leucine with 2% glucose (SD (HL$^-$)) [31].

2. Synthetic Dropout media lacking histidine and leucine with 2% glucose [31] and buffered with potassium phosphate to pH 7 (SD(HL$^-$), pH 7.

3. Synthetic Dropout media lacking histidine with 2% glucose [31] (SD(H$^-$)).

4. Yeast Peptone Dextrose media (YPD) [32].

5. Tris-EDTA (TE) buffer 10×, pH 7.5: 100 mM Tris–HCl, 10 mM EDTA.

6. 1 M Lithium acetate

7. PEG4000 50% (w/v).

8. Solution 1: 8 mL diH2O, 1 mL 10× TE buffer, 1 mL 1 M lithium acetate.

9. Solution 2: 800 μL 50% (w/v) PEG 4000, 100 μL 10× TE buffer, 100 μL 1 M lithium acetate.

10. Salmon Sperm DNA: 1 mg/mL in water, filter sterilized.

11. 1.5 mL Microcentrifuge tubes

12. Petri dishes (100 × 15 mm).

13. Test tubes (25 × 150 mm).

14. Inoculation loops.

15. Flat toothpicks.

16. Parafilm.

2.3 Fluorescent Microscopy

1. Poly-L-lysine slides with coverslip.
2. Calcofluor White Stain.
3. Confocal microscope.
4. 10% KOH in water.

2.4 Luminescent Plate Reader Assay

1. White opaque flat-bottom 96-well plate.
2. Chemical library.
3. Chemical carrier solvent: e.g., Dimethyl Sulfoxide (DMSO).
4. Breathe-Easy® membrane (Electron Microscopy Sciences).
5. Promega Nano-Glo® Assay System.
6. Luminescent enabled plate reader.

2.5 Software

1. Microsoft Excel or similar program.

3 Methods

3.1 Preparation of 5-HTR$_{4B}$-Sensing, Control Sensing, and 5-HTR$_{4B}$-GFP Fusion Strains

1. Streak out PPY140 on a YPD agar plate and incubate at 30 °C for 3 days.
2. Patch a single colony on a fresh YPD agar plate using a sterile flat toothpick or a sterile loop and incubate at 30 °C for 3 days.
3. Inoculate a small loopful of PPY140 into 5 mL YPD in a test tube. Grow overnight at 30 °C at 250 rpm.
4. Add 1 mL of PPY140 overnight culture to 49 mL of fresh YPD and grow at 30 °C, 250 rpm to an $OD_{600} = \sim0.6$–0.8 which takes approximately 3–4 h.
5. Centrifuge the culture for 3 min at $1800 \times g$, decant, and resuspend cells in 50 mL diH_2O.
6. Centrifuge again for 3 min at $1800 \times g$, decant, and resuspend cells in 5 mL Solution 1.
7. Centrifuge again for 3 min at $1800 \times g$, decant, and resuspend in 300 µL Solution 1.
8. To generate the HTR$_{4b}$ sensing strain and empty vector control, combine in a microcentrifuge tube 300 µL of Solution 2, 50 µL of cells in Solution 1 (**step 7**), 5 µL of salmon sperm DNA (denatured at 95 °C for 5 mins, chilled on ice afterward), 500 ng of the GPCR plasmid (replace with No GPCR plasmid for control), 500 ng of the Reporter Plasmids.
9. Incubate at 30 °C, 250 rpm for 30 min. Heat shock the yeast by incubating at 42 °C for 15 min. Spin cells down 1 min at $5000 \times g$.
10. Resuspend the cell pellet in 150 µL of diH_2O and plate on an SD(HL-) agar plate. Incubate plate at 30 °C for 3 days (*see* **Notes 1** and **2**).

11. To generate the HTR$_{4b}$-GFP fusion strain to test GPCR expression in yeast using fluorescent microscopy: use the same protocol as in **step 8** except for replacing the GPCR plasmid with GPCR microscopy plasmid and do not use Reporter Plasmid.

12. Plate on a SD(H$^-$) agar plate. Incubate plate at 30 °C for 3 days (*see* **Notes 1** and **2**).

13. Patch cells from a single colony on a fresh SD(H-) agar plate using a sterile flat toothpick or a sterile loop. Incubate plates at 30 °C for 3 days (*see* **Note 3**).

3.2 Fluorescent Microscopy

1. Start an overnight culture of the HTR$_{4b}$-GFP fusion strain by taking a small loopful of the patch and inoculating 5 mL of SD (H$^-$). Shake at 30 °C and 250 rpm (*see* **Note 4**).

2. The next day, use the overnight culture to inoculate 20 mL of SD(H$^-$) to an OD$_{600}$ of 0.06. Incubate the culture at 15 °C, 150 rpm for 18 h.

3. Centrifuge the culture for 10 min at 1800 × g, decant, and resuspend cells in 200 μL SD(H$^-$).

4. To a microscope slide, layer 2 μL of the cells collected in **step 3**, 2 μL of calcofluor white stain, 2 μL of potassium hydroxide solution. Add a cover slide, making sure to avoid air bubbles.

5. Visualize the yeast on a confocal microscope using a 63× objective lens. The microscope should be equipped with both a 488 and a 405 nm laser lines to excited GFP and calcofluor white, respectively.

3.3 GPCR Luminescence Assay

1. Start an overnight culture of 5-HTR$_{4b}$-sensing strain by taking a small loopful of the patch and inoculating 5 mL of SD(HL$^-$). Shake at 30 °C and 250 rpm (*see* **Note 4**).

2. The next day, use the overnight culture to inoculate 20 mL of SD(H-) to an OD$_{600}$ = 0.06. Incubate the culture at 15 °C, 150 rpm for 18 h.

3. Centrifuge the culture for 10 min at × 2465 × g, decant, and resuspend cells in SD(HL$^-$) to OD$_{600}$ = 1.

4. To a 96-well plate, add 190 μL of fresh SD(HL$^-$) pH 7 (*see* **Note 5**), 8 μL of cell suspension obtained from **step 3**, 2 μL of chemical or chemical carrier solvent as a control (*see* **Notes 6** and **7**). A chemical carrier solvent control and a serotonin positive control should be run in triplicate on the same plate.

5. Cover plate using breathe-easy® membrane (*see* **Note 8**).

6. Incubate for 2 h × 250 rpm at 30 °C.

7. While incubating, allow Promega Nano-Glo® buffer to melt at room temperature. Just before the 2 h incubation is over, create a 1:100 dilution of Nano-Glo® Substrate: Nano-Glo® Buffer.

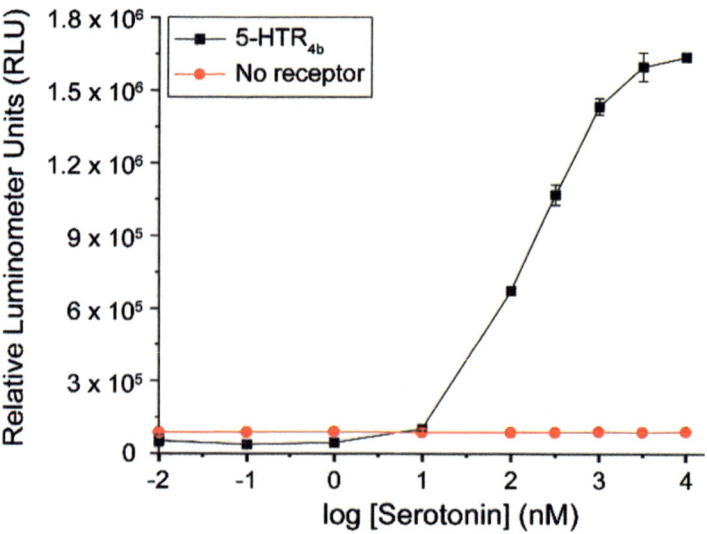

Fig. 1 Representative raw data from the 5-HTR$_4$ assay with serotonin

8. Remove breathe-easy® membrane and pipet 20 μL of Nano-Glo® Substrate:Buffer dilution into each well.

9. Place a new breathe-easy® membrane and incubate at 30 °C, 250 rpm for 30 min.

10. Remove breathe-easy® membrane and read immediately on luminescent plate reader (i.e., Biotek Synergy 2) (*see* **Note 9**).

3.4 Secondary Assay: dose-response Curves and Z-score Analysis

If hits are detected in a screen, follow the same protocol using both the 5-HTR$_{4B}$-sensing and Control sensing strain and run dose-response curves to confirm if luminescence signal is GPCR dependent (*see* **Note 10**). Figure 1 shows representative data.

1. Use the equation below to calculate Z-scores from the screen:

$$Z - score = \frac{(Chemical - mean\ vehicle\ control)}{STDev\ vehicle\ control}$$

Normalize the Z-scores to the serotonin positive control by dividing the chemical Z-score by serotonin's Z-score.

4   Notes

1. Amount of yeast transformation mixture that is plated may need to be varied to see single, well-spaced colonies.

2. Colonies should be pink and approximately 2–5 mm in diameter. Smaller white colonies are indicative of satellite colonies that will not grow well when patched or inoculated overnight.

3. Patches can be stored at 4 °C for up to 1 month wrapped in parafilm.

4. Overnight OD$_{600}$s should be ~6–8.

5. Media pH should be set to 7, as that is the optimal pH for NanoLuc activity [26].

6. Chemical carrier solvent should be chosen based on chemical solubility. When using DMSO, use at ≤1% final concentration. If using other carrier solvents, measure the toxicity to yeast to determine maximal final concentration.

7. The 5-HTR$_{4b}$ assay chemical screen can be run in singlets using all wells in the plate.

8. After placement of breathe-easy® membrane, care should be taken to avoid rough handling to prevent cross-contamination of wells. Shaking at 250 rpm does not cause well cross contamination.

9. On a Biotek Synergy 2, use default luminescence gain settings. Testing different gains should be done on each machine to ensure optimal signal readout and avoid signal saturation.

10. The Control Sensing strain has higher background than the 5-HTR$_{4b}$-sensing strain at low chemical concentrations. The signal of the 5-HTR$_{4b}$-sensing strain should be higher than the Control Sensing strain with increase in chemical concentration.

Acknowledgements

This work funded by an NIH MIRA Award (R35GM124871) to P. P-Y.

References

1. Lv XC, Liu J, Shi Q et al (2016) In vitro expression and analysis of the 826 human G protein-coupled receptors. Protein Cell 7:325–337

2. Dryer L, Berghard A (1999) Odorant receptors: a plethora of G-protein-coupled receptors. Trends Pharmacol Sci 20:413–417

3. Hoon MA, Adler E, Lindemeier J et al (1999) Putative mammalian taste receptors: a class of taste-specific GPCRs with distinct topographic selectivity. Cell 96:541–551

4. Waxham MN (2014) Neurotransmitter receptors. In: Byrne JH, Heidelberger R, Waxham MN (eds) From molecules to networks. Academic Press, Boston, MA, pp 285–321

5. Salazar NC, Chen J, Rockman HA (2007) Cardiac GPCRs: GPCR signaling in healthy and failing hearts. Biochim Biophys Acta Biomembr 1768:1006–1018

6. Hauser AS, Attwood MM, Rask-Andersen M et al (2017) Trends in GPCR drug discovery: new agents targets and indications. Nat Rev Drug Discov 16:829–842

7. Hisahara S, Shimohama S (2011) Dopamine receptors and Parkinson's disease. Int J Med Chem 2011:403039

8. Akin D, Gurdal H (2002) Involvement of 5-HT1B and 5-HT1D receptors in sumatriptan mediated vasocontractile response in rabbit common carotid artery. Br J Pharmacol 136:177–182

9. Prichard BNC (1982) Propranolol and beta-adrenergic-receptor blocking-drugs in the treatment of hypertension. Br J Clin Pharmacol 13:51–60

10. Arendt J, Rajaratnam SMW (2008) Melatonin and its agonists: an update. Br J Psychiatry 193:267–269

11. Zhang R, Xie X (2012) Tools for GPCR drug discovery. Acta Pharmacol Sin 33:372–384

12. Chen Y, Xu Z, Wu D et al (2015) Luciferase reporter gene assay on human 5-HT receptor: which response element should be chosen? Sci Rep 5:8060

13. Cheng Z, Garvin D, Paguio A et al (2010) Luciferase reporter assay system for deciphering GPCR pathways. Curr Chem Genomics 4:84–91

14. Kroeze WK, Sassano MF, Huang XP et al (2015) PRESTO-Tango as an open-source resource for interrogation of the druggable human GPCRome. Nat Struct Mol Biol 22:362–U328

15. Atwood BK, Lopez J, Wager-Miller J et al (2011) Expression of G protein-coupled receptors and related proteins in HEK293 AtT20 BV2 and N18 cell lines as revealed by microarray analysis. BMC Genomics 12:14

16. Versele M, Lemaire K, Thevelein JM (2001) Sex and sugar in yeast: two distinct GPCR systems. EMBO Rep 2:574–579

17. Price LA, Strnad J, Pausch MH, Hadcock JR (1996) Pharmacological characterization of the rat A2a adenosine receptor functionally coupled to the yeast pheromone response pathway. Mol Pharmacol 50:829–837

18. Price LA, Kajkowski EM, Hadcock JR et al (1995) Functional coupling of a mammalian somatostatin receptor to the yeast pheromone response pathway. Mol Cell Biol 15:6188–6195

19. Harashima T, Heitman J (2002) The G alpha protein Gpa2 control yeast differentiation by ineracting with Kelch repeat proteins that mimic G beta subunits. Mol Cell 10:163–173

20. Pausch MH (1997) G-protein-coupled receptors in Saccharomyces cerevisiae: high-throughput screening assays for drug discovery. Trends Biotechnol 15:487–494

21. Pajot-Augy E, Crowe M, Levasseur G et al (2003) Engineered yeasts as reporter systems for odorant detection. J Recep Sig Transd 23:155–171

22. Shaw WM, Yamauchi H, Mead J et al (2019) Engineering a model cell for rational tuning of GPCR signaling. Cell 177:782–796

23. Mukherjee K, Bhattacharyya S, Peralta-Yahya P (2015) GPCR-based chemical biosensors for medium-chain fatty acids. ACS Synth Biol 4:1261–1269

24. Yasi EA, Allen AA, Sugianto W, Peralta-Yahya P (2019) Identification of three antimicrobials activating serotonin receptor 4 in colon cells. ACS Synth Biol 8:2710–2717

25. Laschet C, Dupuis N, Hanson J (2019) A dynamic and screening-compatible nanoluciferase-based complementation assay enables profiling of individual GPCR-G protein interactions. J Biol Chem 294:4079–4090

26. Hall MP, Unch J, Binkowski BF et al (2012) Engineered luciferase reporter from a deep sea shrimp utilizing a novel imidazopyrazinone substrate. ACS Chem Biol 7:1848–1857

27. Ray AM, Kelsell RE, Houp JA et al (2009) Identification of a novel 5-HT4 receptor splice variant (r5-HT4c1) and preliminary characterisation of specific 5-HT4a and 5-HT4b receptor antibodies. Eur J Pharmacol 604:1–11

28. Gershon MD, Tack J (2007) The serotonin signaling system: from basic understanding to drug development for functional GI disorders. Gastroenterology 132:397–414

29. Vijayvargiya P, Camilleri M (2019) Use of prucalopride in adults with chronic idiopathic constipation. Expert Rev Clin Phar 12:579–589

30. Avalos DJ, Sarosiek I, Loganathan P, McCallum RW (2018) Diabetic gastroparesis: current challenges and future prospects. Clin Exp Gastroenter 11:347–363

31. Cold Spring Harbor Protocols (2006) Complete minimal (cm) or synthetic complete (sc) and drop-out media 2006(1):pdbrec8190

32. (2010) YPD media. Cold Spring Harbor Protocols 2010(9):pdbrec12315-pdbrec12315

Chapter 6

Immobilization of Olfactory Receptors Carried by Nanosomes onto a Gold Sensor Surface

Jasmina Vidic and Yanxia Hou

Abstract

Mammalian olfactory receptors (ORs) constitute the largest family of G-protein-coupled receptors, with up to about 1000 different genes per species, each having specific odorant ligands. ORs could be used as sensing elements of highly specific and sensitive bioelectronic hybrid devices such as bioelectronic noses. After optimized immobilization onto the device, natural ORs provide molecular recognition of various odors with their intrinsic sensitivity, discrimination, and detection properties. However, the main difficulties are related to the low expression level of recombinant ORs, their stability and potential loss of activity. Such drawbacks can be successfully overcome in bioelectronic noses integrating nanosomes (nanometric membrane vesicles carrying ORs) that are stably immobilized through a specific antibody. The advantages of such a platform rely on the fact that ORs stay in the natural membrane environment, and thus preserve their full activity. Thanks to their small sizes, nanosomes offer potential for micro- and nano-scale sensor development. In this paper, we summarize the key elements regarding nanosomes production and manipulation and provide an example of their immobilization onto a gold sensor surface. Rat ORI7 is used as a representative OR that can be functionally expressed in *Saccharomyces cerevisiae*. The receptor was not purified but only nanosomes were prepared. Nanosomes were immobilized onto functionalized gold surface using the anti-I7 antibody. Utilization of the antibody provides enrichment of ORI7 on the sensor surface but also uniform and appropriate orientation of the receptors. These features are crucial in optimization of bioelectronic nose' analytical performances.

Key words G-protein-coupled receptors, Olfactory receptors, Nanosomes, Bioelectronic noses, Immobilization, Gold surface

1 Introduction

Olfaction is a chemical sense that is essential for the animals. Various odor molecules in the environment are detected by olfactory receptors (ORs) with exquisite sensitivity [1]. ORs belong to the G-protein-coupled receptors (GPCRs) superfamily that all share a seven transmembrane domains topology [1–3]. They are mainly expressed in the membrane of olfactory sensory neurons in the olfactory epithelium that covers the caudal and upper part of

Sofia Aires M. Martins and Duarte Miguel F. Prazeres (eds.), *G Protein-Coupled Receptor Screening Assays: Methods and Protocols*, Methods in Molecular Biology, vol. 2268, https://doi.org/10.1007/978-1-0716-1221-7_6,
© Springer Science+Business Media, LLC, part of Springer Nature 2021

the nasal cavity. ORs can discriminate between their odorant ligand and similar molecules with slight difference in functional groups, molecular size, or shape. Upon odorant molecule binding to an OR, the conformation of the receptor changes. Such a change triggers the signaling cascade transmission in the olfactory sensory neuron that is conducted to the olfactory bulb, and further to the cortical structures for odor identification. The binding of odorant molecules to ORs is the first event in the perception and discrimination of thousands of different odorants from the external environment.

The intrinsically high sensitivity and selectivity of the animal ORs to detect and discriminate odorant molecules make them very promising as recognition elements in the development of novel analytical tools such as bioelectronic noses. Such bioelectronic hybrid devices with good odor detection performances can be obtained through two main achievements: (i) the reproducible preparation of functional OR sample and (ii) the appropriate immobilization of OR onto a solid sensor surface with maintained recognition activity.

We previously demonstrated the feasibility of bioelectronic noses using the rat I7 OR and human OR1740 [4–14]. To obtain a sufficient amount of the receptor, the ORs were functionally expressed in the budding yeast *Saccharomyces cerevisiae* [15, 16]. For instance, rat olfactory receptor ORI7 (ORL11 in Olfactory data base, OrDB) was heterologously expressed in the yeast *S. cerevisiae* strain MC18 as previously described [15]. For this, yeast cells were transformed with pJH2-I7 plasmid that carries a galactose-inducible GAL1/10 promoter [15]. Other yeast expression plasmids are commercially available. The pJH2-I7 plasmid enables expression of ORs with a cmyc tag fused at its N-terminus and a HA tag at its C-terminus [13, 15, 16]. Tags enable to orientate receptor molecules on the sensor surface by using the specific anti-cmyc or anti-HA antibodies. Alternatively, in case when a specific antibody against the OR of interest exists, it can be used for immobilization purposes.

We successfully demonstrated that olfactory receptors with their seven transmembrane domains conformation localized in the yeast plasma membrane and conserved their odorant recognition activities [4, 15]. Commonly, a high-level expression of an OR in yeast leads to accumulation of the receptor within the cell. However, we showed that inducing receptor expression at 15 °C yields lower expression level but produced receptor molecules are properly addressed to the plasma membrane [4, 12, 14, 15]. Figure 1a shows the ultrastructural visualization of immunogold labeling of OR I7 that confirms adequate receptor cell localization. To preserve OR functional conformation out of the cell, we developed a procedure to prepare membrane lipidic vesicles carrying ORs. These membrane structures, so-called "nanosomes" because of

a)
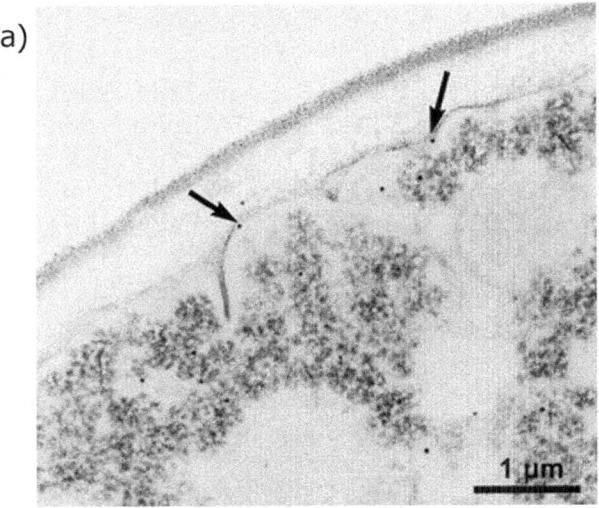

b)
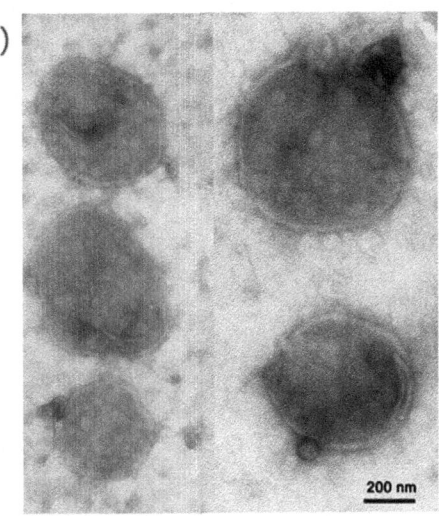

Fig. 1 TEM visualization of a yeast cell expressing ORI7 and its membrane fraction. (**a**) Receptors were immuno-labeled using the primary anti-I7 antibody and the 10 nm gold-conjugated secondary antibody. Arrows show gold grains present on the plasma membrane, which indicates ORI7 localization at the plasma membrane. (**b**) Yeast membrane fraction obtained after cell mechanical disruption and ultracentrifugation in the form of vesicles. (Reproduced from ref. [4] with permission from Elsevier)

their nanometric diameter, stabilize receptor conformation and thereby fully preserve their recognition properties [4, 9, 17–20]. Figure 1b shows membrane vesicles visualized by the transmission electronic microscopy. Next, we elaborated an efficient immobilization of nanosomes onto a gold sensor surface using a specific antibody [6, 7, 13]. Our approach allowed for the development of original bioelectronic noses with good sensitivity and selectivity, which are promising for various applications. Finally, it is noteworthy that the same strategy can be applied for other GPCRs [21, 22].

2 Materials

Prepare all solutions using ultrapure water (18.2 MΩ-cm resistivity, prepared by purifying deionized water) and analytical grade reagents (*see* **Note 1**). Prepare and store all solutions at 4 °C (unless indicated otherwise). Meticulously follow all waste disposal regulations when disposing waste materials.

2.1 Isolation of Membrane Fraction

1. Phenylmethylsulfonyl fluoride (PMSF) stock solution: 1 M. Weigh 0.174 g of PMSF using a high precision balance and transfer it to a 1.5 mL plastic tube. Add 1 mL of ethanol to the tube and vortex (*see* **Note 2**).

2. Glass beads (425–600 μm).

3. Complete™ Protease Inhibitor Cocktail, 1 tablet for 50 mL.

4. Complete Supplement Mixture synthetic drop-out (CSM) media without HIS, LEU, TRP, URA.

5. Lysis buffer: 50 mM Tris–HCl, pH 8.8, 1 mM EDTA, 0.1 mM PMSF, 250 mM sorbitol, and Complete™ Protease Inhibitor Cocktail. Add 100 mL water to a 1 L graduated cylinder or a glass beaker (*see* **Note 3**). Weigh separately: 6.057 g of Tris–HCl, 0.292 g of EDTA, and 45.54 g of sorbitol and transfer them all to the cylinder or the beaker.

6. Add a magnetic flea and place the cylinder or the beaker on a magnetic stirring plate to mix the solution. Adjust pH with HCl (*see* **Note 4**). Make up to 1 L with water. Just before use, take 100 mL of this buffer solution and add 10 μL of 1 M PMSF stock solution and 2 tablets of complete protease inhibitor cocktail.

7. Yeast growing medium: 10× stock solution. Suspend 6.8 g of yeast nitrogen base without amino acids, 5 g of CSM powder, and 5.4 g adenine in 100 mL of ultrapure water. Filter-sterilize this 10× stock solution and store at 2–8 °C.

8. Yeast growing medium: 1×. Right before utilization, aseptically pipette 25 mL of the yeast growing medium 10× stock solution into 225 mL ultrapure water into a 1 L Erlenmeyer. Mix thoroughly by shaking (*see* **Note 5**).

9. Phosphate-buffered saline (PBS) tablet. Prepare 1× PBS.

10. PBS-glycerol solution: 10%. Add 10 mL of pure glycerol to 90 mL PBS.

11. Ultrasonic Bath (*see* **Note 6**).

12. Dounce homogenizer.

13. Pierce™ BCA Protein Assay Reagent.

14. *Saccharomyces cerevisiae* yeast cells transformed with pJH2-I7 plasmid that carries a galactose-inducible GAL1/10 promoter (*see* **Note 7**). Just before inoculation defreeze slowly an aliquot.

15. UV-vis spectrophotometer (*see* **Note 8**).

16. Centrifugation tubes of 250 mL and 50 mL.

2.2 Immobilization

1. Thiol 1 stock solution: 10 mM. Thiol 1: N-(2-{2-[2-(2-{2-[2-(1-mercaptoundec-11-yloxy)-ethoxy]-ethoxy}-ethoxy)-ethoxy]-ethoxy}-ethyl) biotinamide (HS-C11-EG6-Biotin). Add 1 mL absolute ethanol to a 2 mL glass vial with a cap.

2. Weigh 6.94 mg thiol 1 using a high precision balance and transfer to the vial. Mix the solution. Close the cap (*see* **Note 9**).

3. Thiol 2 stock solution: 100 mM. Thiol 2: 2-{2-[2-(2-{2-[2-(1-mercaptoundec-11-yloxy)-ethoxy]-ethoxy}-ethoxy)-ethoxy]-ethoxy]—ethanol (HS-C11-EG6). Add 1 mL

absolute ethanol to a 2 mL glass vial with a cap. Weigh 46.9 mg thiol 2 using a high precision balance and transfer to the vial. Mix the solution. Close the cap (*see* **Note 9**).

4. Mixed thiol solution: 0.1 mM thiol 1 and 1 mM thiol 2. Add 1 mL absolute ethanol to a clean 10 mL glass beaker. Add 50 µL thiol 1 stock solution. Add 50 µL thiol 2 stock solution. Add 3.45 mL absolute ethanol to obtain the final volume of 5 mL. Mix the solution.

5. Phosphate-buffered saline (PBS): 0.01 M phosphate buffer, 0.0027 M potassium chloride, and 0.137 M sodium chloride, pH 7.4, at 25 °C. Add 200 mL ultrapure water to a 250 mL glass beaker.

6. Add one PBS tablet in the beaker. Stir until the complete dissolution of the tablet.

7. Neutravidin stock solution: 0.1 mM. Weigh 6 mg neutravidin using a high precision balance and transfer to a 1.5 mL Eppendorf. Add 1 mL PBS. Mix the solution with a Vortex-Mixer. Store at −20 °C.

8. Rabbit anti-I7 polyclonal antibody raised against its N-terminal 15 amino acids was custom-made by Neosystem (Strasbourg, France). The antibody was biotinylated using a DSB-X™ Biotin Protein Labeling kit.

9. Biotinylated rabbit anti-I7 polyclonal antibody (Biot-anti-I7-Ab) stock solution: 0.1 mM. Weigh 15 mg Biot-anti-I7-Ab using a high precision balance and transfer to a 1.5 mL Eppendorf. Add 1 mL PBS. Mix the solution with a Vortex-Mixer. Store at −20 °C.

10. Solution of nanosomes (0.1 mg/mL) containing OR I7 in PBS (details given later).

3 Methods

Carry out all procedures at room temperature unless otherwise specified.

3.1 Preparation of Membrane Fraction Carrying ORI7

1. Place 250 mL of yeast growing medium 1× (*see* **Note 5**) in 1 L Erlenmeyer. Add sterilely 5 g glucose to the solution.

2. Add 50 µL of transformed *S. cerevisiae* to the medium (*see* **Note 7**) to inoculate the medium very lightly. Incubate the solution at 30 °C under 180 rpm shaking until yeast reaches the exponential growth phase (*see* **Note 8**). It usually takes 24 h.

3. Collect and transfer the yeast suspension into a 250 mL centrifugation tube. Spin at 2500 × g, for 20 min at 4 °C. Remove supernatant by decanting and resuspend pellet in 250 mL of yeast growing medium 1× in an 1 L Erlenmeyer.

4. Add 5 g of galactose to induce ORI7 expression. Incubate solution at 15 °C under shaking (180 rpm) for 60 h (*see* **Note 10**).

5. Collect the solution and transfer it into 250 mL centrifugation tubes. Spin at 2500 × g, for 20 min at 4 °C.

6. Remove supernatant by decanting and resuspend pellet in 50 mL of ice-cold water. Transfer the suspension into a 50 mL centrifugation tube.

7. Vortex tube. Spin at 2500 × g, for 20 min at 4 °C. Repeat this procedure to wash twice yeast cells with ice-cold water.

8. Spin at 2500 × g for 20 min at 4 °C to harvest cells by centrifugation.

9. Resuspend washed cells in 1 mL of ice-cold lysis buffer. Transfer the solution to a 2 mL plastic tube.

10. Add glass beads to the resuspended yeast cells to obtain a total volume of about 1.5 mL.

11. Disturb cells by seven cycles of 1 min of vigorous vortexing / 1 min of cooling on ice.

12. Leave beads to precipitate and pool the supernatant by pipetting.

13. Centrifuge the supernatant at 5000 × g for 10 min at 4 °C to remove by precipitation unbroken cells and cell walls.

14. Pool the second supernatant and ultracentrifuge it at 40,000 × g for 40 min at 4 °C.

15. Discharge supernatant containing cytosolic fraction. Keep pallet containing membrane fraction.

3.2 Nanosomes Preparation

1. Add 200 μL PBS-glycerol solution to the cell pallet enriched in membranes.

2. Resuspend it manually using a Dounce homogenizer. This step allows to obtain small membrane vesicles of different sizes (*see* **Note 11**) that are named microsomes.

3. Determine the total protein concentration in the membrane fraction using the Pierce™ BCA Protein Assay Reagent kit. Usually total protein concentration is in the range between 5 and 15 mg/mL.

4. Immediately before use, sonicate 200 μL of the membrane fraction using ultrasonic bath in ice-cold water for 20 min at 120 W, 42 kHz (*see* **Note 12**) to obtain nanosomes of uniform diameters of about 50 nm [14].

5. Dilute nanosome preparation to 0.1 mg/mL using PBS of total proteins for immobilization purposes. 5 mL of 0.1 mg/mL nanosome solution is used in each immobilization procedure.

3.3 Immobilization of ORs onto the Gold Sensor Surface

1. Treat the gold surface by plasma, generated at 0.6 mbar with 75% oxygen and 25% argon, with a power of 40 W for 3 min, using the Femto plasma cleaner (Diener Electronic, Germany) to remove organic contaminants.

2. Immerse the gold solid substrate into 5 mL mixed thiol solution in the glass beaker. Cover the beaker with Parafilm to avoid solvent evaporation (*see* **Note 5**). Figure 2 shows chemical structures of the used thiols.

3. Leave the gold substrate in the solution for 18 h at room temperature, for the functionalization of the surface by the formation of mixed self-assembled monolayer.

4. Take out the functionalized gold substrate with a tweezer. Immerse it into 5 mL ethanol in a new and clean beaker for 5 min for rinsing.

5. Repeat **step 4** two more times.

6. Take out the gold substrate with the tweezer. Dry the gold surface under nitrogen flow.

7. Add 4995 μL PBS in a new and clean beaker. Inject 5 μL neutravidin stock solution. Mix gently the solution.

8. Put the gold substrate into the neutravidin solution in the glass beaker. Leave for 90 min.

9. Immerse the gold substrate into 5 mL PBS in a new and clean beaker for 5 min for rinsing.

10. Repeat **step 9** two more times.

11. Take out the gold substrate with the tweezer. Dry the gold surface under nitrogen flow.

12. Add 4995 μL PBS in a new and clean beaker. Inject 5 μL Biot-anti-I7-Ab stock solution. Mix gently the solution.

13. Put the gold substrate into Biot-anti-I7-Ab solution in the glass beaker. Leave for 1 h.

14. Immerse the gold substrate into 5 mL PBS in a new and clean beaker for 5 min for rinsing.

15. Repeat **step 14** two more times.

16. Take out the gold substrate with the tweezer. Dry the gold surface under nitrogen flow.

17. Immerse the gold substrate into 5 mL of 0.1 mg/mL nanosome solution in the glass beaker. Incubate for 1 h.

18. Take out the gold substrate with the tweezer. Immerse it into 5 mL PBS in a new and clean beaker for 5 min for rinsing.

19. Repeat **step 16** two more times.

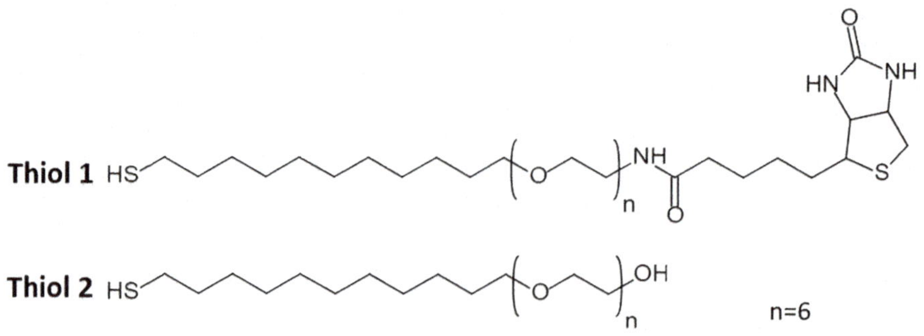

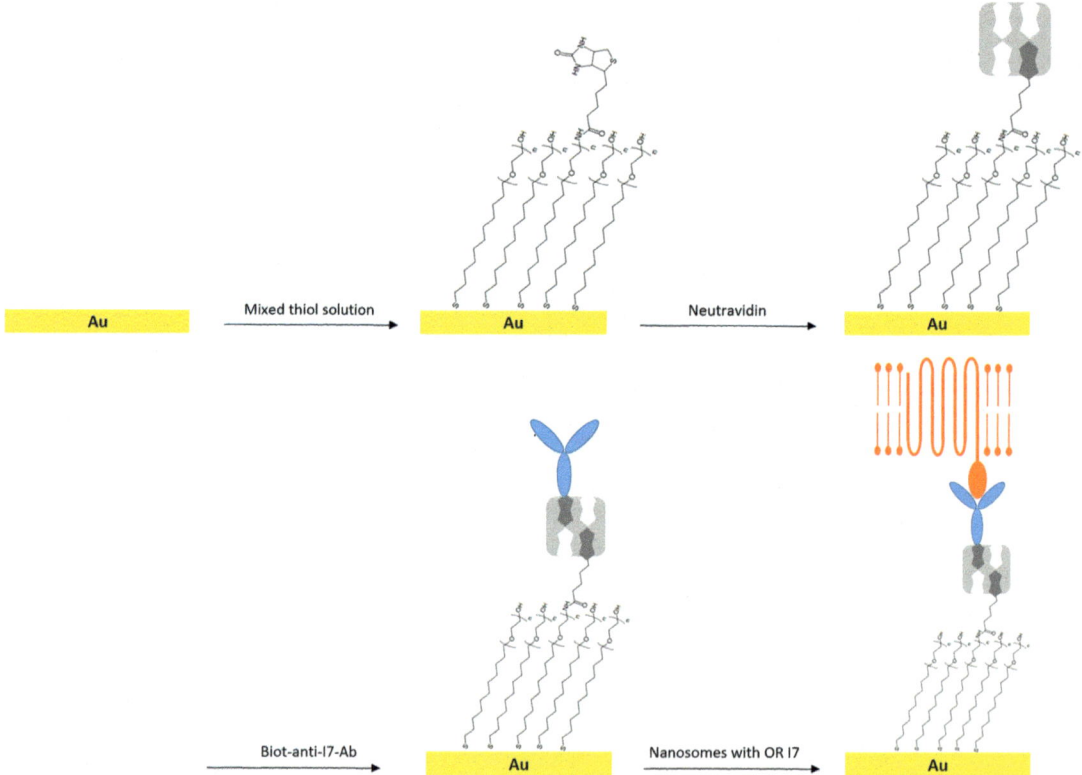

Thiol 1 ... **Thiol 2** ...

Fig. 2 Chemical structures of the thiol 1 and thiol 2

Fig. 3 Schematic illustration of the procedure for the immobilization of ORs carried by nanosomes onto the gold sensor surface

20. Put the gold substrate in 5 mL PBS in a new and clean beaker. Cover the beaker with Parafilm. Store at 4 °C. Figure 3 illustrates the whole immobilization strategy.

4 Notes

1. Wear protective glasses and gloves.

2. Work in the fume hood when manipulating PMSF. PMSF is a serine protease inhibitor that is commonly used when cell lysate are prepared as it prevents protein degradation. However, it cannot inhibit all serine proteases. PMSF is very instable in water solutions. Its half-life is only 35 min at pH 8 at 25 °C. Stock solutions in ethanol are more stable for at least 24 h.

3. Having water at the bottom of the cylinder allows the magnetic stir bar to turn immediately, which helps to dissolve the Tris.

4. At the beginning, concentrated HCl can be used to get closer to the required pH. Then, it would be better to use HCl with lower concentration. When close to the required pH, slowly add HCl solution using a Pasteur pipette, not to add too much at a time, since the pH will change rapidly.

5. Yeast cells and medium should be kept in aseptic conditions to avoid any contamination from foreign bacteria inherent in the environment (like airborne microorganisms or microbes from the lab bench-top or other surfaces). Equipment (Erlenmeyer, glassware, Petri, etc.) has to be sterilized because potential contamination may interfere with the results. Manipulations can be performed near the Bunsen burner. The burner has two adjustments: the first is the knob underneath which adjusts the amount of gas going into the burner tube, and the second is the barrel of the burner, which enables to adjust the amount of air going into the burner. Both air and gas should be adjusted to a minimal open position before lighting a Bunsen burner. After lighting, the air and gas should be adjusted to provide a hissing, small, blue flame within a taller lighter blue/violet flame.

6. Use ear protection because ultrasonic baths create audible sound.

7. Transformed yeast can be stored at −80 °C until OR-nanosome production. For this, 10–50% (v/v) of sterile glycerol is added into yeast culture at exponential growing phase (optical density of 0.4 to 1 at 600 nm) and aliquots within 2 mL plastic vials are frozen at − 80 °C. Before utilization, one aliquot is allowed to defreeze slowly. The plastic vial should be open in aseptic conditions to avoid any contamination from foreign bacteria and fungi inherent in the environment.

8. Exponential growing phase of yeast cells can be determined by measuring absorbance at 600 mn. Because absorbance unit is arbitrary, its value depends on the spectrophotometer used.

The yeast growing kinetic is usually flowed before performing membrane fraction preparation. In our case exponential phase corresponded to absorbance between 0.4 and 1.

9. Work in the fume hood when manipulating thiols and thiol solutions.

10. Various temperatures were used during induction of transformed yeast, and the production of ORs was checked. A low temperature of 15 °C was shown to help receptors to be correctly addressed to the yeast plasma membrane with a preserve natural seven transmembrane conformation.

11. Resuspended pallet contains membrane fraction in a form of microsomes with diameters ranging from 40 to 700 nm. Membrane fraction can be prepared in large amounts, aliquot and stored in aliquots at −80 °C frozen and stored at −80 °C for many weeks without losing biological activity.

12. Preparation is sonicated for 20 min at 120 W, 42 kHz. If ultrasonic bath has low power/frequency, characteristic time of sonication should be increased. During prolonged sonication, ice has to be added to ensure cold environment.

Acknowledgments

This work was financially supported by the SPOT-NOSED Project of the European Community (IST-2001-38739).

References

1. Buck L, Axel R (1991) A novel multigene family may encode odorant receptors: a molecular basis for odor recognition. Cell 65:175–187. https://doi.org/10.1016/0092-8674(91)90418-x

2. Firestein S (2001) How the olfactory system makes sense of scents. Nature 413 (6852):211–218. https://doi.org/10.1038/35093026

3. Persaud K, Dodd G (1982) Analysis of discrimination mechanisms in the mammalian olfactory system using a model nose. Nature 299:352–355. https://doi.org/10.1038/299352a0

4. Gomila G, Casuso I, Errachid A, Ruiz O, Pajot E, Minic J, Gorojankina T, Persuy M, Aioun J, Salesse R (2006) Advances in the production, immobilization, and electrical characterization of olfactory receptors for olfactory nanobiosensor development. Sensors Actuators B Chem 116:66–71. https://doi.org/10.1016/j.snb.2005.11.083

5. Gomila G, Errachid A, Bessueille F, Ruiz O, Pajot E, Minic J, Gorojankina T, Salesse R, Villanueva G, Bausells J (2005) Artificial nose integrating biological olfactory receptors and NEMS. In: International Symposium on Olfaction and Electronic Nose (ISOEN), vol CONF. pp 529–532. https://doi.org/10.1109/SCED.2005.1504506

6. Wade F, Espagne A, Persuy M, Vidic J, Monnerie R, Merola F, Pajot-Augy E, Sanz G (2011) Relationship between homo-oligomerization of a mammalian olfactory receptor and its activation state demonstrated by bioluminescence resonance energy transfer. J Biol Chem 286:15252–15259. https://doi.org/10.1074/jbc.M110.184580

7. Hou Y, Jaffrezic-Renault N, Martelet C, Zhang A, Minic-Vidic J, Gorojankina T, Persuy M-A, Pajot-Augy E, Salesse R, Akimov V (2007) A novel detection strategy for odorant molecules based on controlled bioengineering of rat olfactory receptor I7. Biosens

Bioelectron 22:1550–1555. https://doi.org/10.1016/j.bios.2006.06.018

8. Marrakchi M, Vidic J, Jaffrezic-Renault N, Martelet C, Pajot-Augy E (2007) A new concept of olfactory biosensor based on interdigitated microelectrodes and immobilized yeasts expressing the human receptor OR17-40. Eur Biophys J 36:1015–1018. https://doi.org/10.1007/s00249-007-0187-6

9. Pennetta C, Akimov V, Alfinito E, Reggiani L, Gorojankina T, Minic J, Pajot Augy E, Persuy MA, Salesse R, Casuso I (2006) Towards the realization of nanobiosensors based on G protein-coupled receptors. Nanodevices for the Life Science, Vol.4, in Nanodevices for the Life Science, Edited by Challa S.S.R. Kumar, 1st Edition, Wiley-VCH Verlag GmbH&Co. KGaA, Weinheim, pp 217–240

10. Rodriguez Segui S, Pla M, Minic J, Pajot-Augy E, Salesse R, Hou Y, Jaffrezic-Renault N, Mills C, Samitier J, Errachid A (2006) Detection of olfactory receptor I7 self-assembled multilayer formation and immobilization using a quartz crystal microbalance. Anal Lett 39:1735–1745. https://doi.org/10.1080/00032710600714030

11. Vidic J (2010) Bioelectronic noses based on olfactory receptors. Intell Biosens:377–386. https://doi.org/10.5772/7163

12. Vidic J, Grosclaude J, Monnerie R, Persuy M-A, Badonnel K, Baly C, Caillol M, Briand L, Salesse R, Pajot-Augy E (2008) On a chip demonstration of a functional role for odorant binding protein in the preservation of olfactory receptor activity at high odorant concentration. Lab Chip 8:678–688. https://doi.org/10.1039/B717724K

13. Vidic J, Pla-Roca M, Grosclaude J, Persuy M-A, Monnerie R, Caballero D, Errachid A, Hou Y, Jaffrezic-Renault N, Salesse R (2007) Gold surface functionalization and patterning for specific immobilization of olfactory receptors carried by nanosomes. Anal Chem 79:3280–3290. https://doi.org/10.1021/ac061774m

14. Vidic JM, Grosclaude J, Persuy M-A, Aioun J, Salesse R, Pajot-Augy E (2006) Quantitative assessment of olfactory receptors activity in immobilized nanosomes: a novel concept for bioelectronic nose. Lab Chip 6:1026–1032. https://doi.org/10.1039/B603189G

15. Minic J, Persuy MA, Godel E, Aioun J, Connerton I, Salesse R, Pajot-Augy E (2005) Functional expression of olfactory receptors in yeast and development of a bioassay for odorant screening. FEBS J 272:524–537. https://doi.org/10.1111/j.1742-4658.2004.04494.x

16. Minic J, Sautel M, Salesse R, Pajot-Augy E (2005) Yeast system as a screening tool for pharmacological assessment of g protein coupled receptors. Curr Med Chem 12:961–969. https://doi.org/10.2174/0929867053507261

17. Akimov V, Alfinito E, Bausells J, Benilova I, Paramo IC, Errachid A, Ferrari G, Fumagalli L, Gomila G, Grosclaude J (2008) Nanobiosensors based on individual olfactory receptors. Analog Integr Circ Sig Process 57:197–203

18. Akimov V, Alfinito E, Pennetta C, Reggiani L, Minic J, Gorojankina T, Pajot-Augy E, Salesse R (2006) An impedance network model for the electrical properties of a single-protein nanodevice. In: Nonequilibrium carrier dynamics in semiconductors. Springer US, pp 229–232. https://doi.org/10.1007/s10470-007-9114-0

19. Benilova I, Chegel VI, Ushenin YV, Vidic J, Soldatkin AP, Martelet C, Pajot E, Jaffrezic-Renault N (2008) Stimulation of human olfactory receptor 17-40 with odorants probed by surface plasmon resonance. Eur Biophys J 37:807–814. https://doi.org/10.1007/s00249-008-0272-5

20. Benilova I, Vidic JM, Pajot-Augy E, Soldatkin A, Martelet C, Jaffrezic-Renault N (2008) Electrochemical study of human olfactory receptor OR 17–40 stimulation by odorants in solution. Mater Sci Eng C 28:633–639. https://doi.org/10.1016/j.msec.2007.10.040

21. Hou Y, Helali S, Zhang A, Jaffrezic-Renault N, Martelet C, Minic J, Gorojankina T, Persuy M-A, Pajot-Augy E, Salesse R (2006) Immobilization of rhodopsin on a self-assembled multilayer and its specific detection by electrochemical impedance spectroscopy. Biosens Bioelectron 21:1393–1402. https://doi.org/10.1016/j.bios.2005.06.002

22. Minic J, Grosclaude J, Aioun J, Persuy M-A, Gorojankina T, Salesse R, Pajot-Augy E, Hou Y, Helali S, Jaffrezic-Renault N (2005) Immobilization of native membrane-bound rhodopsin on biosensor surfaces. Biochim Biophys Acta 1724:324–332. https://doi.org/10.1016/j.bbagen.2005.04.017

Chapter 7

Screening Methods for Cell-Free Synthesized GPCR/Nanoparticle Samples

Zoe Köck, Volker Dötsch, and Frank Bernhard

Abstract

Cell-free protein expression systems and lipid nanoparticle technologies are core platforms for membrane protein synthesis. The implementation of preassembled nanodiscs allows the co-translational insertion of membrane proteins into tailored lipid bilayers in the absence of any artificial hydrophobic compounds. This strategy is particularly interesting for detergent sensitive or otherwise critical membrane proteins such as G-protein-coupled receptors (GPCRs). Cell-free expression reactions are completed within a day and the formed GPCR/nanodisc particles can be purified directly out of the reaction mixture by affinity tags and without any further manipulation. The streamlined procedure reduces risk of GPCR denaturation and the sample quality can further be supported by supplying chaperones or other beneficial compounds directly into the expression reactions.

GPCRs inserted into nanoparticle membranes are excellent tools for a variety of applications such as ligand screening, engineering or even structural characterization. In this chapter, we provide protocols for the reaction set-up and efficient cell-free production of functionally folded GPCRs reaching µM concentrations in the final expression reactions. We further exemplify the tuning of GPCR sample quality and discuss their application for throughput ligand screening and for the analysis of ligand-binding characteristics.

Key words G-protein-coupled receptors, Cell-free protein expression, Nanodiscs, GPCR ligand screening, Preformed membranes, Synthetic biology

1 Introduction

Cell-free (CF) protein expression systems have emerged as alternative tools for the production of diverse types of membrane proteins including GPCRs [1–3]. Advantages of CF systems are reduced biological complexity, open accessibility, and in particular high versatility of the reaction environment. The addition of preformed nanodiscs (NDs) into CF reactions is a new approach that allows the co-translational insertion of membrane proteins into membranes of defined lipid composition [2–5]. This strategy enables the completely detergent-free production of membrane proteins for functional and structural characterization [2, 3, 5].

Sofia Aires M. Martins and Duarte Miguel F. Prazeres (eds.), *G Protein-Coupled Receptor Screening Assays: Methods and Protocols*, Methods in Molecular Biology, vol. 2268, https://doi.org/10.1007/978-1-0716-1221-7_7,
© Springer Science+Business Media, LLC, part of Springer Nature 2021

NDs are stable particles and can be assembled in different sizes and with a multitude of different lipids and lipid mixtures [4]. The planar discs are accessible from both sides making them an ideal tool to analyze ligand-binding mechanisms of membrane proteins. The co-translational insertion of GPCRs and other membrane proteins into NDs takes a few hours to finally obtain μM concentrations of the synthesized proteins in the CF reaction mixture [2, 3, 5]. A single milliliter of CF reaction can therefore be sufficient to perform thousands of subsequent ligand-binding studies in microplates. GPCR/ND particles are purified directly from the CF reaction by conventional affinity chromatography via small purification tags [3]. Critical steps such as membrane extraction, exchange of hydrophobic environments, and contact with detergents are not required. CF expression reactions are performed in several formats such as microplates or dialysis devices and in volumes ranging from nano/microliters for throughput screening applications up to several milliliters for preparative scale preparations [6].

This chapter gives a detailed protocol for the co-translational insertion of GPCRs into preformed NDs. We exemplify the strategy with the human endothelin B receptor (ETB) and the turkey β1-adrenergic receptor (β1AR). Specific requirements of ETB and β1AR for improved folding, such as redox conditions, lipid composition of the ND membranes, or specific CF lysate preparations, are highlighted [2, 3, 7]. Full-length wild-type as well as engineered and truncated GPCRs can be synthesized with similar CF expression protocols. Described applications are (i) the tuning of GPCR folding, (ii) the screening of ligand libraries, and (iii) the characterization of GPCR ligand-binding characteristics (Fig. 1).

2 Materials

2.1 S30 and Heat-Shocked (HS-S30) Lysate Preparation

1. Fermenter (5–10 L).

2. French press or similar mechanic cell disruptor.

3. Thermoshaker.

4. Common centrifuges and rotors.

5. LB agar plates.

6. Dialysis tubes, 12–14 kDa MWCO.

7. Antifoam.

8. *E. coli* strain A19, D10, or BL21.

9. 1× YTPG medium: 10 g/L yeast extract, 16 g/L peptone, 5 g/L NaCl, 100 mM glucose, 22 mM KH_2PO_4, 40 mM K_2HPO_4.

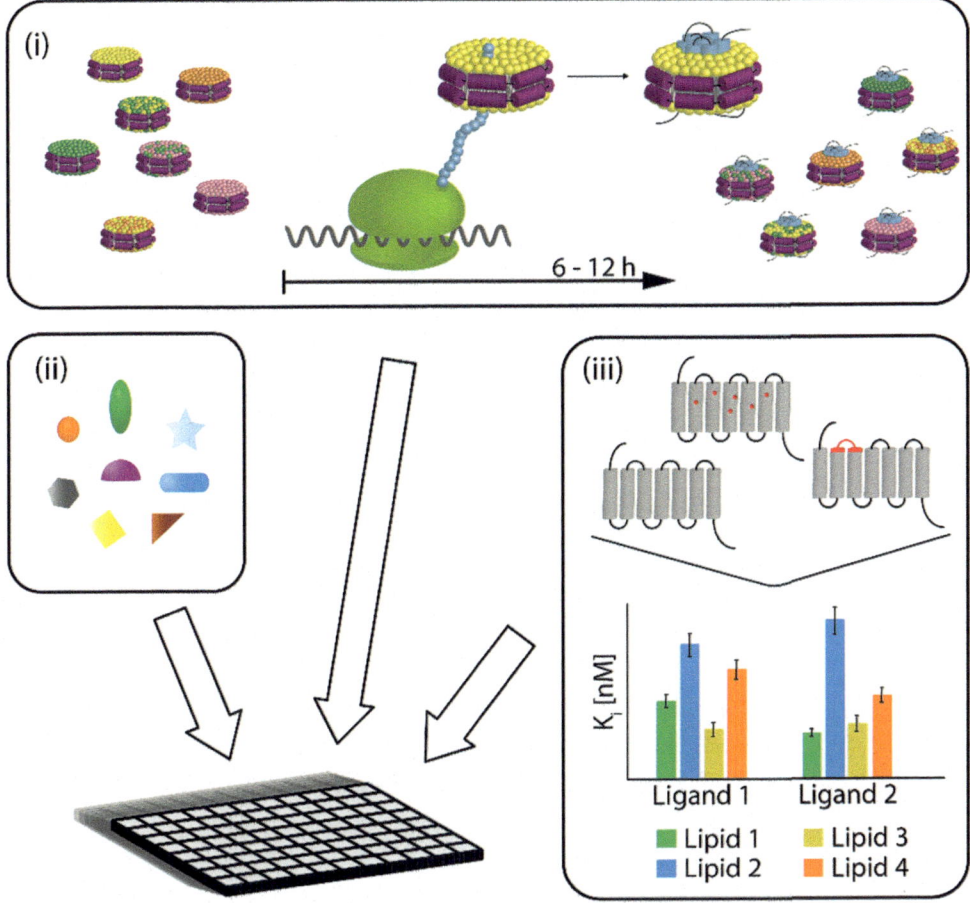

Fig. 1 Applications for CF-expressed GPCR/ND complexes in screening assays. (i) Lipid screening. The GPCR is co-translationally inserted in a set of NDs containing different lipid compositions. Subsequent quantitative ligand-binding assays will reveal preferred lipid environments for functional GPCR folding and stability. (ii) Ligand screening. Ligand libraries can be screened in microplate formats by using CF-synthesized GPCR/ND samples. (iii) GPCR functional tuning. Libraries of engineered GPCRs can be CF synthesized and effects on ligand binding can be assessed in different lipid environments by determination of K_D, B_{max}, or IC_{50} values

10. LB medium: 10 g/L peptone, 5 g/L yeast extract, 5 g/L NaCl.

11. S30 buffer A: 10 mM Tris-acetate pH 8.2, 14 mM Mg(OAc)$_2$, 60 mM KCl, 6 mM β-mercaptoethanol.

12. S30 buffer B: 10 mM Tris-acetate pH 8.2, 14 mM Mg(OAc)$_2$, 60 mM KCl, 1 mM DTT, 1 mM PMSF.

13. S30 buffer C: 10 mM Tris-acetate pH 8.2, 14 mM Mg(OAc)$_2$, 60 mM KOAc, 0.5 mM DTT.

14. 5 M NaCl.

15. Ethanol absolute.

2.2 Membrane Scaffold Protein (MSP) Expression, Purification, and ND Assembly

1. Chromatographic system (e.g., ÄKTA purifier).
2. Immobilized metal affinity chromatography column.
3. Centriprep filter, 10 kDa MWCO.
4. Dialysis tube, 12–14 kDa MWCO.
5. Slide-A-lyzer, 10 kDa MWCO.
6. pET-28-MSP1E3D1 vector.
7. T7 express cells (NEB).
8. LB: 10 g/L peptone, 5 g/L yeast extract, 5 g/L NaCl, in liquid medium and agar plates.
9. $1000\times$ kanamycin stock: 30 mg/mL in H_2O.
10. 10% (w/v) glucose stock.
11. 1 M IPTG stock.
12. cOmplete EDTA-free protease inhibitor.
13. 10% (v/v) Triton X-100 stock.
14. MSP buffer 1: 40 mM Tris–HCl pH 8.0, 300 mM NaCl, 1% Triton X-100.
15. MSP buffer 2: 40 mM Tris–HCl pH 8.9, 300 mM NaCl, 50 mM cholic acid.
16. MSP buffer 3b: 40 mM Tris–HCl pH 8.9, 300 mM NaCl.
17. MSP buffer 3: 40 mM Tris–HCl pH 8.0, 300 mM NaCl.
18. MSP buffer 4: 40 mM Tris–HCl pH 8.0, 300 mM NaCl, 50 mM imidazole.
19. MSP elution buffer: 40 mM Tris–HCl pH 8.0, 300 mM NaCl, 300 mM imidazole.
20. MSP dialysis buffer: 40 mM Tris–HCl pH 8.0, 300 mM NaCl, 10% glycerol.
21. Lipid stocks: 50 mM lipid in 100 mM sodium cholate.
22. DF buffer: 10 mM Tris–HCl pH 8.0, 100 mM NaCl.
23. 10% Dodecylphosphocholine (DPC) stock in H_2O.

2.3 Cell-Free Expression

1. CF expression container, e.g., MD100 dialysis cartridges.
2. 96-Deep-well microplates as feeding mix containers.
3. 96-Well plate with V-shape bottom.
4. Dialysis tube, 12–14 kDa MWCO.
5. Fluorescence spectrophotometer for detection of green fluorescent protein (GFP).
6. CF stock solutions (Table 1).
7. T7RNA-polymerase (T7RNAP).
8. DNA plasmids in vectors with T7 promoter.

Table 1
CF compound stocks and final concentrations[a]

3× Mastermix with common compounds	Concentration		Volume [μL][b]	Remark
	Stock	Final		
Acetyl phosphate (Li+, K+), pH 7.0	1 M	20 mM	378	NTP (nucleotide triphosphate) regeneration / Stock not fully soluble; mix well before pipetting
Phosphoenolpyruvic acid (K+), pH 7.0	1 M	20 mM	378	NTP regeneration
NTP mix, pH 7.0	75×	1×	252	
Folinic acid	10 mg/mL	0.1 mg/mL	189	Formyl-methionine formation
Tris-acetate, pH 8.0	2.4 M	100 mM	788	
20 Amino acid mix[c]	25 mM[d]	1 mM	757	Stock not fully soluble; mix well before pipetting
Redox system[e]: GSH:GSSG = 3:1	100 mM[f]	3/1 mM	189/63	Reducing conditions / Disulfide formation
$Mg(OAc)_2$[g]	1 M	14 mM	265	
KOAc[g]	10 M	181 mM	342	
H_2O			2700	
RM only				
RiboLock	40 U/μL	0.3 U/μL	1.5	RNAse inhibitor
Pyruvate kinase	10 mg/mL	0.04 mg/mL	0.8	NTP regeneration
Template DNA (plasmid)	0.5 mg/mL	5–15 μg/mL	6.0	
T7RNAP[h]	> 20 U/μL	2–6 U/μL	≥ 10	
S30 lysate	30–40 mg/mL	10–15 mg/mL	70	
Optional				

(continued)

Table 1
(continued)

3× Mastermix with common compounds	Concentration		Volume [μL][b]	Remark
	Stock	Final		
Nanodiscs[i]	1 mM	20–100 μM		Added in RM only
PEG 8000[j]	40%	2%		Crowding reagent
NaN_3[k]	10%	0.05%		Preservative
Complete cocktail[l]	50×	1×		Protease inhibitor

[a]Compounds subject to optimization are indicated in yellow

[b]Volumes of compound stocks needed for the exemplified screening of 5 Mg^{2+} concentrations. For common compounds the volumes for the 3× mastermix are given. For RM compounds the volumes for one Mg^{2+} concentration (= one duplicate reaction) are given. Volumes for common compounds are rounded

[c]Adjusting the amino acid mixture according to the composition of the target protein may increase expression

[d]Each amino acid in the amino acid mix has a final concentration of 25 mM

[e]Selection of a redox systems depends on the presence of disulfide bridges in the target protein. For GPCRs we recommend the indicated GSH (glutathione)/GSSG (glutathione disulfide) system. GSH stocks should be prepared fresh

[f]For both, GSH and GSSG, a 100 mM stock is prepared

[g]Mg^{2+} and potentially also K^+ concentrations should be adjusted to each new DNA template as described in Subheading 3.4

[h]Commercial T7RNAP might need to be concentrated to obtain a suitable stock concentration. We recommend an initial screen to identify the best final concentration. Alternatively, T7RNAP can efficiently be synthesized in *E. coli* and prepared in-house [8]

[i]Nanodiscs are used as hydrophobic environment for the co-translational solubilization of GPCRs in the exemplified protocols. Depending on the intended applications, liposomes or detergents might replace them

[j]PEG could increase expression efficiency but might also support aggregation of proteins

[k]In case of prolonged incubation times, adding a preservative is recommended

[l]Recommended additive in case of proteolytic degradation of the target protein

2.4 Radioligand-Binding Assay

1. Glass fiber filter plate.

2. Shaker.

3. Vacuum manifold.

4. Gamma counter or scintillation counter.

5. 96-Well plates.

6. RBA buffer: 50 mM HEPES-NaOH pH 7.5, 1 mM $CaCl_2$, 5 mM $MgCl_2$.

7. 0.3% Polyethyleneimine (PEI).

8. Radiolabeled ligand (*see* **Note 1**).

9. Cold ligand.

3 Methods

3.1 E. coli *Extract* Preparation

We recommend lysate preparation out of the *E. coli* A19 strain, from which we typically obtain soluble GFP expression with yields of 3–5 mg/mL (*see* **Note 2**) [8]. The documented standard S30 lysate preparation protocol is tuned for high protein expression efficiency. The optional HS-S30 lysate preparation for chaperone enrichment results in higher folding competence, but also reduces expression yields. However, using blends of S30 and HS-S30 lysates restores high expression efficiencies and combines benefits of both lysate types [7]. Lysate preparation takes 2.5 days in total. Starting with a 10 L fermenter, we routinely obtain 60 to 80 mL S30 lysate sufficient for 180–240 mL of final CF reaction. Lysates can be stored in aliquots at −80 °C for at least 1 year.

1. On day 1, prepare a preculture by inoculating the *E. coli* A19 strain in 200 mL LB media and incubate overnight at 37 °C with shaking (*see* **Note 3**).

2. For fermentation, 10 L 1× YPTG media are needed. Dissolve yeast extract, peptone, and NaCl in 8 L H_2O and autoclave in the fermenter. Prepare 1 L 10× KH_2PO_4/K_2HPO_4 (220 mM and 400 mM, respectively) in H_2O and autoclave separately. Prepare 1 L 10× (1 M) glucose in H_2O and sterile filtrate.

3. On day 2, add the 1 L 10× KH_2PO_4/K_2HPO_4 and 1 L 10× glucose to the rest of the media in the fermenter.

4. Add an appropriate amount of antifoam to the supplemented media (*see* **Note 4**).

5. Inoculate the fermenter with 100 mL preculture and start fermentation with vigorous stirring (~500 rpm) and good aeration at 37 °C.

6. Record bacterial growth by measuring OD_{600} every 30 min.

7. For standard S30: At an OD_{600} of 3.8 to 4.5 (mid-log phase), chill the broth to ~20 °C (*see* **Note 5**).

8. Optional for HS-S30: At an OD_{600} of 3.5 to 4.0, add 300 mL 100% ethanol and increase the temperature to 42 °C for 45 min to induce a heat-shock response. Then chill the broth down to ~20 °C.

9. Harvest the cells by centrifugation at 8000 × g and 4 °C for 15 min.

10. Discard the supernatant and resuspend the pellets in a total of 300 mL S30A buffer. Centrifuge at 8000 × g and 4 °C for 10 min.

11. Repeat **step 10** twice with prolonged centrifugation for 30 min in the last centrifugation step.

12. Weigh the pellet and resuspend it in 110% (w/v) of S30B buffer.

13. Disrupt the cells by using a French press at 1000 psi (*see* **Note 6**).

14. Centrifuge disrupted cells at 30,000 × g and 4 °C for 30 min. Transfer the supernatant into fresh centrifugation tubes and repeat the centrifugation.

15. Carefully transfer the cleared supernatant into a fresh tube (*see* **Note 7**).

16. Adjust the supernatant to a final concentration of 400 mM NaCl from a 5 M NaCl stock. Mix the solution by inverting the tube a couple of times and incubate it at 42 °C for 45 min. The solution will become turbid.

17. Dialyze the turbid solution twice against 5 L of S30C buffer at 4 °C. The first dialysis buffer should be exchanged after ~3 h, followed by dialysis overnight.

18. On day 3, centrifuge the dialyzed lysate at 30,000 × g and 4 °C for 30 min and collect the cleared supernatant.

19. Aliquot the lysate in suitable volumes, directly shock-freeze it in liquid nitrogen, and store it at −80 °C (*see* **Note 8**).

3.2 MSP Expression and Purification

We express MSP derivatives from pET28 vectors in T7 express cells cultivated in 2 L baffled Erlenmeyer-flasks with LB media containing 0.5% (w/v) glucose and the appropriate antibiotic. The following protocol describes MSP expression and purification from 9.6 L LB medium, yielding approximately 60 to 150 mg of protein. Other strains suitable for T7RNAP-based protein production or other volumes may be used.

1. Prepare two 600 mL precultures. Inoculate single colonies from a fresh agar plate in each 600 mL LB media containing the appropriate antibiotic, but no glucose. Incubate overnight at 37 °C with shaking.

2. For expression, 16 baffled flasks containing 600 mL LB media each are supplemented with kanamycin and glucose to final concentrations of 30 mg/L and 10%, respectively.

3. Inoculate each flask 1:12 with the preculture and incubate at 37 °C with shaking.

4. At an OD_{600} of 1.0, induce protein expression by addition of IPTG to a final concentration of 1 mM.

5. First, cultures are incubated at 37 °C and shaking for 1 h, then temperature is decreased to 28 °C and cultures are incubated for further 4 h.

6. Harvest the cells by centrifugation at 6000 × g and 4 °C for 15 min and combine the pellets of 3 flasks in a 50 mL falcon tube. The combined pellets should weigh approximately 18 g.

7. Store the pellets at −20 °C or directly proceed with MSP purification.

8. Resuspend one pellet in 45 mL MSP buffer 3 supplemented with one tablet of cOmplete protease inhibitor.

9. Add 5 mL 10% (v/v) Triton-X-100 to a final concentration of 1% (v/v) (see **Note 9**).

10. Sonicate the resuspended cells 3× for 60 s and 3× for 45 s with rest periods on ice for 60 s.

11. Centrifuge sonicated cells at 30,000 × g and 4 °C for 30 min and filter the supernatant through sterile 0.45 μm filter.

12. Equilibrate a Ni^{2+}-IMAC column with 5 CVs of MSP buffer 1.

13. Load the filtered supernatant on the column, collect the flow through, and reload twice.

14. Wash the column with 5 CVs of MSP buffer 1, 2 CVs of MSP buffer 3b, and 5 CVs of each MSP buffers 2, −3, and −4, respectively.

15. Elute MSP with 5 CVs of MSP elution buffer in 1 mL fractions.

16. Pool MSP-containing fractions, add glycerol to a final concentration of 10% (v/v) and dialyze 2× 12 h against 5 L MSP dialysis buffer at 4 °C.

17. Determine protein concentration, shock-freeze MSP in liquid nitrogen and store at −80 °C.

3.3 Nanodisc Assembly

NDs (see **Note 10**), self-assemble by mixing purified MSP derivates with detergent solubilized lipids, followed by subsequent dialysis to remove the detergent. Important for ND quality is to carefully adjust the appropriate MSP:lipid ratio, which depends on the selected MSP derivate and lipid (Table 2; see **Note 11**). We generally prepare ND assembly mixtures with volumes of 8–10 mL. The yield of preassembled NDs then depends on the MSP concentration added into the mixture. We usually add MSP in final concentrations of 1.5–4 mg/mL.

Table 2
Recommended MSP:lipid ratios for ND assembly

lipid[a]	MSP1E3D1: lipid (1: ×)	MSP1: lipid (1: ×)
Aso-PC	50	40
DEPG	85	–
DMPA	110	–
DMPC	115	80
DMPG	110	70
DOPA	90	–
DOPE	80	30
DOPC	80	30
DOPG	80	30
DOPS	90	–
POPC	85	55
POPG	90	–
POPS	90	–
SOPG	80	–

[a]*Aso-PC* L-α–phosphocholine, *DEPG* 1,2-dielaidoyl-sn-glycero-3-phospho-(1′-rac-glycerol), *DMPA* 1,2-dimyristoyl-sn-glycero-3-phosphate, *DMPC* 1,2-dimyristoyl-sn-glycero-3-phos-phocholine, *DMPG* 1,2-dimyristoyl-sn-glycero-3-phospho-(1′-rac-glycerol), *DOPA* 1,2-dioleoyl-sn-glycero-3-phosphate, *DOPE* 1,2-dioleoyl-sn-glycero-3-phosphoetha-nolamine, *DOPC* 1,2-dioleoyl-sn-glycero-3-phosphocholine, *DOPG* 1,2-dioleoyl-sn-glycero-3-phospho-(1′-rac-glycerol), *DOPS* 1,2-dioleoyl-sn-glycero-3-phospho-L-serine, *POPC* 1–1,2-dioleoyl-sn-glycero-3-phospho-L-serine, *POPG* 1-palmitoyl-2-oleoyl-sn-glycero-3-phospho-(1′-rac-glycerol), *POPS* 1-palmitoyl-2-oleoyl-sn-glycero-3-phospho-L-serine, *SOPG* 1-stearoyl-2-oleoyl-sn-glycero-3-phospho-(1′-rac-glycerol)

1. Prepare a 50 mM stock of the desired lipid in 100 mM sodium cholate. For an 8–10 mL assembly mixture, 2–3 mL lipid stock should be sufficient. However, the volume is dependent on the final MSP concentration.

2. Mix lipid and MSP in the appropriate concentration (Table 2), add 10% (w/v) DPC to a final concentration of 0.1% (v/v) (*see* **Note 12**), and fill up to the final desired volume with DF buffer.

3. Incubate the mixture at RT and shaking for 1 h.

4. Dialyze the ND mix in a 12 mL Slide-A-lyzer (10 kDa MWCO) 3× for 12 h against 5 L DF buffer at RT.

5. Transfer the dialyzed NDs into fresh tubes and centrifuge at 30,000 × g and 4 °C for 20 min to remove potential precipitates.

6. Transfer the supernatant into a fresh tube and concentrate to a final concentration of 500 to 700 μM (~0.5 mL), using a Centriprep concentrating unit (10 kDa MWCO).

7. Concentrated NDs can be analyzed for homogeneity by size exclusion chromatography.

8. Freeze appropriate ND aliquots in liquid nitrogen and store at −80 °C.

3.4 Basic Cell-Free Expression Protocol

For CE-CF expression (*see* **Note 13**), a reaction mix (RM) is separated from a feeding mix (FM) via a semipermeable membrane, allowing diffusion of inhibitory by-products from the RM into the FM, as well as continuous supply of fresh energy sources for a certain period. The expression reaction is therefore extended to approx. 12 h. We usually work with a ratio of RM:FM of 1:15–1:17.

The optimal Mg^{2+} concentration is essential for the expression of the target. The default Mg^{2+} concentration of the S30 lysate is approximately 5 mM. Depending on the protocol and on selected additives, optimal Mg^{2+} concentration of a reaction usually varies between 14 and 28 mM. The following protocol exemplifies a screen with 5 Mg^{2+} concentrations (14, 16, 18, 20, and 22 mM) in a CE-CF expression setup with a RM volume of 100 μL and a FM volume of 1700 μL (1:17). Each concentration is tested in duplicates, making 10 reactions in total. RM and FM of duplicate reactions are prepared in common tubes (= 200 μL and 3400 μL). Stock solutions and final concentrations of each compound are given in Table 1. The compounds listed as common are of low molecular weight and have identical concentrations in RM and FM. For these compounds, a 3× mastermix is prepared and the individual volumes required to reach the threefold concentration in the mastermix are already given in Table 1. Suitable aliquots from the completed 3× mastermix are then transferred to the RMs and FMs (*see* **Note 14**).

1. Calculate the total final volumes needed for RM and FM (10 × 100 μL = 1000 μL and 10 × 1700 μL = 17,000 μL, respectively).

2. Prepare one 15 mL falcon for the 3 × mastermix, 5 × 1.5 mL reaction tubes for the RMs and 5 × 2 mL reaction tubes for the FMs.

3. Prepare the 3× mastermix according to the final compound concentrations given in Table 1, *see* **Note 15**. The total volume of the 3× mastermix is 6301 μL (for RM 334 μL and for FM 5667 μL = 6001 μL + 5% (= 300 μL) excess volume = 6301 μL). Take into account that the final concentrations refer to the final reaction volume. For the 3× mastermix, final concentrations for the common compounds have to be multiplied by 3. The example volumes for the individual common compounds given

in Table 1 already refer to the volumes needed for the $3\times$ mastermix. The $3\times$ mastermix is completed with 2700 μL H_2O to the final volume of 6301 μL. Mix thoroughly after completion.

4. Transfer 67 μL of the $3\times$ mastermix into each RM tube and 1134 μL into each FM tube.

5. Adjust the FMs to final Mg^{2+} concentrations of 14, 16, 18, 22, and 22 mM from a 100 mM $Mg(OAc)_2$ stock. Then, fill up with H_2O to final volumes of 3400 μL.

6. Adjust the RMs to final Mg^{2+} concentrations of 14, 16, 18, 22, and 22 mM from a 100 mM $Mg(OAc)_2$ stock.

7. Add all high molecular weight compounds to the RMs and fill up to final volumes of 200 μL (Table 1). The example volumes given in Table 1 indicate the volumes needed for each RM tube.

8. Fill 1700 μL of the FMs into each cavity of a 96-deep-well plate.

9. Fill 100 μL of the RMs into MD100 cartridges and place them into the cavities of the 96-deep-well plate containing the appropriate FMs (*see* **Note 16**).

10. Incubate the reactions overnight at 30 °C with slight shaking.

11. After incubation, transfer the RMs into fresh tubes. If the screen is performed with GFP or a GFP fusion protein, determine protein expression by measuring the GFP fluorescence at 495 nm. In case of precipitates, centrifuge the RM at $18,000 \times g$ and 4 °C for 10 min and separate the supernatant from the precipitate, before fluorescence measurement.

3.5 Tuning of Co-translational GPCR Folding

In the following, we discuss applications for tuning of GPCR folding in CF reactions. CF expression is an artificial system, and depending on the selected hydrophobic environment, the functional solubilization of the synthesized GPCR is not quantitative. A primary task is therefore to increase the fraction of functionally folded protein by adjusting the CF expression environment according to the specific requirements of a selected GPCR [2, 3] (*see* **Notes 17** and **18**). For the described co-translational insertion of GPCRs into preformed NDs, membrane protein solubilization is mainly dependent on three determinants: ND size, concentration, and composition. For GPCR expression, the larger MSP1E3D1 NDs proved to be most suitable. Nevertheless, for some applications, e.g., NMR, smaller NDs would be favorable. Lipids are available with different head group charges, fatty acid chain lengths, and saturation states that could significantly affect membrane protein stability. Further parameters relevant for protein stability can be the addition of ligands, allosteric modulators, or cofactors.

Depending on the expression efficiency, screening for quality tuning is operated in the two-compartment CE-CF or in the single compartment batch configurations. Screening in the batch configuration can be performed in microplate formats with automated pipetting devices that allow also the correlated screening of several compounds. We describe a batch configuration screening setup of NDs with various lipid compositions in 96-well plates with V-shape bottom. The effects of lipid composition on GPCR folding and stability can be evaluated by measuring specific ligand binding. Measuring GFP fluorescence by using GPCR-GFP fusions will further indicate lipid environments that improve the membrane insertion of the receptor.

1. Prepare a set of NDs with different lipid compositions according to Subheading 3.3 (*see* **Note 19**).

2. Express the GPCR in the presence of the different NDs according to a CF batch configuration protocol (*see* **Note 20**).

3. Determine receptor concentration via GFP fluorescence.

4. Dilute the receptor in RBA buffer, so that the final concentration of active receptor in 30 µL is approximately $1/10$ K_D of the selected radioligand (*see* **Note 21**).

5. For determination of nonspecific binding, preincubate control reactions of the receptor with a large excess of unlabeled ligand at RT with shaking, until equilibrium (*see* **Note 22**).

6. Titrate the labeled ligand to the receptor and incubate at RT with shaking, until equilibration. To reach saturation, ligands should be titrated up to $10 \times K_D$.

7. Add suitable aliquots of an appropriate resin into the microplate cavities (e.g., Ni^{2+} resin for His-tagged GPCRs or streptactin beads for strep-tagged GPCRs).

8. Incubate the plate with shaking for 20 min to allow coupling of the proteins to the resin.

9. Centrifuge the microplate at $1 \times g$ for 1 min to pellet down the resin.

10. Discard the supernatants and wash each cavity with RBA buffer.

11. Repeat **steps 9** and **10** for 2–5 times.

12. Measure the remaining counts per minute (CPM) in a suitable counter according to the type of isotope.

13. Subtract nonspecific CPM, determined with the control reactions, from total CPM to calculate specific binding. To plot the specific CPM against the corresponding radioligand concentration, enter the datasets into a standard calculation program, e.g., Origin or GraphPadPrism.

The programs will determine K_D and B_{max} values by using the formula: $(CPM \text{ specific}) = B_{max} \times \frac{[\text{radioligand}]}{K_D + [\text{radioligand}]}$ and nonlinear regression.

14. For calculation of the percental ligand-binding activity, use the datapoint where B_{max} is reached. Determine the amount of bound ligand [fmol] by using the specific activity [CPM/fmol] of the ligand. From (i) the amount of bound ligand [fmol], (ii) the final receptor concentration [mol/L] in the assay, and (iii) the assay volume [L], the percental binding activity of the GPCR can be determined.

15. Comparison of GPCR-specific ligand-binding activity and GPCR concentration of the expression reactions with the different NDs will reveal the most suitable lipid environment.

3.6 Screening Ligand Libraries

Competitive binding assays can be used to screen large compound libraries by displacing the labeled ligands. Radioligand-binding assays are highly specific and can be performed with non-purified GPCR samples directly from the CF RM. Samples of ETB and β1AR are typically diluted 500–1000×. 1 mL of CE-CF expression is therefore sufficient for >10,000 assay reactions. Alternatively, ligand-receptor interactions can be determined via fluorescence anisotropy or SPR applications. However, these methods require higher receptor amounts and protein purification.

We exemplify a competition screen based on radioligand binding performed in 96-well microplates and starting from a 1 mL CE-CF reaction. Appropriate GPCR-GFP fusions allow determination of receptor concentration directly from the RM. The exemplified assay volume is 30 μL; however much smaller volumes are possible (Fig. 2).

1. Express the GPCR-GFP construct in 1 mL RM according to Subheading 3.4 with the optional addition of suitable pre-formed NDs as determined in Subheading 3.5 (*see* **Note 23**). Set the final Mg^{2+} concentration to the optimal concentration determined previously for your expression protocol.

2. After synthesis, centrifuge the RM at $18,000 \times g$ and 4 °C for 10 min and separate the supernatant from the precipitate.

3. Determine receptor concentration via GFP fluorescence.

4. Dilute the receptor with RBA buffer in microplate cavities, so that its final concentration in 30 μL is approximately $1/10$ K_D of the selected radioligand (*see* **Note 21**).

5. Add an excess or a set of suitable concentrations of the unlabeled library compounds to the receptor and incubate with shaking at 10–20 °C.

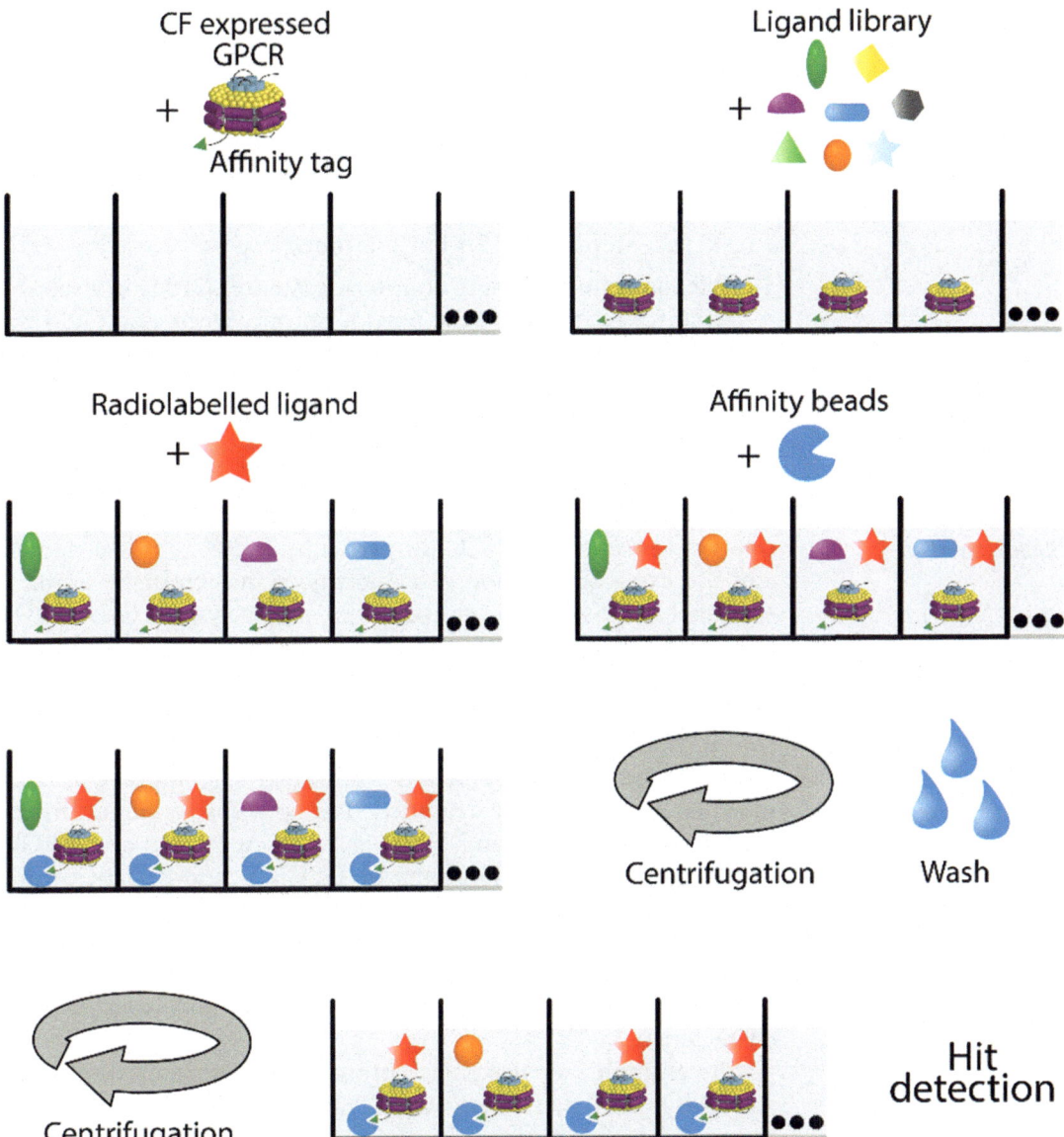

Fig. 2 Illustration of a ligand library screen based on competition binding with a radiolabeled ligand. Steps are (i) aliquoting the CF synthesized GPCR/ND samples into microplate cavities; (ii) addition of the ligand library compounds; (iii) addition of the radiolabeled competitor ligand; (iv) addition of affinity beads for immobilization of the GPCR/NDs; (v) incubation, centrifugation, and washing steps to remove unbound ligands; (vi) hit detection by measuring residual radioactivity in the microplate wells

6. Add the labeled ligand at a final concentration of approximately its K_D and incubate with shaking, until equilibrium (*see* **Note 24**).

7. Add suitable aliquots of an appropriate resin into the cavities (e.g., Ni^{2+} resin for His-tagged proteins or strep-tactin beads for strep-tagged proteins).

8. Incubate the plate with shaking for 20 min to allow coupling of the proteins to the resin.

9. Centrifuge the microplate at $1 \times g$ for 1 min to pellet down the resin.

10. Discard the supernatants and wash each cavity with RBA buffer.

11. Repeat **steps 9** and **10** for 2–5 times.

12. Measure the remaining counts per minute (CPM) in a suitable counter according to the type of isotope. Decreased CPM in a cavity indicate interaction between the compound and the receptors' ligand-binding site.

3.7 Characterization of GPCR Ligand-Binding Characteristics

Ligand-binding characteristics such as selectivity, dissociation constants, or IC_{50} values are of prime importance for pharmaceutical applications. Key determinants for ligand selectivity are the extracellular domains of GPCRs, in particular the N-terminal domain and the extracellular loop 2 connecting transmembrane domains 4 and 5 [10]. Systematic construction and analysis of truncated or chimeric derivatives of a wild-type GPCR can be a straightforward approach to localize residues involved in ligand selectivity [11]. Most derivatives can be synthesized with similar CF expression protocols. This is often in contrast to in vivo expression systems, where engineered derivatives of wild-type membrane proteins may be miss-targeted or degraded. The lipid environment furthermore modulates ligand-binding characteristics of GPCRs [3]. Observed tissue variations of GPCR ligand selectivity may therefore account to specific membrane lipid compositions. GPCRs inserted into nanoparticles containing different and defined membrane compositions offer a convenient and rapid tool to analyze such lipid-dependent variations in ligand-binding characteristics.

As example, we describe a protocol to determine the ligand selectivity of two derivatives, a thermostabilized and a non-stabilized construct, of the turkey β1-adrenergic receptor inserted into NDs containing different lipid compositions (Fig. 3) [3]. The strategy is based on competitive binding of non-labeled antagonists by replacement of the radiolabeled antagonist [^{3}H] alprenolol. Expression of the GPCR-GFP constructs can be performed either in 96-well microplates (batch configuration) or in MD100 devices (CE-CF configuration).

1. Prepare a set of NDs with different lipid compositions according to Subheading 3.3 (*see* **Note 19**).

2. Express the GPCRs in the presence of the various NDs either in CE-CF or in batch configuration [9] (*see* **Note 20**).

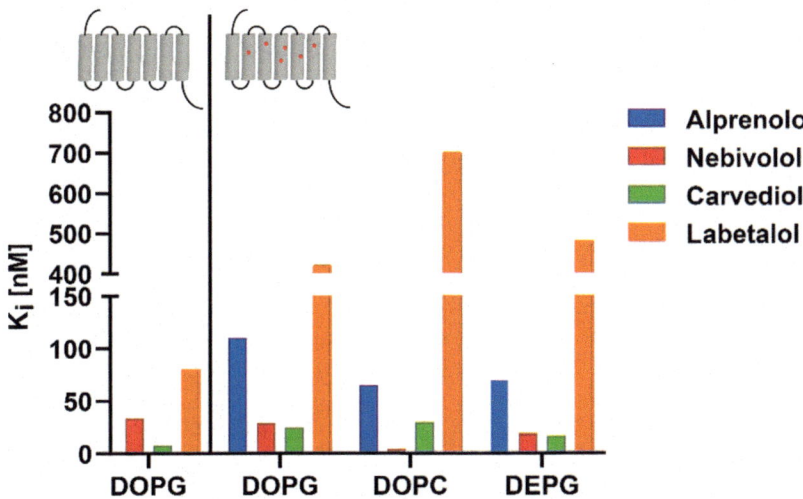

Fig. 3 Ligand-binding properties of engineered GPCRs. In the examples, wild-type and an engineered derivative of the turkey ß1-adrenergic receptor are analyzed for binding of a variety of antagonists within the context of different nanodisc membranes. GPCR engineering as well as the GPCR membrane environment significantly affect the affinity of the analyzed ligands

3. After synthesis, determine receptor concentration via GFP fluorescence.

4. Dilute the receptors in RBA buffer, so that the final concentration in 30 μL is approximately $1/10$ K_D of the selected radioligand (*see* **Note 21**).

5. Titrate the unlabeled antagonists to be analyzed to the receptor and incubate at RT with shaking, until equilibration (*see* **Note 22**).

6. Add the labeled ligand at a final concentration of approximately $1\times$ K_D and incubate with shaking, until equilibrium (*see* **Note 24**).

7. Add a suitable amount of an appropriate resin into the reactions (e.g., Ni^{2+} resin for His-tagged proteins or strep-tactin beads for strep-tagged proteins).

8. Incubate the plate with shaking for 20 min to allow coupling of the proteins to the resin.

9. Centrifuge the microplate at $1 \times g$ for 1 min to pellet down the resin.

10. Discard the supernatants and wash each cavity with RBA buffer.

11. Repeat **steps 9** and **10** for 2–5 times.

12. Measure the remaining CPM in a suitable counter according to the type of isotope.

13. To plot the CPM against the corresponding unlabeled compound concentration, enter the datasets into a standard

calculation program, e.g., Origin or GraphPadPrism. The curve should have two plateaus. The first plateau describes the total binding (TB) of the radioligand. The second plateau equals nonspecific binding (NS).

14. The programs will determine IC_{50} values using the formula:

$$(CPM) = NS + \frac{TB - NS}{1 + 10^{\log[\text{unlabelled ligand}] - \log(IC_{50})}}$$

and nonlinear regression.

15. If necessary, convert IC_{50} into K_i using the formula:

$$K_i = \frac{IC_{50}}{1 + \frac{[\text{radioligand}]}{K_D}}$$

16. For concluding remarks *see* **Note 25**.

4 Notes

1. For ETB: $[^{125}I]$endothelin-1 (e.g. PerkinElmer); for β1AR: $[^{3}H]$alprenolol (e.g., American radiolabeled chemicals).

2. The GFP expression efficiency is given for a CE-CF configuration (1 mL RM and 14 mL FM) with a final S30 lysate concentration of 35%. Other *E. coli* strains such as D10 or BL21 derivatives might result in lysates of similar efficiency.

3. Only fresh liquid precultures should be used for fermenter inoculation. Avoid inoculation directly from glycerol stocks, agar plates, or old precultures.

4. The amount of antifoam depends on the selected fermentation conditions; we use 4 mL antifoam for the described protocol.

5. The initially recording of a growth curve of the selected strain in the local fermentation device to determine the optimal time (mid-log phase) for harvesting is recommended.

6. We do not recommend sonification for this step.

7. The pellet might contain a loose upper layer containing polysaccharides or membrane fractions. Avoid transferring parts of this layer.

8. Appropriate concentrations of T7RNAP may be added prior freezing. Alternatively, T7RNAP can be added into the individual reactions.

9. MSP is partial hydrophobic. Addition of Triton-X-100 is therefore essential to release MSP attached to membrane fractions.

10. NDs provide an artificial lipid environment for membrane proteins and are highly tolerated by CF expression reactions.

They consist of a certain amount of lipids and two copies of a membrane scaffold protein (MSP), which wrap around the lipids. Different MSP derivates result in NDs with varying diameters (e.g. MSP1E3D1: 12 nm, MSP1: 10 nm, MSP1D1ΔH5: 8–9 nm). Moreover, a large variety of lipids, as well as mixtures of different lipids, can be used to form NDs, thus allowing functional expression of membrane proteins with different lipid preferences.

11. The optimal MSP:lipid ratio depends on the selected compounds and should be considered upon ND preparation (Table 2). When using lipids not listed, optimal MSP:lipid ratios should be screened and analyzed by size exclusion chromatography.

12. Sodium cholate has a high critical micellar concentration and might become too diluted upon addition of MSP. Addition of DPC therefore keeps the lipids solubilized and prevents liposome formation and precipitation.

13. CF expression can be performed in two expression modes: the two-compartment continuous-exchange (CE)-CF expression and the single compartment batch expression [8, 9]. In our hands, the efficiency of batch reactions is reduced to approx. 10% compared to CE-CF reactions and the reaction is terminated after 5–6 h. However, batch reactions can easily be automatized for throughput screening applications.

14. When calculating reaction volumes, a small excess volume may be added to compensate for volume loss during mixing.

15. The 3× mastermix may remain turbid, as several stocks are oversaturated. In this case, aliquots should be removed instantly after vortexing.

16. Other RM and FM volumes may be used. However, the liquid levels of RM and FM after placing the MD100 container into a cavity of the deep-well plate should be equal to ensure optimal substance exchange.

17. For each membrane protein target, the basic CF reaction protocol summarized in Table 1 may need to be adjusted according to individual requirements. For co-translational insertion of membrane proteins into preformed NDs, ND size, concentration, and lipid composition are crucial. For GPCR expression, the larger MSP1E3D1 NDs appear to be most suitable. However, optimal lipid composition may change from receptor to receptor.

18. For β1AR, utilization of NDs assembled with MSP1E3D1 and DOPG showed significantly improved receptor folding [3]. Modifications beneficial for the production of ETB included (i) using a blend of standard S30 lysate and the

chaperone enriched S30-HS lysate in a ratio of 70:30; (ii) replacement of DTT with a GSH/GSSG redox system (GSH:GSSG = 3:1, with final concentrations of 3 mM and 1 mM); and (iii) using NDs assembled with MSP1E3D1 and DEPG [2]. In our hands, final concentrations of 10–20 μM for ETB and β1AR can be obtained in CE-CF reactions. Despite optimized protocols, only a fraction (approx. 10–50%) of the synthesized GPCR is ligand-binding active for both ETB and β1AR. Structural approaches would therefore require affinity purification steps with immobilized ligands. However, biochemical studies or screening applications can be performed directly with the non-purified GPCRs in the CF RM or with GPCRs purified via terminal affinity tags.

19. Initial screens may be performed with lipids similar to the natural environment of the protein. In our case, the so far analyzed GPCRs favor PG lipids over PC lipids.

20. For a batch configuration, the compounds listed in Table 1 with the same final concentrations can be used. The reaction setup according to Subheading 3.4 is modified by leaving out the FM and incubation time can be shortened to 5–6 h. Reaction volumes may be reduced to 20–30 μL. Optimal Mg^{2+} concentration can vary between CE-CF and batch reactions and may be determined again.

21. We use final assay concentrations of 5–10 nM for both ETB and β1AR to take into account that only a fraction of the GPCR sample is ligand-binding active.

22. The time until equilibrium is adjusted should be determined beforehand; 1 h incubation might be a good starting point.

23. For CE-CF expressions ≥1 mL several MD100 container may be used. Alternatively, reaction volumes could be upscaled by using commercial Slide-A-lyzers as RM container (Pierce, Thermo Scientific) and appropriate FM containers.

24. Example: We add [^{125}I]endothelin-1 in a final concentration of 0.5 nM and [^{3}H]alprenolol in a final concentration of 50 nM. These concentrations are in the range of previously reported K_D measurements [2, 3].

25. In eukaryotic cells, many GPCRs undergo posttranslational modifications, such as glycosylation or lipidation. Such modifications are completely absent in samples produced in the described *E. coli* S30 lysates, while they still might affect ligand-binding characteristics. CF-synthesized GPCRs in eukaryotic cell lysates may contain posttranslational modification to a certain extent, while the overall expression efficiency is usually significantly lower [12, 13]. In a more comprehensive approach, the highly efficient and robust CF systems based on *E. coli* lysates may initially be used for the screening of large

compound libraries. Once a short list of potential binders is selected, more complex CF systems based on eukaryotic lysates may be implemented to analyze the role of posttranslational modifications.

References

1. Shelby ML, He H, Coleman MA (2019) Cell-free co-translational approaches for producing mammalian receptors: expanding the cell-free expression toolbox using nanolipoproteins. Front Pharmacol 10:744. https://doi.org/10.3389/fphar.2019.00744

2. Rues R-B, Dong F, Dötsch V, Bernhard F (2018) Systematic optimization of cell-free synthesized human endothelin B receptor folding. Methods San Diego Calif 147:73–83. https://doi.org/10.1016/j.ymeth.2018.01.012

3. Rues R-B, Dötsch V, Bernhard F (2016) Co-translational formation and pharmacological characterization of beta1-adrenergic receptor/nanodisc complexes with different lipid environments. Biochim Biophys Acta BBA - Biomembr 1858:1306–1316. https://doi.org/10.1016/j.bbamem.2016.02.031

4. Denisov IG, Sligar SG (2017) Nanodiscs in membrane biochemistry and biophysics. Chem Rev 117:4669–4713. https://doi.org/10.1021/acs.chemrev.6b00690

5. Henrich E, Peetz O, Hein C, et al (2017) Analyzing native membrane protein assembly in nanodiscs by combined non-covalent mass spectrometry and synthetic biology. eLife 6. doi:https://doi.org/10.7554/eLife.20954

6. Kai L, Roos C, Haberstock S et al (2012) Systems for the cell-free synthesis of proteins. Methods Mol Biol Clifton NJ 800:201–225. https://doi.org/10.1007/978-1-61779-349-3_14

7. Foshag D, Henrich E, Hiller E et al (2018) The E. coli S30 lysate proteome: a prototype for cell-free protein production. New Biotechnol 40:245–260. https://doi.org/10.1016/j.nbt.2017.09.005

8. Schwarz D, Junge F, Durst F et al (2007) Preparative scale expression of membrane proteins in Escherichia coli-based continuous exchange cell-free systems. Nat Protoc 2:2945–2957. https://doi.org/10.1038/nprot.2007.426

9. Voloshin AM, Swartz JR (2005) Efficient and scalable method for scalling up cell free protein synthesis in batch mode. Biotechnol Bioeng 91:516–521

10. Wheatley M, Wootten D, Conner M et al (2012) Lifting the lid on GPCRs: the role of extracellular loops. Br J Pharmacol 165:1688–1703. https://doi.org/10.1111/j.1476-5381.2011.01629.x

11. Dong F, Rues RB, Kazemi S et al (2018) Molecular determinants for ligand selectivity of the cell-free synthesized human endothelin B receptor. J Mol Biol 430:5105–5119. https://doi.org/10.1016/j.jmb.2018.10.006

12. Sonnabend A, Spahn V, Stech M et al (2017) Production of G protein-coupled receptors in an insect -based cell-free system. Biotechnol Bioeng 114:2328–2338. https://doi.org/10.1002/bit.26346

13. Merk H, Rues RB, Gless C et al (2015) Biosynthesis of membrane dependent proteins in insect cell lysates: identification of limiting parameters for folding and processing. Biol Chem 396:1097–1107. https://doi.org/10.1515/hsz-2015-0105

Chapter 8

Fluorescence Anisotropy-Based Assay for Characterization of Ligand Binding Dynamics to GPCRs: The Case of Cy3B-Labeled Ligands Binding to MC₄ Receptors in Budded Baculoviruses

Santa Veiksina, Maris-Johanna Tahk, Tõnis Laasfeld, Reet Link, Sergei Kopanchuk, and Ago Rinken

Abstract

During the past decade, fluorescence methods have become valuable tools for characterizing ligand binding to G protein-coupled receptors (GPCRs). However, only a few of the assays enable studying wild-type receptors and monitor the ligand binding in real time. One of the approaches that is inherently suitable for this purpose is the fluorescence anisotropy (FA) assay. In the FA assay, the change of ligand's rotational freedom connected with its binding to the receptor can be monitored with a conventional fluorescence plate reader equipped with suitable optical filters. To achieve the high receptor concentration required for the assay and the low autofluorescence levels essential for reliable results, budded baculoviruses that display GPCRs on their surfaces can be used. The monitoring process generates a substantial amount of kinetic data, which is usually stored as a proprietary file format limiting the flexibility of data analysis. To solve this problem, we propose the use of the data curation software *Aparecium* (http://gpcr.ut.ee/aparecium.html), which integrates experimental data with metadata in a Minimum Information for Data Analysis in Systems Biology (MIDAS) format. *Aparecium* enables data export to different software packages for fitting to suitable kinetic or equilibrium models. A combination of the FA assay with the novel data analysis strategy is suitable for screening new active compounds, but also for modeling complex systems of ligand binding to GPCRs. We present the proposed approach using different fluorescent probes and assay types to characterize ligand binding to melanocortin 4 (MC₄) receptor.

Key words Budded baculoviruses, Fluorescence anisotropy, Melanocortin 4 receptor, Global analysis, Binding kinetics

1 Introduction

G-protein coupled receptors (GPCRs) are responsible for signal transduction from the extracellular to the intracellular environment. This is a multistep process, where the ligand binding to the receptor plays a key role, and the majority of drugs target this

Sofia Aires M. Martins and Duarte Miguel F. Prazeres (eds.), *G Protein-Coupled Receptor Screening Assays: Methods and Protocols*, Methods in Molecular Biology, vol. 2268, https://doi.org/10.1007/978-1-0716-1221-7_8,
© Springer Science+Business Media, LLC, part of Springer Nature 2021

particular step. The development of efficient fluorescent dyes coupled to high-affinity pharmacophores is the basis for the implementation of novel fluorescence-based methods to characterize ligand binding to GPCRs [1, 2]. Of these methods, the fluorescence anisotropy (FA) assay is one of the few that enables the monitoring of reactions in real time, works with wild-type receptors, and is compatible with high-throughput screening systems. FA is based on the phenomenon that the fluorophores emit light with a certain degree of polarization when excited by plane-polarized light. The polarization of the emitted light depends on the ratio of fluorophore's rotational freedom and its fluorescence lifetime [3]. The binding of fluorescent ligands to a relatively large receptor protein with slow rotational diffusion limits the rotational movement of the fluorophore and causes the polarization of the emitted fluorescence to increase. This can be detected by measuring fluorescence intensities parallel ($I_{\parallel}$) and perpendicular ($I_{\perp}$) to the plane of excitation. The process does not require any separation step, and the reaction can be monitored in real time with a conventional fluorimeter equipped with suitable optical filters.

Structure-guided design as well as trial and error experimentation have resulted in an ever-increasing number of fluorescent ligands for GPCRs [4], which retain their high affinity and specificity after coupling with relatively bulky fluorophores. However, not all such ligands are suitable for FA measurements. There is a quite long list of requirements for fluorophores for FA assays [5]. For example, for optimal results the fluorescence lifetime has to be suitable for this assay and excitation in the red part of the spectrum is preferred (autofluorescence from biomolecules is significantly lowered by using excitation beyond 500 nm).

As the FA method is ratiometric, the detected signal depends on the relative concentrations of both the free and the bound states of the fluorescent ligand. For reliable measurements, the binding affinity of the ligand and the receptor's concentration should be sufficiently high to cause significant depletion of the probe during binding [6]. This is usually difficult to achieve with native tissue preparations. Although membranes prepared from cells that over-express receptors can solve this problem, the high autofluorescence of cellular components still complicates the measurements. Therefore, finding a suitable receptor display system, where receptors are in their natural environment, and their concentration is sufficiently high, has become a critical parameter for successful FA assay performance. The first solution was the preparation of membranes of transfected Sf9 cells [7]. Since then, budded baculoviruses (BBVs) that display GPCRs on their surfaces have led to a better signal-to-noise ratio [8].

Baculoviruses have been widely used as a gene expression tool based on their ability to express high levels of proteins in insect cells [9]. BBVs are rod-shaped viruses (250–300 nm in length and

30–60 nm in diameter) [10], surrounded by a lipid-bilayer envelope that is derived directly from the plasma membrane of insect cells and hence have captured their membrane proteins in the correct orientation, including GPCRs. Therefore, BBV technology is a universal system for gene delivery, and both protein expression and subsequent protein display. The small size of BBVs precludes their fast sedimentation, and homogeneous FA assays can be conducted for up to 12 h essentially without loss of signal [8]. Furthermore, baculoviruses can be handled in Biosafety Level 1 conditions (baculoviruses are neither hazardous for the environment nor dangerous for humans), which considerably simplifies the maintenance of baculovirus/insect cell systems [11].

Although the procedures for FA assay may seem straight forward, several important aspects have to be considered during the interpretation of obtained data. As these experiments are usually carried out in second-order kinetics conditions, the probe becomes depleted within the course of the reaction, a fact that has to be carefully taken into account in data analysis [8, 12]. Moreover, if the quantum yield of the probe changes during the binding, the data analysis becomes even more complex [13] and requires global data analysis.

To efficiently manage large kinetic datasets and quickly perform global data analysis, we have developed the data curation software *Aparecium* (http://gpcr.ut.ee/aparecium.html). *Aparecium* enables easy handling of large datasets measured by plate readers, including FA experiments, and is integrated with several global analysis software. This allows fitting of ordinary differential equation (ODE) based kinetic models to experimental data. Such models can efficiently simulate the effects of nonspecific binding, ligand and receptor depletion, changes in quantum yield and bleaching, but also receptors' cooperative effects. These models allow quantitative estimation of model parameters such as total receptor concentration in baculovirus preparations, as well as kinetic and equilibrium constants of fluorescent probes and unlabeled competitive ligands. The assay system has already been successfully implemented for the characterization of melanocortin 4 (MC_4) [14], serotonin 1A ($5\text{-}HT_{1A}$) [15], and dopamine 1 (DA_1) [16] receptors, and adaptation for muscarinic and neuropeptide Y receptors are in progress. This is not limited to the GPCR field as FA assay is also used for characterization of peptide binding to multi-ligand endocytic receptors [17], and inhibitor binding to protein kinases [18]. Our earlier studies have revealed that in the case of MC_4 receptors, apparent potencies of the same ligand depend on the kinetic properties of the probe and may differ up to three orders of magnitude [19]. Therefore, probes with different kinetic properties have to be used depending on the aim of the experiments [20].

Thus, the developed BBVs/FA-based assay system can provide experimental data at a level that enables to solve complex models of ligand-receptor interactions and would become a valuable tool for the screening of pharmacologically active compounds.

2 Materials

1. Aliquots of fluorescent ligands Cy3B-NDP-α-MSH, UTBC101, and UTBC102 [20] 1 µM stock solutions in dimethyl sulfoxide (DMSO) were stored at −90 °C. Dilute aliquots with incubation buffer (IB) before fluorescence anisotropy (FA) experiment (*see* **Note 1**). Always use ultrapure water (Millipore Milli-Q, 18 MΩ/cm) for the preparation of all solutions needed.

2. Microplate spectrofluorimeter compatible with FA measurements (*see* **Note 2**).

3. 1 M $CaCl_2$ stock solution: for preparing 50 mL of solution, weigh 7.35 g $CaCl_2 \cdot 2H_2O$, and add water to a volume 50 mL. Store at 4 °C.

4. 1 M Na-HEPES stock solution, pH 7.4: for preparing 100 mL of solution, weigh 23.83 g HEPES (free acid), adjust pH with 4 M NaOH, and make up to 100 mL with water. Store at 4 °C.

5. 4 M NaOH stock solution: for preparing 100 mL of solution, weigh 16 g NaOH, and add water up to 100 mL. Store at 4 °C.

6. 2% Pluronic F-127 solution: for preparing 5 mL of solution, weigh 20 mg Pluronic F-127 and add 980 µL water and dissolve by extensive shaking/vortexing (*see* **Note 3**). Store at 4 °C.

7. 1 M $MgCl_2$ stock solution: for preparing 50 mL of solution, weigh 10.17 g $MgCl_2 \cdot 6H_2O$, and add water to a volume 50 mL. Store at 4 °C.

8. 4 M NaCl stock solution: for preparing 50 mL of solution, weigh 11.7 g NaCl, and add water to a volume 50 mL. Store at 4 °C.

9. 1 M KCl stock solution: for preparing 50 mL of solution, weigh 3.73 g KCl, and add water to a volume 50 mL. Store at 4 °C.

10. Incubation buffer (IB) for MC_4 receptors: 135 mM NaCl, 5 mM KCl, 2 mM $CaCl_2$, 1 mM $MgCl_2$, 11 mM Na-HEPES, 0.1% Pluronic F-127, and CompleteEDTA-Free Protease Inhibitor Cocktail according to the manufacturer's description (Roche Applied Science), pH 7.4. Store buffer component stock solutions at 4 °C and always use freshly prepared IB (*see* **Note 4**).

11. 96-Well black microplates for FA assays (*see* **Note 5**).

12. Sf9 cells: insect cells isolated from pupal ovarian tissue of the fall armyworm *Spodoptera frugiperda*.

13. Serum-free insect cell growth medium (*see* **Note 6**).

14. The main components necessary for molecular biology procedures and construction of recombinant baculoviruses: donor vector containing cDNA of human MC_4 receptor: pcDNA3.1 (+)-hMC4; acceptor vector: pFastBac1; competent cells for recombinant bacmid DNA generation: DH10Bac; ligase; restrictases (in our case EcoRI and XhoI).

15. Cell imaging reader for image-based cell-size estimation (ICSE) assay (*see* **Note 7**).

3   Methods

3.1 Recombinant Baculovirus Construction

1. Subclone the cDNA of your GPCR of interest (commonly in pcDNA3.1 vector) into the appropriate restriction site of the pFastBac1 vector under the control of the polyhedrin promoter and transform DH10Bac-competent cells with it. We subcloned the cDNA of human MC_4 receptor from a pcDNA3.1(+)-hMC4 vector into the EcoRI-XhoI restriction site of pFastBac1 vector (*see* **Note 8**).

2. Purify the recombinant bacmid DNA and use it for the generation of recombinant baculovirus via transfection of Sf9 cells (*see* **Note 9**). Culture Sf9 cells in serum-free insect cell growth medium as a suspension culture in a non-humidified environment at 27 °C (*see* **Note 10**).

3. Look for signs of infection and harvest your generated recombinant baculovirus (typically 3–5 days post-infection)—centrifuge the cell suspension at $1600 \times g$ for 10–15 min, collect the supernatant fraction that contains recombinant baculoviruses, and store at 4 °C as an initial transfection virus stock (P0). For long-term storage, aliquot supernatant (500–1000 µL) and store at −90 °C (*see* **Note 11**).

3.2 Virus Amplification and Virus Titer Determination

Initial transfection virus stock (P0) should be amplified to prepare high-titer baculovirus working stocks for production of budded baculoviruses that will be used as a receptor source in FA assay. If the titer of the virus stock is unknown for P0, then 0.1–1% of the total end volume (virus + cells) of the virus should be added.

1. Infect the cell culture at low multiplicity of infection (MOI 0.01–0.1).

2. Harvest virus supernatant $\approx$ 72–96 h post infection when cell viability has dropped down to 30–40%. Centrifuge the cell suspension for 10–15 min at 1600 $\times$ g and filtrate the supernatant using the 0.45 μm syringe filter (*see* **Note 12**).

3. Determine the titer of the amplified baculovirus. In our laboratory, we use the image-based cell-size estimation (ICSE) assay, which is a microscopy-based method for quantification of baculoviruses based on the characteristic increase in the size of infected Sf9 cells (*see* **Note 13**) [21]. Shortly:
   (a) Seed 2×10^5 Sf9 cells in 250 μL/well on 24-well cell culture plate and allow 20–60 min to adhere.

   (b) Make threefold serial dilutions of virus supernatant in cell growth medium and add 250 μL of each virus dilution (in duplicates) to wells.

   (c) Incubate cells for 24 h and thereafter determine the average cell diameter by image-based cell-size estimation (ICSE) assay (or using cell counter that allows measuring of cell diameters) (*see* **Notes 14** and **15**).

   (d) Plot the average cell diameters versus virus dilutions (in log scale) and calculate virus concentration (infectious virus particles per mL; ivp/mL) from a sigmoidal dose-response curve using Eq. (1):

$$\text{Virus concentration (ivp/mL)} = \frac{\frac{1}{ED_{50}} \times 50\% \text{ of infected cells}}{V} \quad (1)$$

where V is the sample volume in wells (here 0.5 mL), ED_{50} – 50% effective virus dilution corresponding to dilution at which the average cell diameter has changed by 50%, and 50% of infected cells – 50% of the cells in wells at the time of infection (here 1×10^5 cells).

4. Store the amplified baculovirus supernatants at 4 °C in tubes protected from light. For long-term storage, the preparation can be aliquoted and stored at −90 °C (*see* **Note 11**).

3.3 Production of GPCR Displaying Budded Baculovirus Particles

1. Infect the cell suspension at a density 2×10^6 viable cells/mL with a high-titer baculovirus supernatant encoding your GPCR of interest at an MOI typically 3–5 and grow the cells for the next 3–5 days by monitoring the cell viability and infection's efficiency (*see* **Note 16**).

2. Centrifuge the cell suspension at 1600 $\times$ g for 10–15 min, collect the supernatant fraction that contains budded baculoviruses, and use it for the budded baculovirus preparation.

3. Centrifuge the collected supernatant at 48,000 $\times$ g for 40–45 min. Carefully wash the obtained pellet that contains budded baculoviruses with IB.

4. Resuspend the obtained budded baculovirus pellet in IB at a volumetric ratio of 1:30–1:50 with regard to the initial cell suspension volume (*see* **Note 17**). Choose the concentration factor based on the desired final receptor concentration in the preparation and the expression level achieved.

5. Pool the resuspended baculovirus preparations together and homogenize it using a syringe and a needle with an inner diameter of 0.3 mm.

6. Aliquot suitable volumes of the pooled baculovirus preparation and store at −90 °C until used in FA measurements (*see* **Note 18**).

3.4 FA Measurements in Multiwell Microplates: Determination of Receptor Concentration

All pre-measurement procedures should be carried out at room temperature unless otherwise mentioned. All FA measurements are carried out in the kinetic mode at 27 °C (optimal temperature for maintenance of Sf9 cells). The total volume per well is 100 μL, and reactions are started by the addition of the baculovirus preparation (e.g., 70 μL (*see* **Notes 19**) to the microplate wells that contains the fluorescent ligand (e.g., 20 μL) with or without competing ligands (*see* **Notes 20**), and fluorescence intensities are measured at the appropriate time points. The FA signal at each time point t after the initiation of the binding reaction is calculated as parameter $FA(t)$ according to the Eq. (2):

$$FA(t) = \frac{I(t)_{\mathrm{II}} - I(t)_{\perp}}{I(t)_{\mathrm{II}} + 2 \cdot I(t)_{\perp}} \tag{2}$$

The ligand-specific effects are measured in the presence (non-specific binding) or absence (total binding) of an excess of non-labeled ligand (e.g., 3 μM NDP-α-MSH or SHU9119 for MC_4 receptors), and specific binding is defined as the difference between these values. In addition, the background fluorescence of the assay (caused by cellular components and buffer components, competing ligands, detector noise, etc.) should also be taken into account—these values are measured in the absence of fluorescent ligand (wells contain only IB and the appropriate amount of baculovirus preparation) and subsequently subtracted separately from all channels of all the total and nonspecific binding data, resulting in background-corrected values (*see* **Note 21**).

1. The fluorescent ligand's binding saturation to the receptor is determined by varying the receptor concentration and keeping the ligand's concentration constant. Using two fluorescent ligand concentrations in parallel for this type of experiment helps to achieve more accurate results for the receptor stock concentration determination. We varied the concentration of MC_4 receptor in the 0–38 nM range and kept the UTBC102 concentration constant (0.5 nM or 7 nM). For UTBC101 the

concentrations were 0.4 nM and 4 nM and the receptor concentration was varied from 0 to 90 nM and for Cy3B-NDP-α-MSH the concentrations were 0.5 nM and 1 nM and the receptor concentration was varied from 0 to 1.38 nM (*see* **Note 22**). Total volume per well is 100 μL, and the total, nonspecific, and background binding parameters are measured in duplicates. For total binding measurements, each well contains 10 μL IB + 20 μL fluorescent ligand +70 μL baculovirus receptor preparation, for nonspecific binding measurement each well contains 10 μL excess concentration of non-labeled ligand +20 μL fluorescent ligand +70 μL baculovirus preparation, and for background measurement each well contains 30 μL IB + 70 μL baculovirus preparation.

2. Prepare seven serial dilutions of budded baculovirus receptor preparation in IB in vials of appropriate volume, considering that 70 μL of each baculovirus dilution will be required for each of the 10 wells + solution for reserve. The 10 wells for each baculovirus dilution consist of total and nonspecific binding for two concentrations of fluorescent ligand plus background, and all these performed in duplicates. Make seven serial dilutions and leave the eighth microplate row as a "zero" point, where no virus will be carried. For example, the dilution step for the serial dilutions can be about twofold (*see* **Note 23**).

3. Dispense 10 μL IB in all total binding duplicate wells.

4. Dispense 10 μL excess concentration of non-labeled ligand solution in all nonspecific binding duplicate wells (in our case, 3 μM SHU9119).

5. Dispense 30 μL IB in all background binding duplicate wells.

6. Add 20 μL of fluorescent ligand solution to the total and nonspecific binding duplicate wells (in our case, 0.4 nM and 4 nM for UTBC101 or 0.5 nM and 7 nM for UTBC102 or 0.5 nM and 1 nM for Cy3B-NDP-α-MSH) (*see* **Note 24**).

7. If the baculovirus stock has visible aggregates after thawing, homogenize the preparation using a syringe and a needle with an inner diameter of 0.3 mm.

8. Start binding reactions by adding 70 μL of previously prepared dilutions of budded baculovirus receptor preparation to corresponding wells. Work quickly, definitely make sure to use a multichannel pipette (preferably electronic ones, that enable multiple dispensing steps within one "fill" step); do not forget to register the time of the start of reactions and time to the first anisotropy measurement. Take it into account while constructing the reaction kinetics curves! (*see* **Note 25**).

9. If the equilibrium for the association is reached, prepare a more concentrated nonspecific ligand solution, which is able to induce full dissociation of the fluorescent ligand (in our case, 150 µM for SHU9119 resulting in 3 µM concentration in the well). To start the dissociation, add 2 µL of the nonspecific ligand to one of the duplicates and add 2 µL of IB to the other.

10. Plot the data points (Fig. 1c, e) and calculate binding parameters by using the global data analysis with the simultaneous fitting of two data surfaces (total and nonspecific binding) as a function of both, baculovirus preparation volume and time (*see* Global data analysis for more details).

3.5 FA Measurements in Multiwell Microplates: Competition Binding Experiments

Total reaction's volume per well is 100 µL (60 µL competitor solution (or 60 µL IB in case of "zero" point), 20 µL fluorescent ligand, and 20 µL budded baculovirus preparation) (*see* **Note 26**).

1. Prepare 10 serial dilutions of competitors (in duplicates, typically in the concentration range between 0.1 nM and 10 µM (*see* **Note 27**)) in 60 µL of IB on the multiwell microplate (considering that the total volume of the well will be 100 µL and that the final concentration of competitor's should be calculated according to the total volume).

2. Dispense 60 µL IB in 11-th duplicate—this will be "zero" point (without competitor).

3. Dispense 80 µL IB in 12-th duplicate—these wells will be used for detection of background fluorescence (without fluorescent ligand).

4. Add 20 µL of fluorescent ligand to duplicates 1–11 (our final concentrations were for UTBC101, and UTBC102 was 1.3 nM, and for Cy3B-NDP-α-MSH 0.5 nM (*see* **Note 28**).

5. Start the competition reactions by adding 20 µL of budded baculovirus preparation to all wells and immediately insert the microplate in the plate reader to start anisotropy measurements. Work quickly, definitely use a multichannel pipette (preferably electronic pipette that enables multiple dispensing steps within one "fill" step); do not forget to check the time from the start of reactions to the first anisotropy measurement and use it for construction of reaction kinetics curves!

6. Plot the data points on a graph (Fig. 1b, d, f) and calculate apparent IC_{50} values according to the logistic function with a variable slope. Alternatively, you can use global modeling to calculate the K_i values. For details on Global analysis using GraphPad Prism, *see* **Note 29**. For details on Global analysis using IQMTools, *see* **Note 30**.

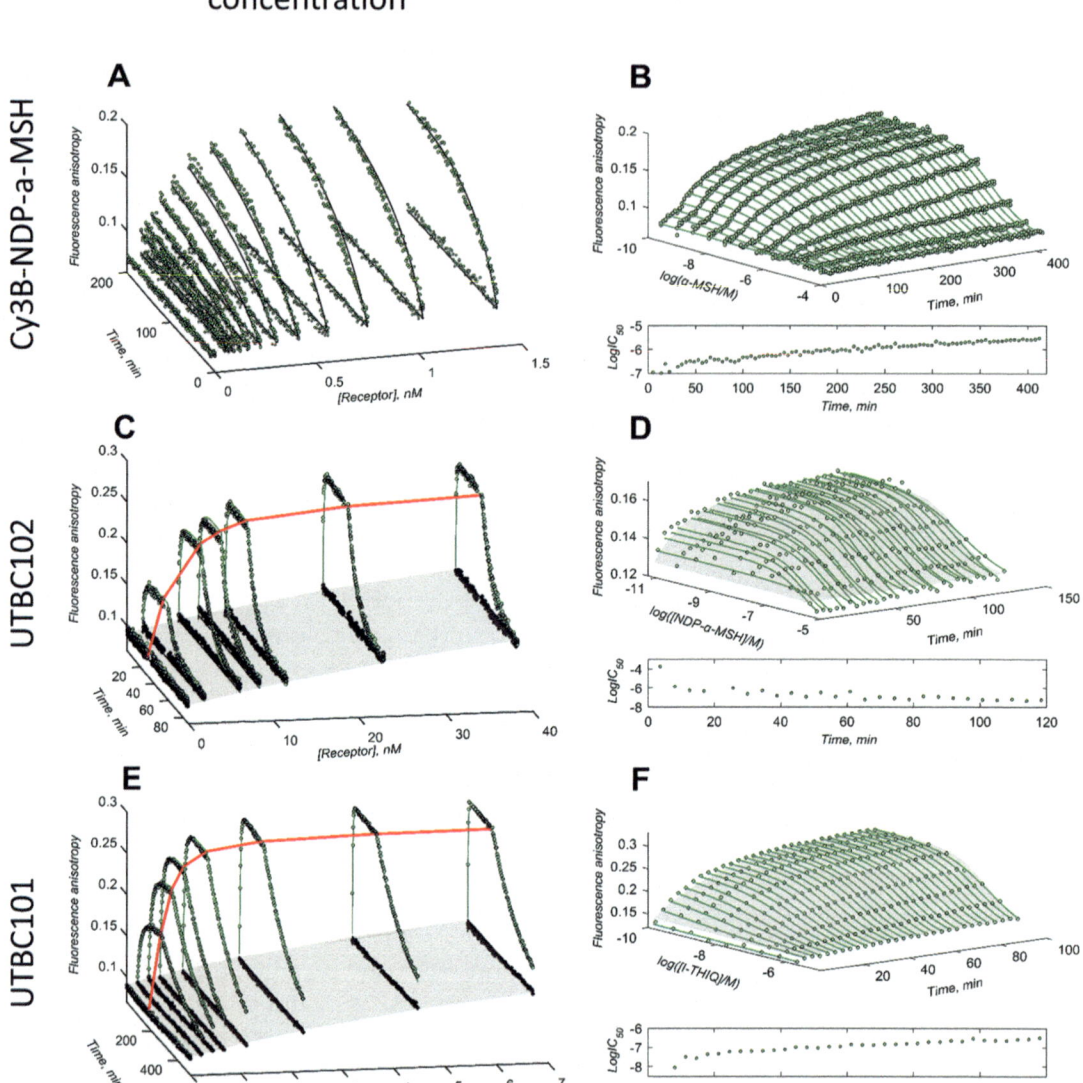

Fig. 1 Time courses of fluorescence anisotropy changes caused by binding of Cy3B-NDP-α-MSH (**a**, **b**), UTBC102 (**c**, **d**), and UTBC101 (**e**, **f**) to MC$_4$ receptors. Dependence of total and nonspecific binding on the receptor concentration (**a**, **c**, **e**). Displacement of fluorescent ligand by competitive ligand (**b**, **d**, **f**). Data were transformed with an *Aparecium* 2.0 software. Subsequent fitting was performed with Graphpad Prism to global Model 2 (**a**) or to sigmoidal dose-response at different time points (lines on **b**, **d**, **f**) or with a modified version of IQMTools V1.2.2.2 software by a single-site binding model (surfaces on **c**, **d**, **e**, **f**). With event definition, even non-continuous assay steps such as initiation of dissociation (red lines **c**, **e**) can be included within a single fitting procedure, as indicated in the receptor concentration determination experiments for UTBC102 and UTBC101 (**c** and **e**)

4   Notes

1. The fluorescence lifetime, τ, of red-shifted Cy3B-labeled fluorescent ligand UTBC102 has been determined to be 2.8 ns and for Cy3B-NDP-α-MSH to be 2.7 ns. These ligands demonstrated high photostability, insensitivity to buffer ionic strength, and their fluorescence characteristics are not pH-dependent. The concentrations of the ligands were confirmed by the absorbance reading of Cy3B ($\varepsilon_{558} = 130{,}000$ M^{-1} cm^{-1}).

2. We have used Synergy NEO (Biotek) and PheraSTAR (BMG Labtech) with dual emission detection mode that allow the simultaneous recording of intensities that are parallel ($I_{\parallel}$) and perpendicular ($I_{\perp}$) to the plane of excitation light. For Cy3B-labeled compounds, the optical module used in PheraSTAR has excitation and emission filters of 540 nm (slit 20 nm) and 590 nm (slit 20 nm), and optical module in Synergy NEO has excitation and emission filters of 530 nm (slit 25 nm) and 590 nm (slit 35 nm). Sensitivities of channels (G factor) were corrected with a gain adjustment of the photomultiplier tubes (PMTs) using erythrosine B as a standard [22].

3. Pluronic F-127 is hard to dissolve and foams extensively during the shaking/vortexing. It is better to prepare always a fresh 2% solution for the experiment.

4. The incubation buffer needs to be optimized for every reporter ligand, and receptor studied. Here we have used a minimal buffer for ligand binding to the MC$_4$ receptor, but buffer composition can differ depending on the fluorescent ligand and/or GPCR used—e.g., our used ligands have been found to be more sensitive to the ionic strength of buffer solution [7], and MC$_4$ receptor to Ca^{2+}, Zn^{2+}, and Cu^{2+}ions [23]. Additionally, the buffer solution should be as little autofluorescent as possible in the range of fluorescence readings used. The 0.1% Pluronic F-127 has been found to be essential for stabilizing the anisotropy signal during measurements and has no significant influence on the properties of the used fluorescent ligands binding to the MC$_4$ receptor [7].

5. Among several microplates tested, Corning half area black flat-bottom polystyrene non-binding-surface microplates were found to give optimal results for our FA assays as a result of low background fluorescence, low adsorption of ligands onto the plastic surface, and decreased evaporation due to the smaller well surface that contacts with air.

6. We have used EX-CELL 420 (Sigma-Aldrich GmbH), but other serum-free insect cell growth mediums can be used as well.

7. For the determination of average cell diameter, we used a Cytation 5 (Biotek) with 4× objective, but any other equipment allowing cell diameter measurements can be used as well.

8. All plasmid constructs should be sequenced. Design your GPCR plasmid constructs according to your interests, possibilities, and needs—other promoters than polyhedrin (e.g., dual promoter systems) and other restriction sites can be used.

9. Use standard plasmid DNA extraction MiniPrep Kit for recombinant bacmid DNA purification. For transfection of Sf9 cells with bacmid DNA, we used a PolyEthyleneImidine (PEI)-based transfection reagent ExGen500 according to the manufacturer's description, but other transfection methods can be used as well.

10. The usage of a serum-free medium is preferred as it can be considered as a "cleaner" environment since it does not contain different proteins and/or other components present in serum that would become concentrated within budded baculovirus preparation (see Subheading 3.3) and influence FA measurements. Antibiotics and antimycotics are not recommended for serum-free cell cultures, but if necessary, use reduced concentrations (e.g., 25 U/mL penicillin and 25 μg/mL streptomycin [7]).

11. You can add a small amount of FBS or DMSO to viral stocks to freeze them for a longer time at −90 °C. Depending on the virus, this addition is not always required (MC$_4$ receptor baculoviruses in EX-CELL 420 TM medium have remained active for years when stored at 90 °C). It is advisable to determine the titer of the stock solution after one freeze-thaw cycle; the obtained titer value can be subsequently used for MOI calculation in future experiments.

12. We have used 0.45 μm filters with Surfactant-Free Cellulose Acetate (SFCA) Membrane because this material is hydrophilic and has low protein binding properties.

13. The cell counter-based determination of cell diameter has also been used for this purpose [24]. The assay has been successfully used also with Sf9 cell-derived MIMIC™ Sf9 cells (Thermo-Fisher Scientific).

14. The detailed protocol for the assay with Cytation 5 can be downloaded from http://gpcr.ut.ee/icse%2D%2D-baculovirus-quantification-assay.html

15. Analyze the images using *Aparecium* software, with the support of the software manual: http://gpcr.ut.ee/uploads/ApareciumHelp/Aparecium.html?Welcome1.html

16. We have optimized the infection and virus collection conditions by varying the MOI from 3 to 30 and the infection time from 48 to 120 h. In the case of MC$_4$ receptors, the optimal

conditions were found to be MOI 3–5 and virus collection time 72–96 h after infection, but these conditions were not optimal for other GPCRs used (e.g., for Y_1—neuropeptide Y receptor and 5-HT_{1A}—serotonin receptor 1A).

17. In our case, a 50-fold concentration of budded baculovirus suspension gave an MC_4 receptor concentration in the stock of about 94 nM, as estimated after kinetic saturation binding experiments and global data fitting [8].

18. According to our experience, the pharmacological properties of MC_4 receptor in the budded baculovirus preparation was not affected by single freezing and thawing (but we tried to avoid multiple freezing/thawing cycles), and the receptor concentration in the preparation did not significantly change during storage at −90 °C for at least 2 years. For aliquoting, choose volumes that are convenient for further experiments.

19. As the receptor preparation is the component that starts the reaction, its volume should be as large as possible to enable the extensive mixing of solutions during the addition.

20. The total volume per well as well as the volumes of each component (e.g., fluorescent ligand, competitor, baculovirus preparation) can be different depending on the receptor concentration in the budded baculovirus preparation used, fluorescent ligand's properties, and practical pipetting possibilities offered by your multichannel pipettes (e.g., we used Thermo Scientific E1-ClipTip electronic multichannel pipettes that have numerous mixing, diluting, and dispensing possibilities, which enable flexible assay design, moreover). The usage of small compound amounts and hydrophobicity of fluorescent ligands sets high requirements to pipette tips (in our assay, ClipTip pipette tips had very small nonspecific binding as opposed to other pipette tips tested).

21. Background correction can easily be applied by using Export-Tools in *Aparecium* software.

22. If the receptor concentration in budded baculovirus preparation is high enough and assay sensitivity enables it, it is possible to perform saturation binding experiments by varying the concentration of the fluorescent ligand and keeping the receptor concentration constant [8].

23. As the receptor concentration in baculovirus preparation is unknown, the optimal dilution factor should be chosen after analyzing the results of a pilot experiment. In general, dilutions have to be done until all receptors become saturated with the ligand, plus one additional dilution. Ligand binding affinity and initial receptor concentration in baculovirus preparation are factors that influence the choice of dilution's factor.

24. The usage of an automated multichannel pipette considerably facilitates and accelerates these dispensing steps.

25. To obtain more accurate kinetic data for models, a stopwatch can be used to register reaction start times (i.e., the time that baculovirus was added to the well) for each row/column/well. *Aparecium* can directly import registered stopwatch times for data analysis (http://gpcr.ut.ee/uploads/ApareciumHelp/Aparecium.html?Welcome1.html).

26. The optimal concentrations of particular baculovirus preparation and fluorescent ligand depend on the binding affinity and available stock concentration. This can be decided after performing receptor concentration determination assay (*see* Subheading 3.5).

27. When results from a pilot experiment with the unknown ligand are obtained, you can optimize/individualize the used concentration range for each of the displacers.

28. Reactions are carried out under second-order conditions where the ligand and receptor concentrations should be usually comparable and close to the fluorescent ligand's K_d value.

29. Global data analysis has proven to be a useful tool for data analysis. However, here we provide a general description as the software and specific instructions are continuously being developed and updated. As fluorescence anisotropy experiments in kinetic mode usually generate large data matrices, special attention must be paid to data management and analysis. Although most plate readers provide the possibility to perform data analysis to some extent, this is often inconvenient if devices from several different manufacturers are used in a single study. Additionally, reliance on proprietary software can lead to difficulties in long-term data management and sharing. Finally, only the most common data analysis algorithms are usually covered, and hence the possibilities for global data analysis are very limited. Fortunately, the plate readers' software usually provides the possibility of data export as Excel files. As the data layout is not standardized, it requires further data transposition for compatibility with different data analysis software. These manual operations in Excel can lead to different errors such as misalignment, missing data, and unwanted changes in values. These data management issues can be addressed by using *Aparecium* software (http://gpcr.ut.ee/aparecium.html), which is designed to handle kinetic datasets from plate readers. It offers possibilities to integrate experimental measurement data with experiment metadata and save this integrated dataset as a MIDAS file. Thereafter, MIDAS files can be exported to different data analysis platforms such as MATLAB and Excel for custom workflows or IQMTools and GraphPad Prism for

global kinetic analysis. *Aparecium* is distributed as an open-source software and works in the MATLAB environment. Documentation, downloads, help, and most recent tutorials for *Aparecium* and global fitting can be found at www.gpcr.ut. ee/Aparecium.

For data analysis of fluorescence anisotropy, both Graphpad Prism and IQMTools can be used. While Graphpad Prism is often easier to use, it is restricted to nonlinear fitting of analytically solvable mathematical models. IQMTools, however, can solve almost any system of ordinary differential equations. The choice of software depends on both the user experience and the complexity of the mathematical model used to describe the system. We have proposed three analytically solved mathematical models for the above-described assay types:

Equilibrium analysis for receptor concentration determination experiments (**Model 1**).

Kinetic analysis of ligand binding from receptor concentration determination experiments (**Model 2**).

Equilibrium analysis for competition binding experiments (**Model 3**).

All three models have been described previously [8]. Briefly, all models assume the possibility of three binding reactions:

$$R + L \rightleftarrows RL$$

$$R + C \rightleftarrows RC$$

$$NS + L \rightleftarrows NSL$$

Here, R is the receptor, L is the fluorescently labeled ligand, C is the competitive unlabeled ligand, and NS is the nonspecific binding site associated with the baculovirus. The right-hand sides of equations represent the corresponding complexes. These underlying chemical reactions can be mathematically linked to measurable properties such as the fluorescence anisotropy [25]. The measured FA described by Eq. (2) is related to the concentrations of each fluorescence ligand species by Eq. (3).

$$FA(t) = \frac{\sum_i QY_i * C_i(t) * FA_i}{\sum_i QY_i * C_i(t)} \tag{3}$$

where $FA(t)$ indicates the predicted fluorescence anisotropy value at time t, QY_i represents the quantum yield of the fluorescence ligand in state i, $C_i(t)$ represents the thermodynamic concentration of fluorescence ligand in state i at time t, and FA_i represents the intrinsic fluorescence anisotropy of fluorescence ligand state i. For this particular system, assuming equal quantum yields for all states, Eq. 3 simplifies to Eq. (4).

$$FA_t = \frac{[L]_t * FA_L + [RL]_t * FA_{RL} + [NSL]_t * FA_{NSL}}{[L]_t + [RL]_t + [NSL]_t}$$ (4)

Each different state of the fluorescence ligand is described by the intrinsic anisotropy of the particular state (FA_L, FA_{NSL}, FA_{RL}). **Model 1** and **Model 2** also assign intrinsic relative fluorescence quantum yield to RL and NSL states with respect to L. Additionally, **Model 2**, as it is based on association kinetics, also requires the pre-knowledge of k_{off} of the fluorescent ligand. The value can easily be obtained by adding a small volume of high concentration competitive ligand to the total binding rows of receptor concentration determination experiments, but the value must be fitted separately using a one-phase exponential decay. **Model 1** and **Model 2** can both predict the ligand K_d value, intrinsic fluorescence anisotropy, and quantum yield values of all three possible states of the fluorescent ligand (L, RL, and NSL), as well as receptor stock concentration. In addition, **Model 2** can directly predict the k_{on} value of the fluorescent ligand. While **Model 1** is limited to a single time point, the global fitting of **Model 2** is limited to only a single fluorescence ligand concentration. Due to the model definition syntax of GraphPad Prism, the possibility to create generalized and assay type independent global models is not always possible. More detailed descriptions of all models have been presented [8] previously.

30. An alternative approach to using analytically defined equations is to use a system of ordinary differential equations (ODEs). Each ODE corresponds to a reaction velocity definition for a particular chemical or pharmacological species. In this case, the equation system is solved numerically. Fitting is performed by simulating the model and adjusting parameter values to achieve a minimal optimization function value. For this purpose, several specialized global optimum search algorithms can be employed, such as the Nelder-Mead method and particle swarm. A significant benefit of this approach is that a single model file can be used for all different experimental setups as long as all the components remain exactly the same. Additionally, all data points can be fitted together, including association and dissociation kinetics, since non-continuous events such as the addition of the competitive ligand can also be defined and simulated. Although GPCRs are, in some cases, already known to be described by much more complex binding mechanisms, ODE based global fitting is not limited to a single binding site model, and models can be created to describe any reaction mechanism.

We have used IQMTools previously distributed as SBTool-box2 available at https://iqmtools.intiquan.com/ to perform the ODE based global fitting. For fitting experimental data obtained from receptor concentration determination experiments as well as competition binding experiments, we have used a single model based on the same assumptions as described for GraphPad Prism **Model 2**. All available data, including both fluorescent ligand concentrations, receptor, and competitive ligand concentrations as well as association and dissociation kinetics, were used for fitting. As a result of such analysis, a single value for each model parameter, such as intrinsic fluorescence anisotropy values, is obtained. This is especially advantageous for the determination of K_i values for competitors since K_i value calculation from IC_{50} values using Cheng-Pursoff equation does not take ligand and receptor depletion effect into account and ignores most of the kinetic data, resulting in a less accurate K_i value estimate compared to the global analysis approach. The fact that fluorescent ligand and displacer may have different kinetic characteristics and that kinetic properties of displacer are *a priori* unknown should not be ignored when results from competition binding experiments are interpreted—otherwise, it may lead to over- or underestimation of a compound's apparent potency. This aspect is evident as $\log IC_{50}$ is increasing in time for the competition between UTBC101 and I-THIQ indicating that the fluorescent ligand has lower k_{on} compared to the competitor (Fig. 1f) while the $\log IC_{50}$ is decreasing in time for the competition between UTBC102 and NDP-α-MSH indicating that fluorescent ligand has higher k_{on} compared to the competitor (Fig. 1d). Additionally, global fitting can indicate fundamental flaws in the models' assumptions by a clear lack-of-fit while single-curve fitting procedures overfit the local values and can, therefore, hide the discrepancies between the data and the model.

Acknowledgments

The work has been financed by the Estonian Ministry of Education and Science (PSG230) and by the NATO (SPS 985261).

References

1. Hoffmann C, Castro M, Rinken A, Leurs R, Hill SJ, Vischer HF (2015) Ligand residence time at G-protein-coupled receptors - why we should take our time to study it. Mol Pharmacol 88:552–560

2. Soave M, Briddon SJ, Hill SJ, Stoddart LA (2020) Fluorescent ligands: bringing light to emerging GPCR paradigms. Br J Pharmacol 177:978–991

3. Perrin F (1926) Polarisation de la lumičre de fluorescence. Vie moyenne des molécules dans l'etat excité. J Phys Radium 7:390–410

4. Vernall AJ, Hill SJ, Kellam B (2014) The evolving small-molecule fluorescent-conjugate toolbox for Class A GPCRs. Br J Pharmacol 171:1073–1084

5. Rinken A, Lavogina D, Kopanchuk S (2018) Assays with detection of fluorescence anisotropy: challenges and possibilities to characterize ligand binding to GPCRs. Trends Pharmacol Sci 39:187–199

6. Nosjean O, Souchaud S, Deniau C, Geneste O, Cauquil N, Boutin JA (2006) A simple theoretical model for fluorescence polarization binding assay development. J Biomol Screen 11:949–958

7. Veiksina S, Kopanchuk S, Rinken A (2010) Fluorescence anisotropy assay for pharmacological characterization of ligand binding dynamics to melanocortin 4 receptors. Anal Biochem 402:32–39

8. Veiksina S, Kopanchuk S, Rinken A (2014) Budded baculoviruses as a tool for a homogeneous fluorescence anisotropy-based assay of ligand binding to G protein-coupled receptors: the case of melanocortin 4 receptors. Biochim Biophys Acta 1838:372–381

9. Kost TA, Condreay JP, Jarvis DL (2005) Baculovirus as versatile vectors for protein expression in insect and mammalian cells. Nat Biotechnol 23:567–575

10. Wang Q, Bosch BJ, Vlak JM, van Oers MM, Rottier PJ, van Lent JWM (2016) Budded baculovirus particle structure revisited. J Invert Pathol 134:15–22

11. Kost TA, Condreay JP, Ames RS (2010) Baculovirus gene delivery: a flexible assay development tool. Curr Gene Ther 10:168–173

12. Roehrl MH, Wang JY, Wagner G (2004) A general framework for development and data analysis of competitive high-throughput screens for small-molecule inhibitors of protein-protein interactions by fluorescence polarization. Biochemistry 43:16056–16066

13. Jameson DM, Mocz G (2005) Fluorescence polarization/anisotropy approaches to study protein-ligand interactions: effects of errors and uncertainties. Methods Mol Biol 305:301–322

14. Rinken A, Veiksina S, Kopanchuk S (2016) Dynamics of ligand binding to GPCR: Residence time of melanocortins and its modulation. Pharmacol Res 113:747–753

15. Tõntson L, Kopanchuk S, Rinken A (2014) Characterization of 5-HT$_{1A}$ receptors and their complexes with G-proteins in budded baculovirus particles using fluorescence anisotropy of Bodipy-FL-NAN-190. Neurochem Int 67:32–38

16. Allikalt A, Kopanchuk S, Rinken A (2018) Implementation of fluorescence anisotropy-based assay for the characterization of ligand binding to dopamine D1 receptors. Eur J Pharmacol 839:40–46

17. Scodeller P, Simon-Gracia L, Kopanchuk S, Tobi A, Kilk K, Säälik P, Kurm K, Squadrito ML, Kotamraju VR, Rinken A, De Palma M, Ruoslahti E, Teesalu T (2017) Precision targeting of tumor macrophages with a CD206 binding peptide. Sci Rep 7:14655

18. Sormus T, Lavogina D, Enkvist E, Uri A, Viht K (2019) Efficient photocaging of a tight-binding bisubstrate inhibitor of cAMP-dependent protein kinase. Chem Commun (Camb) 55:11147–11150

19. Kopanchuk S, Veiksina S, Mutulis F, Mutule I, Yahorava S, Mandrika I, Petrovska R, Rinken A, Wikberg JES (2006) Kinetic evidence for tandemly arranged ligand binding sites in melanocortin 4 receptor complexes. Neurochem Int 49:533–542

20. Link R, Veiksina S, Rinken A, Kopanchuk S (2017) Characterization of ligand binding to melanocortin 4 receptors using fluorescent peptides with improved kinetic properties. Eur J Pharmacol 799:58–66

21. Laasfeld T, Kopanchuk S, Rinken A (2017) Image-based cell-size estimation for baculovirus quantification. BioTechniques 63:161–168

22. Thompson RB, Gryczynski I, Malicka J (2002) Fluorescence polarization standards for high-throughput screening and imaging. BioTechniques 32:34–42

23. Link R, Veiksina S, Tahk MJ, Laasfeld T, Paiste P, Kopanchuk S, Rinken A (2020) The constitutive activity of melanocortin-4 receptors in cAMP pathway is allosterically modulated by zinc and copper ions. J Neurochem 153:346–361

24. Veiksina S, Kopanchuk S, Mazina O, Link R, Lille A, Rinken A (2015) Homogeneous fluorescence anisotropy-based assay for characterization of ligand binding dynamics to gpcrs in budded baculoviruses: the case of Cy3B-NDP-alpha-MSH binding to MC4 receptors. Methods Mol Biol 1272:37–50

25. Beechem JM, Gratton E, Ameloot M, Knutson JR, Brand L (2002) The global analysis of fluorescence intensity and anisotropy decay data: second-generation theory and programs. In: Lakowicz JR (ed) Topics in fluorescence spectroscopy: principles. Springer, Boston, MA, pp 241–305

Chapter 9

Bioluminescence in G Protein-Coupled Receptors Drug Screening Using Nanoluciferase and Halo-Tag Technology

Hannes Schihada, Katarina Nemec, Martin J. Lohse, and Isabella Maiellaro

Abstract

Here we describe the stepwise application of bioluminescence resonance energy transfer (BRET)-based conformational receptor biosensors to study GPCR activation in intact cells. This technology can be easily adopted to various plate reader devices and microtiter plate formats. Due to the high sensitivity of these BRET-based receptor biosensors and their ability to quantify simultaneously receptor activation/deactivation kinetics as well as compound efficacy and potency, these optical tools provide the most direct and unbiased approach to monitor GPCR activity in a high-throughput-compatible assay format, representing a novel promising tool for the discovery of potential GPCR therapeutics.

Key words Bioluminescence resonance energy transfer, G protein-coupled receptors, Conformational biosensors, Activation dynamics, Drug screening, Z-factor

1 Introduction

G protein-coupled receptors (GPCRs) are transmembrane proteins whose primary function is to transduce extracellular stimuli into intracellular function, thereby regulating a multiplicity of cellular processes.

Due to their physiological importance and their role as drug targets, effort by the scientific community has been put in place to develop sensitive methods to study receptor pharmacology and push forward their use as therapeutic targets.

In this chapter, we describe the principle of bioluminescence resonance energy transfer (BRET) and specify how this phenomenon can be exploited to quantify efficacy and potency of GPCR ligands in intact cells in a high-throughput screening (HTS) format. We analyze advantages and limitations of BRET methods compared to other established GPCR screening assays and provide guidance in transferring this sensor design to other GPCRs.

Sofia Aires M. Martins and Duarte Miguel F. Prazeres (eds.), *G Protein-Coupled Receptor Screening Assays: Methods and Protocols*, Methods in Molecular Biology, vol. 2268, https://doi.org/10.1007/978-1-0716-1221-7_9,
© Springer Science+Business Media, LLC, part of Springer Nature 2021

1.1 Conformational GPCR BRET-Based Biosensors

Bioluminescence describes a naturally occurring phenomenon that utilizes luciferases, a family of oxidative enzymes, that emit light upon catalytic oxidation of a substrate. Bioluminescence resonance energy transfer (BRET) relies on radiationless energy transfer from a luminescent donor (the luciferase) to a fluorescent acceptor. The amount of energy transferred between donor and acceptor depends on the inter-fluorophore distance and relative orientation [1, 2].

GPCRs contain seven transmembrane helices. It is known that receptor activation upon ligand binding induces a reorganization of the helices involving, most prominently, an outward movement of helix 6 (14Å) in the class A β_2AR [3] and even up to 19Å for class B GPCRs [4, 5], which is then propagated to the third intracellular loop connecting helix 5 and 6. By tagging these intracellular conformationally sensitive sites with BRET donor and acceptor molecules, ligand-induced receptor activation can be visualized as a change in energy transfer between them (Fig. 1) [6].

In our previous work [7, 8], we described a universal BRET biosensor design that employs Nanoluciferase (Nluc), an engineered catalytic subunit of *Oplophorus gracilirostris* luciferase [9], as an energy donor and the red fluorescent dye HaloTag® Nano-BRET™ 618 Ligand (HaloTag(618)) covalently bound to the self-labeling protein tag HaloTag [10] as the energy acceptor. This sensor design allows the discrimination of GPCR ligands of distinct efficacies and potencies (Fig. 2a).

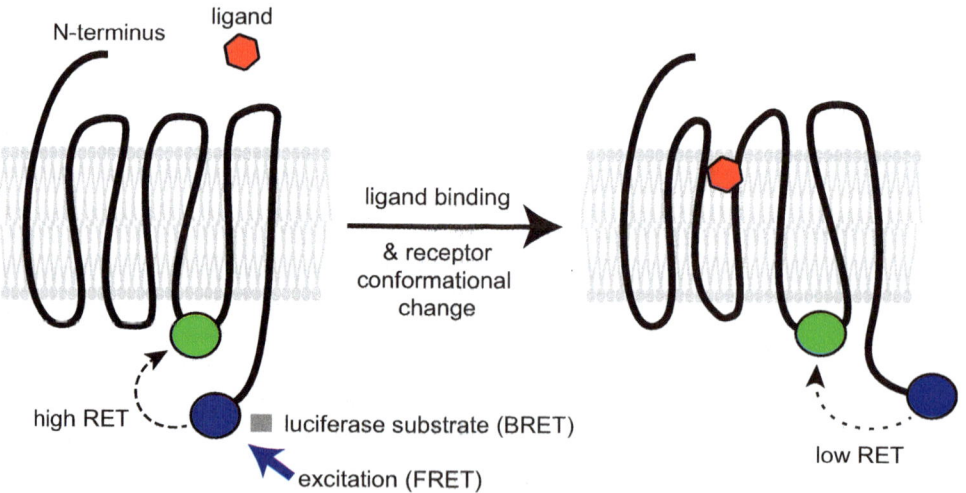

Fig. 1 Principle of BRET-based GPCR conformational biosensors. The receptor of interest is labeled with a BRET donor/acceptor pair at the C-terminus and the third intracellular loop allowing for resonance energy transfer in the ligand-free GPCR state. Ligand-receptor interaction induces a conformational transition of the receptor affecting the relative distance and/or dipole-dipole orientation of the energy partners and results in a change in BRET

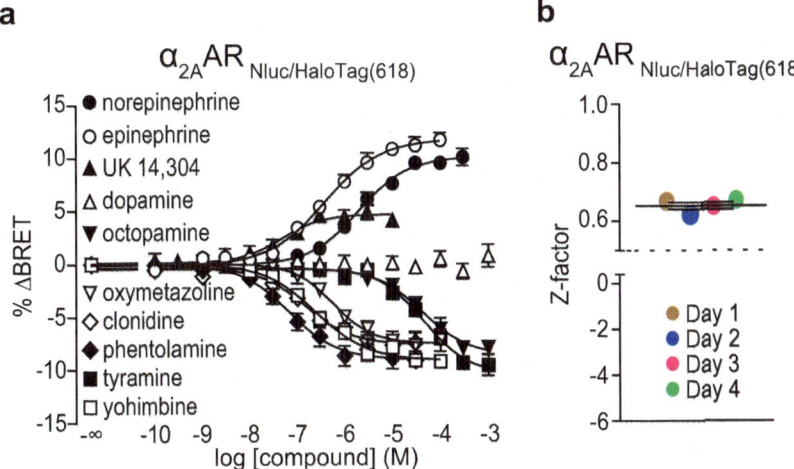

Fig. 2 GPCR_{Nluc/HaloTag(618)} biosensors for high-throughput GPCR compound screening and ligand characterization. (**a**) Concentration-response curves generated with $\alpha_{2A}AR_{Nluc/HaloTag(618)}$ enable the evaluation of ligand efficacy (maximum BRET response) and potency (EC_{50}) at the level of GPCR conformation. (**b**) The $\alpha_{2A}AR_{Nluc/HaloTag(618)}$ biosensor shows excellent sensitivity for application in high-throughput screenings (Z-factor > 05) and low inter-day variability. Figures have been adapted from Schihada et al., 2018 [7] according to Creative Commons Attribution 40 International License (http:/creativecommonsorg/licenses/by/40/)

1.2 HTS-Suitability

The BRET-based receptor biosensor offers the unprecedented possibility to study the receptor's conformational dynamics in a high-throughput format. The assay suitability for HTS format is quantified through the standard Z-factor determination [11]. The ideal assay without signal variation would yield the maximum Z-factor of 1.0. Assays with Z-factors ≥0.5 are described as excellent, whereas assays with $0 \leq Z < 0.5$ require more replicates to be tested to allow for a reliable interpretation of the screening data. $Z < 0$ defines methods that cannot be used for HTS. Additionally, comparing Z-factors measured on different days enables to evaluate the robustness and reproducibility of the assay (Fig. 2b).

1.3 Advantages and Limitations of BRET-Based Assays

Since BRET-based assays do not require external illumination and are therefore not susceptible to cellular autofluorescence, light scattering, or fluorophore photobleaching, BRET often results in higher sensitivity than FRET-based assays [12]. Implementing a novel BRET-based assay is often straightforward since no specialized microscopes are required and luminescent plate readers are commonly used in biochemical labs. Most of these devices enable luminescence measurements based either on monochromator or filter optics. Table 1 lists the different plate readers and respective reading settings we deployed to measure the conformational dynamics of Nluc/HaloTag(618)-based GPCR sensors.

BRET screening data may still be affected when compounds are tested that (i) significantly absorb light of 400–650 nm wavelength and therefore interfere with the donor's or acceptor's

Table 1
Technical specifications of different plate readers used for GPCR$_{Nluc/HaloTag(618)}$ biosensors

Plate reader	Filter sets/monochromator settings	Properties
Synergy Neo2 (BioTek Instruments)	460/40 nm + 620/20 nm *or* 450/50, 610 LP | DM 550 (for dual PMT detectors) *or* EM 450/50, 610 LP (one PMT detector)	Filter optics
GloMax® Discover Multimode Microplate Reader (Promega)	495 SP + 600 LP	Filter optics
CLARIOstar (BMG Labtech)	450/80 nm + 630/60 nm	Monochromator optics
PHERAstar (BMG Labtech)	460/80 + 610 LP	Filter optics
Mithras (Berthold Technologies)	480/20 + 620 LP	Filter optics
EnSight® Multimode Plate Reader (PerkinElmer®)	NanoBRET Blue 460/80 + BRET deep red *or* ET645/75 nm	Filter optics
Spark® Cyto (Tecan)	460 − 485 nm + 610 − 650 nm	Filter optics

photophysical properties or (ii) if the tested compounds alter the catalytic activity of the luciferase [13]. These compounds are often identified as false positive hits in compound screening campaigns and can be excluded through chemical evaluation of a screening library, negative counter screens based on the same donor/acceptor combination (e.g., Nluc-rigid linker-HaloTag(618)), or through subsequent secondary screens for hit validation [14].

1.4 Transferability of the Sensor Design to Other GPCRs

Below, we outline the detailed workflow to monitor the conformational GPCR dynamics for drug screening using the genetically encoded BRET-based biosensor $\alpha_{2A}AR_{Nluc/Halo(618)}$ (Fig. 3). This biosensor design was successfully employed on other GPCRs including β_2AR and PTHR1 [7], as well as the chemokine CXCR4, the angiotensin AT$_1$, and the histamine H$_3$ receptor [8]. The selection of appropriate insertion sites for Nluc and HaloTag represents the most critical step in the transfer of this sensor design to other G protein-coupled receptors.

All functional Nluc/HaloTag(618)-based GPCR sensors are labeled with Nluc at the full-length or truncated receptor C-terminus and HaloTag is inserted in the third intracellular loop. Unfortunately, the exact positioning of the tags within these receptor domains often needs to be found through a trial and error procedure. However, sequence-based alignments with validated receptor sensors as well as 3D structural information and insights from site-directed mutagenesis studies about the receptor of

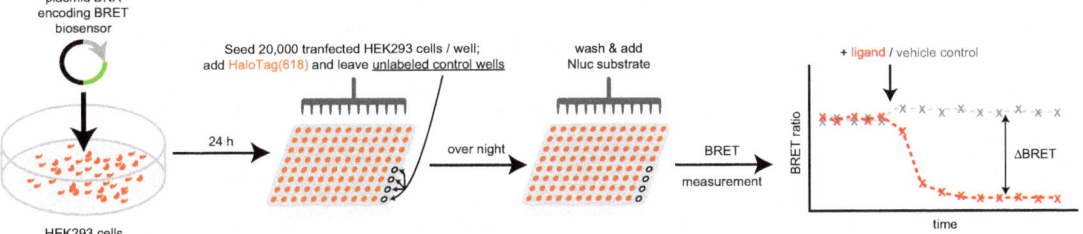

Fig. 3 Workflow of GPCR$_{Nluc/HaloTag(618)}$ experiments

interest can be very helpful to select tolerated insertion sites and increase the chances that the new fusion protein is (i) trafficked to the cell surface and (ii) maintains wild-type-like ligand binding properties and conformational dynamics.

2   Materials

Prepare all solutions using ultrapure water and analytical grade reagents.

2.1   Cell Culture

1. Human Embryonic Kidney (HEK)-293 cells, HEK-293A or HEK-293 T cells (ThermoFisher, ATCC).

2. Cell culture medium: Dulbecco's Modified Eagle's Medium (DMEM) supplemented with 10% fetal bovine serum (FBS), 1% L-Glutamine (200 mM), 1% Penicillin/Streptomycin (100 U/mL Penicillin; 0.1 mg/mL Streptomycin).

3. Dulbecco's Phosphate Buffered Saline (DPBS).

4. Solution of trypsin (0.25%) and ethylenediaminetetraacetic acid (EDTA, 0.53 mM).

5. 100 mm and 55 mm tissue culture dishes or T75 and T25 culture flasks.

2.2   Transfection Materials

1. Biosensor plasmid DNA. We used plasmid α_{2A}-adrenergic receptor α_{2A}-AR$_{Nluc/Halo618}$.

2. Effectene Transfection Reagent or Lipofectamine 2000.

2.3   Counting and Plating

1. Solution of trypsin (0.25%) and EDTA (0.53 mM).

2. Hemocytometer.

3. Trypan blue.

4. White-wall and white-bottomed 96-well plates.

5. Poly-D-lysine: 0.1 mg/mL.

6. HaloTag® NanoBRET™ 618 Ligand.

2.4 Measurement Materials and Devices

1. Experimental buffer: 2 mM HEPES, 28 mM NaCl, 1.08 mM KCl, 0.2 mM $MgCl_2$, 0.4 mM $CaCl_2$, pH 7.3 or Hank's Balanced Salt Solution (HBSS).

2. Furimazine (NanoBRET™ Nano-Glo® Substrate (*see* **Note 1**).

3. Plate reader Synergy Neo2 equipped with filter optics (emission filters: 460/40 nm, 620/20 nm or similar, set integration time to 0.3 s) (*see* **Note 2**).

4. Compound library and reference ligand(s) serving as positive control (confirmed agonist / inverse agonist of the targeted GPCR).

2.5 Data Analysis

1. Plate reader control software (e.g., Gen5 Data Analysis Software).

2. Spreadsheet application (e.g., Microsoft Office Excel).

3 Methods

All steps are performed at room temperature unless specified otherwise.

3.1 Cell Culture, Counting, and Transfection

1. Grow HEK-293 T cells in cell culture medium at 37 °C and 5% CO_2 Passage cells every 2–3 days when they reach confluency of 80–100% (*see* **Note 3**). Aspirate medium and wash cells carefully with 3 mL of DPBS. Subsequently, incubate cells with 1 mL trypsin-EDTA solution for 1 min and resuspend in 6–10 mL of fresh medium. Transfer an aliquot (1 mL) of resuspended cells to a new dish / flask with fresh cell culture medium. Use cells up to passage 30.

2. To determine cell density, mix 10 μL of cell suspension with 10 μL of Trypan blue and transfer 10 μL of the new suspension to a hemocytometer. Determine the number of living unstained cells in one large square. Calculate the cell number/mL suspension according to the following equation:
   Cell number/mL = counted cells × 2 (Dilution Factor) × 10^4.

3. To prepare cells for transient expression, seed 1.5×10^6 HEK-293 T cells onto a 5.5 cm dish or T25 culture flask (*see* **Note 4**). After 24 h, transfect 2 μg plasmid DNA using Effectene transfection reagent or 1 μg plasmid DNA using Lipofectamine 2000 according to the manufacturer's protocol. Add DNA solution dropwise to the cells and grow overnight at 37 °C, 5% CO_2.

3.2 Labeling and Plating

1. Precoat (*see* **Note 5**) whitewall, white-bottomed 96-well plates with 50 μL/well of a 0.1 mg/mL poly-D-lysine and incubate for ≥20 min. Aspirate the coating solution and wash twice with

DPBS. Allow the plate to dry for ≥ 20 min in the laminar flow hood before seeding cells.

2. 24 h after transfection, wash cells with 2 mL DPBS. Incubate with 1 mL trypsin-EDTA solution for 1 min, remove trypsin-EDTA, and resuspend cells in 5 mL of fresh cell culture medium.

3. Count cells as described above.

4. To seed 20,000 cells into each well of a whole 96-well plate, prepare 11 mL of cell suspension with a density of 200,000 cells per mL (Fig. 3). Seed 100 μL into four wells each, without the addition of HaloTag(618) to correct for donor bleed-through into the BRET acceptor channel (unlabeled control). Add HaloTag(618) to the residual cell suspension to a final concentration of 100 nM (*see* **Note 6**).

5. Incubate overnight at 37 °C, 5% CO_2.

3.3 BRET Measurement

1. Turn on the plate reader and the computer.

2. Open the plate reader control software and, if possible, warm up the reader to 37 °C.

3. Set up a read protocol that allows for well-by-well luminescence intensity measurements of donor and acceptor emission intensities. Set integration time to 0.3 s per data point. Program five repeats for basal and 15 repeats for stimulated readings for the entire plate (*see* **Note 7**).

4. Wash the cells once with 100 μL of pre-warmed experimental buffer or HBSS.

5. Mix 10 μL of furimazine stock solution with 10 mL experimental buffer/HBSS and add 90 μL of this solution to each well carefully and slowly to not create bubbles (*see* **Note 8**). Incubate for 2–5 min at 37 °C to allow for substrate diffusion.

6. Insert the plate into the microplate reader and establish a five-point baseline reading.

7. Add 10 μL per well of a tenfold compound solution or vehicle and positive control—again take care to not create bubbles. Use each compound in duplicates to quadruplicates (Fig. 4).

8. Record 15 BRET reads after compound / vehicle / positive control addition (Fig. 3).

3.4 Data Analysis and Graphing

1. Open data in Excel or similar spreadsheet application. Calculate the raw BRET ratio for each well at each individual time point according to the following equation:

$$\text{raw BRET} = \frac{\text{emission intensity in the BRET acceptor channel}}{\text{emission intensity in the BRET donor channel}}$$

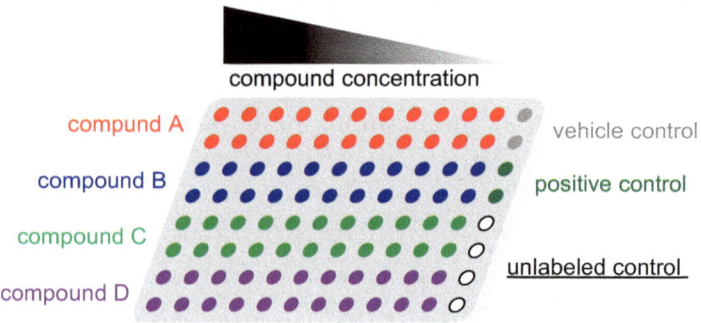

Fig. 4 Schematic of a typical plate stimulation layout for experiments with GPCR$_{Nluc/HaloTag(618)}$ biosensors. In primary screenings, the number of compounds to be tested on one plate can be substantially increased by testing only one concentration per compound

2. To correct for donor bleedthrough into the acceptor emission channel, calculate the average raw BRET ratio of all HaloTag (618)-unlabeled control wells for each time point (*see* **Note 9**). Subsequently, subtract these average values from the raw BRET ratios of all HaloTag(618)-labeled wells of the same time point to obtain the bleedthrough-corrected BRET ratio (corr BRET):

$$\text{corr BRET} = \text{raw BRET (labeled wells)} - \text{average raw BRET (unlabeled wells)}$$

3. Plot the corr BRET values over time to generate a BRET time course (*see* **Note 7**).

4. To quantify compound-induced BRET changes, first average the five corr BRET values before (BRET$_{basal}$) and after (BRET$_{stim}$) compound / vehicle addition for each well to calculate the raw ΔBRET according to the following equation:

$$\text{raw } \Delta\text{BRET } [\%] = 100 * \frac{(\text{BRET}_{stim} - \text{BRET}_{basal})}{\text{BRET}_{basal}}$$

Subsequently calculate corrected ΔBRET (corr ΔBRET) as a percent over basal to correct for non-compound-induced BRET signals (e.g., due to increases in well volume). Therefore, subtract the average raw ΔBRET of the vehicle control from the raw ΔBRET of each compound-treated well:

$$\text{corr } \Delta\text{BRET} = \text{raw } \Delta\text{BRET (compound)} - \text{average raw } \Delta\text{BRET (vehicle)}$$

5. To determine the Z-factor as a measure of the assay's HTS-suitability, calculate the average (μ) and standard deviation (σ) of corr. ΔBRET of positive control and vehicle treated wells (*see* **Note 10**).

$$Z - \text{factor} = 1 - \frac{3*(\sigma[\text{positive control}] + \sigma[\text{vehicle}])}{|\mu[\text{positive control}] - \mu[\text{vehicle.}]|}$$

If the positive control induces a decrease in BRET, the denominator of the equation has to be substituted with $|\mu[\text{vehicle}] - \mu[\text{positive control}]|$

6. To identify hits, calculate the background of the experiment defined as $\mu \pm 3*\sigma$ of corr. ΔBRET of the vehicle control. Screening hits evoke corr. ΔBRET signals above or below this background (*see* **Note 11**).

4   Notes

1. Promega provides distinct furimazine preparations for different assays. In our hands, the furimazine stock solutions provided with Promega products #N157A (NanoBRET™ Nano-Glo® Substrate) and #G980A (NanoBRET™ Nano-Glo® Detection System; kit containing the fluorescent dye HaloTag® Nano-BRET™ 618 Ligand and furimazine) yielded high assay sensitivity when applied in a final 1/1000 dilution.

2. Our assays were optimized using the Synergy Neo2 plate reader from BioTek Instruments with a 460/40 nm and 620/20 nm filter set. Additionally, we successfully conducted experiments with BRET-based conformational GPCR sensors using the plate readers listed in Table 2 with 0.3 s integration time per data point and the indicated filter / monochromator settings.

3. All specifications apply to HEK-293 T cells. The use of other cell lines may require optimization of cell culture, transfection techniques, and plating density.

4. We employed the optimized specifications in the table below for application of BRET conformational GPCR biosensors in a 384-well microtiter format.

5. Precoating may not be needed for cell lines with improved adhesive properties.

6. The HaloTag(618) labeling concentration required for optimal labeling efficiency depends on several factors including cell number, expression level of the protein of interest, and accessibility of HaloTag within the fusion protein. In our experiments with conformational GPCR biosensors, reducing the labeling concentration to 50 nM has never negatively affected assay sensitivity.

7. The optimal length of the protocol can differ between different biosensors and compounds tested depending on the kinetics of, e.g., the receptor-ligand binding process. A BRET time-course experiment can be conducted to determine the optimal

Table 2
Optimized assay parameters for 384-well microtiter plates

Step	Optimized parameter	Microtiter plate format	
		96-well	384-well
Cell culture	Recommended dish/flask size	55 mm/T25	100 mm/T75
	Seeding density (10^6 cells/dish or flask)	1.5	4
Transfection	Amount of DNA (μg/culture dish or flask)	1.5	4
Plating	Seeding density (cells/well)	20,000 – 40,000	5000–20,000
Measurement	Basal measurement well volume (μL)	90	45
	Compound addition	10 μL of a 10× solution	5 μL of a 10× solution

time window. In our experience, reading and averaging at least three independent time points for $BRET_{basal}$ and $BRET_{stim}$ (after saturation of the BRET signal upon compound addition) significantly improved assay sensitivity.

8. Higher dilutions than 1/1000 of furimazine can be tested for their effect on the dynamic range of the conformational sensors. For instance, a 1/4000 furimazine dilution yields an even higher dynamic range for $\alpha_{2A}AR_{Nluc/Halo(618)}$, but lower signal-to-noise ratios for $\beta_2AR_{Nluc/Halo(618)}$ and $PTHR1_{Nluc/Halo(618)}$.

9. A statistical comparison of the basal raw BRET ratios of labeled vs. unlabeled wells indicates whether donor-to-acceptor resonance energy transfer occurs in the ligand-free state of the receptor biosensor. In our experience, all functional GPCR conformational biosensors showed significant energy transfer in the basal receptor state. However, this might not be in the case in receptors with long C-termini (≥ 200 amino acids) due to increased distance between the Nluc and HaloTag(618). In such a scenario, optimization of the sensor design (e.g., truncation of the C-terminus) may be required to allow for basal BRET and to increase the dynamic range of the biosensor.

10. A sufficient Z-factor is essential for the reproducibility and robustness of the assay and can be improved by using a cell line that stably expresses the conformational biosensor.

11. After the primary screen, a confirmation screen with different concentrations of the primary hits should be considered. Furthermore, a Nluc/HaloTag(618)-based counter-screen can be employed to identify compounds that interfere with the assay principle.

Acknowledgments

This work was supported by the Federal Ministry of Research (BMBF; 03 V0830), National Institutes of Health (NIH, 0255-8521-4609), SFB/TR166 to M.J.L and by the Deutsche Forschungsgemeinschaft (DFG, German Research Foundation – 427840891), Nottingham Research Anne McLaren fellowship to IM.

Competing interests: The authors declare that The University of Würzburg holds a patent on this technology: WO2004057333 A1.

References

1. Forster T (1948) Zwischenmolekulare Energiewanderung Und Fluoreszenz. Annalen Der Physik 2:55–75. https://doi.org/10.1002/andp19484370105

2. Wu P, Brand L (1994) Resonance energy transfer: methods and applications. Anal Biochem 218:1–13. https://doi.org/10.1006/abio19941134

3. Rasmussen SG et al (2011) Crystal structure of the beta2 adrenergic receptor-Gs protein complex. Nature 477:549–555. https://doi.org/10.1038/nature10361

4. Zhao LH et al (2019) Structure and dynamics of the active human parathyroid hormone receptor-1. Science 364:148–153. https://doi.org/10.1126/scienceaav7942

5. Qiao A et al (2020) Structural basis of Gs and Gi recognition by the human glucagon receptor. Science 367:1346–1352. https://doi.org/10.1126/scienceaaz5346

6. Kauk M, Hoffmann C (2018) Intramolecular and intermolecular FRET sensors for GPCRs - monitoring conformational changes and beyond. Trends Pharmacol Sci 39:123–135. https://doi.org/10.1016/jtips201710011

7. Schihada H et al (2018) A universal bioluminescence resonance energy transfer sensor design enables high-sensitivity screening of GPCR activation dynamics. Commun Biol 1:105. https://doi.org/10.1038/s42003-018-0072-0

8. Schihada H et al (2020) Development of a conformational histamine H3 receptor biosensor for the synchronous screening of agonists and inverse agonist. ACS Sens Jun 26;5(6):1734–1742. https://doi.org/10.1021/acssensors.0c00397. Epub 2020 May 28

9. Hall MP et al (2012) Engineered luciferase reporter from a deep sea shrimp utilizing a novel imidazopyrazinone substrate. ACS Chem Biol 7:1848–1857. https://doi.org/10.1021/cb600135w

10. Los GV et al (2008) HaloTag: a novel protein labeling technology for cell imaging and protein analysis. ACS Chem Biol 3:373–382. https://doi.org/10.1021/cb800025k

11. Zhang JH, Chung TD, Oldenburg KR (1999) A simple statistical parameter for use in evaluation and validation of high throughput screening assays. J Biomol Screen 4:67–73. https://doi.org/10.1177/108705719900400206

12. Boute N, Jockers R, Issad T (2002) The use of resonance energy transfer in high-throughput screening: BRET versus FRET. Trends Pharmacol Sci 23:351–354. https://doi.org/10.1016/S0165-6147(02)02062-X

13. Braeuning A (2015) Firefly luciferase inhibition: a widely neglected problem. Arch Toxicol 89:141–142. https://doi.org/10.1021/jm701302v

14. Thorne N, Auld DS, Inglese J (2010) Apparent activity in high-throughput screening: origins of compound-dependent assay interference. Curr Opin Chem Biol 14:315–324. https://doi.org/10.1016/jcbpa201003020

Chapter 10

NanoluOciferase-Based Complementation Assay to Detect GPCR-G Protein Interaction

Céline Laschet and Julien Hanson

Abstract

G protein-coupled receptors (GPCR) are one of the principal class of membrane proteins and around 30% of the currently marketed drugs act on one of them. The efficacious detection of ligands with the desired pharmacological profile remains a challenge of paramount importance in the GPCR drug discovery and pharmacological research. Recent evidences demonstrate that GPCR ligands can stabilize distinct receptor conformation and trigger various signaling pathways with different efficacies and/or potencies. This phenomenon called functional selectivity or biased signaling may lead to improved drugs with fewer side effects. Most receptors are promiscuous and can couple to more than one G protein family. To enable the discovery of biased ligands able to selectively trigger one G protein pathway over another, simple and efficient screening procedures are needed. The traditional assays aiming at detecting G protein activation monitor the generation of second messengers ($[Ca^{2+}]_i$, cAMP, IP_1) or active G proteins (with GTP-g-S for instance). While these approaches have proven sensitive and robust, they are not suited for the detection of a single GPCR-G protein interaction. Here, we present in detail a method to assess directly the interaction between the receptor and the G protein. It permits the profiling of a receptor or a ligand toward G protein interactions and is compatible with high-throughput screening.

Key words GPCR, G protein, Complementation, Nanoluciferase, NanoBit, Biased signaling, Functional selectivity

1 Introduction

G protein-coupled receptors (GPCR) are one of the principal class of membrane proteins and around 30% of the currently marketed drugs act on one of them [1, 2]. The efficacious detection of ligands with the desired pharmacological profile remains a challenge of paramount importance in the GPCR drug discovery and pharmacological research. Recent advances in our understanding of GPCR pharmacology have added new layers of complexity that revealed the importance of the choice of the assay when designing high-

Sofia Aires M. Martins and Duarte Miguel F. Prazeres (eds.), *G Protein-Coupled Receptor Screening Assays:*
Methods and Protocols, Methods in Molecular Biology, vol. 2268, https://doi.org/10.1007/978-1-0716-1221-7_10,
© Springer Science+Business Media, LLC, part of Springer Nature 2021

throughput screening (HTS) campaigns [3]. Indeed, these receptors are not simple on-off switches but rather elaborated transmitters that can adopt multiple conformational states [4]. The current paradigm states that discrete conformations may have different affinities for intracellular transducers. Consequently, chemically different ligands may stabilize distinct receptors entities and thus trigger various pathways with varying potency and efficacy. This property is called functional selectivity or biased signaling and opens new avenues in drug discovery [5].

For the ligand hunter, the choice of the technology used for a given screening will have an impact on the profile of ligands found. Thus, if the goal of the HTS is to identify ligands with a defined bias toward a given pathway, more exquisite strategies to monitor receptor activation must be developed. A purposely designed assay will directly select the right subset of ligands possessing the desired properties and speed up the discovery of drugs with an improved therapeutic profile.

An important characteristic of GPCRs is their capacity to bind more than one G protein, a feature that is more widespread than initially thought [6, 7]. Actually, most of the GPCR are "promiscuous" and show some degree of nonselectivity between G protein families. In the frame of the functional selectivity theory, some ligands should in principle be able to select distinct GPCR conformations that in turn will only bind a subset of G proteins or even a single one. The traditional assays aiming at detecting G protein activation monitor the generation of second messengers ($[Ca^{2+}]_i$, cAMP, IP_1, among others) or active G proteins (with GTP-γ-S for instance) [8]. While these approaches have proven sensitive and robust, they are not suited for the detection of a single GPCR-G protein interaction. Other strategies exist to detect the interaction between a given receptor and its G proteins, principally using BRET and FRET. However, they are usually not sensitive or simple enough to be used in the context of HTS.

Here, we present a method to assess directly the interaction between the receptor and the G protein. It permits the profiling of a receptor or a ligand toward G protein interactions but, in addition, is compatible with HTS [9]. The technology is based on the complementation of two parts of nanoluciferase (NLuc) [10, 11]. The initial report by Dixon *et al.* suggested the use of a modified small stretch of the enzyme called small BiT (SmBiT) to label one partner and the large part (LgBiT) to label the other partner. The advantage of the SmBiT is that it has a very weak affinity for the LgBiT (Fig. 1). We modified this approach by using the unmodified small part of the NLuc (called "Natural peptide" or NP, *see* Fig. 1)

Peptide	sequence												K_D (M)
SmBiT	-	V	T	G	Y	R	L	F	E	E	I	L	-	1.9×10^{-4}
Natual peptide (NP)	G	V	T	G	W	R	L	C	E	R	I	L	A	0.9×10^{-6}

Fig. 1 Amino acid sequence and K_D value of two different small peptides of the NLuc (SmBiT and NP). The amino acids of the NP that have been mutated to generate the SmBiT sequence are labeled in red

because we demonstrated having some residual affinity between the two complementing partners was more adapted to detect the highly transient interaction between GPCR and G proteins [9].

2   Materials

2.1   Cell Culture and Transfection

1. HEK293 cells (American Type Culture Collection).
2. T-175 cell culture flask.
3. Culture medium: Dulbecco's modified Eagle medium (DMEM) supplemented with 10% fetal bovine serum (FBS), 2 mM L-glutamine, and penicillin/streptomycin (50 units/mL).
4. 100-mm Cell culture dishes.
5. Opti-MEM (*see* **Note 1**).
6. X-tremeGENE 9 DNA transfection reagent.
7. GPCR-NP and Gα-LgBiT expression constructs (*see* Subheading 2.3).

2.2   Cell Preparation and Measurement for Nanoluciferase Complementation Assay

1. PBS (phosphate-buffered saline).
2. Trypsin-Versene.
3. 15-ml Tube.
4. HBSS:120 mM NaCl, 5.4 mM KCl, 0.8 mM $MgSO_4$, 10 mM HEPES, pH 7.4. Weigh 5.96 g of NaCl, 0.4 g of KCl, 0.197 g of $MgSO_4.7H_2O$, and 2.38 g of HEPES and transfer to 1 L graduated cylinder. Add 900 mL of ultrapure water, mix, and adjust pH with NaOH at 7.4. Make up to 1 L with ultrapure water. Store at room temperature.
5. Nano-Glo® Live Cell Reagent: Mix 1 volume of Nano-Glo® Live Cell Substrate with 19 volumes of Nano-Glo® LCS Dilution Buffer (a 20-fold dilution).
6. White flat-bottom 96-well plates (*see* **Note 2**).
7. Centro XS^3 LB960 plate reader.

2.3 Design of GPCRs and Gα Subunit Expression Constructs

2.3.1 GPCR

1. Clone GPCR coding sequence into an expression vector (e.g., pcDNA3.1). A signal sequence (example: KTIIALSYIFCLVFA) can be added upfront to enhance the expression of the receptor at the cell surface [12]. To facilitate further detection of the receptor, a tag epitope (e.g., Flag: DYKDDDDK [13]) may be added at the N-terminus of the receptor.

2. SmBiT (VTGYRLFEEIL) or NP (GVTGWRLCERILA) are added with a flexible linker (GNSGSSGGGGSGGGGSSG) in frame at the C terminus of the receptor by polymerase chain reaction (PCR) (Fig. 2) [9–11] (*see* **Note 3**).

2.3.2 Gα Subunit

1. Clone Gα subunit coding sequence into a vector (e.g., pcDNA3.1). A tag epitope (e.g., HA) may be added at the N terminus of the Gα subunit to facilitate the detection.

A

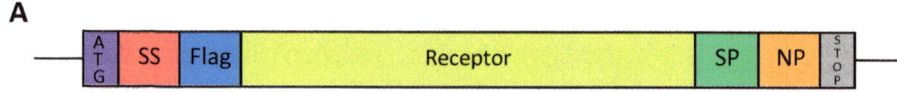

B

ATGAAGACGATCATCGCCCTGAGCTACATCTTCTGCCTGGTATTCGCCGATTATAAA
GATGATGATGATAAAGATCCACTGAATCTGTCCTGGTATGATGATGATCTGGAGAG
GCAGAACTGGAGCCGGCCCTTCAACGGGTCAGACGGGAAGGCGGACAGACCCCA
CTACAACTACTATGCCACACTGCTCACCCTGCTCATCGCTGTCATCGTCTTCGGCAA
CGTGCTGGTGTGCATGGCTGTGTCCCGCGAGAAGGCGCTGCAGACCACCACCAAC
TACCTGATCGTCAGCTCCGCAGTGGCCGACCTCCTCGTCGCCACACTGGTCATGCC
CTGGGTTGTCTACCTGGAGGTGGTAGGTGAGTGGAAATTCAGCAGGATTCACTGTG
ACATCTTCGTCACTCTGGACGTCATGATGTGCACGGCGAGCATCCTGAACTTGTGT
GCCATCAGCATCGACAGGTACACAGCTGTGGCCATGCCCATGCTGTACAATACGCG
CTACAGCTCCAAGCGCCGGGTCACCGTCATGATCTCCATCGTCTGGGTCCTGTCCT
TCACCATCTCCTGCCCACTCCTCTTCGGACTCAATAACGCAGACCAGAACGAGTGC
ATCATTGCCAACCCGGCCTTCGTGGTCTACTCCTCCATCGTCTCCTTCTACGTGCCC
TTCATTGTCACCCTGCTGGTCTACATCAAGATCTACATTGTCCTCCGCAGACGCCGC
AAGCGAGTCAACACCAAACGCAGCAGCCGAGCTTTCAGGGCCCACCTGAGGGCTC
CACTAAAGGGCAACTGTACTCACCCCGAGGACATGAAACTCTGCACCGTTATCATG
AAGTCTAATGGGAGTTTCCCAGTGAACAGGCGGAGAGTGGAGGCTGCCCGGCGAG
CCCAGGAGCTGGAGATGGAGATGCTCTCCAGCACCAGCCCACCCGAGAGGACCCG
GTACAGCCCCATCCCACCCAGCCACCACCAGCTGACTCTCCCCGACCCGTCCCACC
ATGGTCTCCACAGCACTCCCGACAGCCCCGCCAAACCAGAGAAGAATGGGCCATGC
CAAAGACCACCCCAAGATTGCCAAGATCTTTGAGATCCAGACCATGCCCAATGGCA
AAACCCGGACCTCCCTCAAGACCATGAGCCGTAGGAAGCTCTCCCAGCAGAAGGA
GAAGAAAGCCACTCAGATGCTCGCCATTGTTCTCGGCGTGTTCATCATCTGCTGGCT
GCCCTTCTTCATCACACACATCCTGAACATACACTGTGACTGCAACATCCCGCCTGT
CCTGTACAGCGCCTTCACGTGGCTGGGCTATGTCAACAGCGCCGTGAACCCCATCA
TCTACACCACCTTCAACATTGAGTTCCGCAAGGCCTTCCTGAAGATCCTCCACTGCG
GGAATTCTGGCTCGAGCGGTGGTGGCGGGAGCGGAGGTGGAGGGTCGTCAGGTG
GTGTGACCGGCTGGCGGCTGTGTGAGCGAATTCTGGCATAA

Fig. 2 (**a**) Typical construction of a GPCR for the NLuc-based complementation assay. SS: signal sequence, SP: spacer, NP: natural peptide. (**b**) Example for the dopamine D$_2$ receptor. Nucleotide sequence of the D$_2$ fused with the NP (D$_2$-NP). Start codon (purple), signal sequence (orange), Flag epitope (blue), D$_2$ (black), linker (green), NP (red), and stop codon (gray)

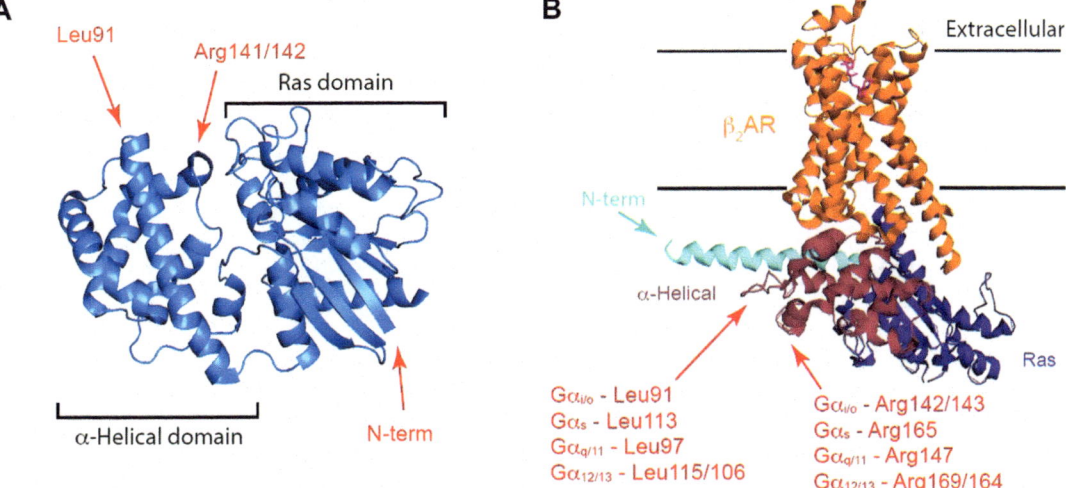

Fig. 3 (**a**) Overall architecture of Gα$_{i1}$ (Coleman et al. 1994; Weng et al. 1998). PDB number 1gfi. Red arrows indicate positions for LgBiT insertion that gave the best results in our hands. (**b**) Crystal structure of the β$_2$AR-Gαs complex (PDB number 3SN6; (Rasmussen et al., 2011)). Red arrows indicate positions of LgBiT insertion that are topologically identical between Gα subunit. Residues number for Gαs and Gαi/o are indicated as examples

2. Add two different restriction sites by targeted mutagenesis at the location where you want to insert the coding sequence of the LgBiT (Fig. 3) (*see* **Note 4**).

3. Amplify the coding sequence of the LgBiT flanked by a flexible linker (SGGGGS) and the respective restriction sites by PCR and insert the sequence into the Gα subunit sequence by cloning (Fig. 4).

3   Methods

3.1   Cell Culture and Transfection

Cell culture and transfection should be performed in sterile conditions under a laminar flow hood.

1. Maintain HEK293 cells in T-175 cell culture flask with 25 mL of culture medium at 37 °C with 5% CO_2.

2. Seed cells at 70% confluence into 100-mm cell culture dishes with 8 mL of culture medium.

3. Incubate the cells at 37 °C with 5% CO_2 for 24 h.

4. Transfect the cells with the plasmids coding for GPCR-NP and Gα-LgBiT expression constructs. Dilute 12 μL of X-tremeGENE 9 DNA transfection reagent with 400 μL of Opti-MEM. Add 4 μg of DNA to diluted X-tremeGENE 9 DNA transfection reagent and mix gently (*see* **Note 5**).

A

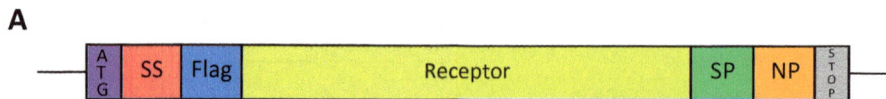

B

ATGTACCCATACGATGTTCCAGATTACGCTGGCTGCACGCTGAGCGCCGAGGACAA
GGCGGCGGTGGAGCGGAGTAAGATGATCGACCGCAACCTCCGTGAGGACGGCGA
GAAGGCGGCGCGCGAGGTCAAGCTGCTGCTGCTCGGTGCTGGTGAATCTGGTAAA
AGTACAATTGTGAAGCAGATGAAAATTATCCATGAAGCTGGTTATTCAGAAGAGGA
GTGTAAACAATACAAAGCAGTGGTCTACAGTAACACCATCCAGTCAATTATTGCTAT
CATTAGGGCTATGGGGAGGTTGGATATCAGCGGAGGCGGAGGAAGCGTCTTCACA
CTCGAAGATTTCGTTGGGGACTGGGAACAGACAGCCGCCTACAACCTGGACCAAG
TCCTTGAACAGGGAGGTGTGTCCAGTTTGCTGCAGAATCTCGCCGTGTCCGTAACT
CCGATCCAAAGGATTGTCCGGAGCGGTGAAAATGCCCTGAAGATCGACATCCATGT
CATCATCCCGTATGAAGGTCTGAGCGCCGACCAAATGGCCCAGATCGAAGAGGTG
TTTAAGGTGGTGTACCCTGTGGATGATCATCACTTTAAGGTGATCCTGCCCTATGGC
ACACTGGTAATCGACGGGGTTACGCCGAACATGCTGAACTATTTCGGACGGCCGTA
TGAAGGCATCGCCGTGTTCGACGGCAAAAAGATCACTGTAACAGGGACCCTGTGG
AACGGCAACAAAATTATCGACGAGCGCCTGATCACCCCCGACGGCTCCATGCTGTT
CCGAGTAACCATCAACAGTAGCGGAGGCGGAGGAAGCCCGCGGAAGATAGACTTT
GGTGACTCAGCCCGGGCGGATGATGCACGCCAACTCTTTGTGCTAGCTGGAGCTGC
TGAAGAAGGCTTTATGACTGCAGAACTTGCTGGAGTTATAAAGAGATTGTGGAAAG
ATAGTGGTGTACAAGCCTGTTTCAACAGATCCCGAGAGTACCAGCTTAATGATTCTG
CAGCATACTATTTGAATGACTTGGACAGAATAGCTCAACCAAATTACATCCCGACTC
AACAAGATGTTCTCAGAACTAGAGTGAAAACTACAGGAATTGTTGAAACCCATTTTA
CTTTCAAAGATCTTCATTTTAAAATGTTTGATGTGGGAGGTCAGAGATCTGAGCGGA
AGAAGTGGATTCATTGCTTCGAAGGAGTGACGGCGATCATCTTCTGTGTAGCACTG
AGTGACTACGACCTGGTTCTAGCTGAAGATGAAGAAATGAACCGAATGCATGAAAG
CATGAAATTGTTTGACAGCATATGTAACAACAAGTGGTTTACAGATACATCCATTAT
ACTTTTTCTAAACAAGAAGGATCTCTTTGAAGAAAAAATCAAAAAGAGCCCTCTCAC
TATATGCTATCCAGAATATGCAGGATCAAACACATATGAAGAGGCAGCTGCATATA
TTCAATGTCAGTTTGAAGACCTCAATAAAAGAAAGGACACAAAGGAAATATACACC
CACTTCACATGTGCCACAGATACTAAGAATGTGCAGTTTGTTTTTGATGCTGTAACA
GATGTCATCATAAAAAATAATCTAAAAGATTGTGGTCTCTTTTAA

Fig. 4 (**a**) Example of construction for the G protein, given for Gαᵢ₁. HA = hemagglutinin tag; RS = restriction site, left: EcoRV, right: SacII; LK = linker; LgBiT = truncated Nanoluciferase (first 156 AA). (**b**) Nucleotide sequence of Gαᵢ₁ with the insertion of the LgBiT between residues 91 and 92 of Gαᵢ₁. Start codon (purple), HA epitope (blue), Gαᵢ₁ (black), restriction site (yellow), linker (green), LgBiT (red), and stop codon (gray)

5. Incubate the transfection reagent-DNA complex for 15 min at RT.

6. Add the transfection complex to the cells in a dropwise manner (*see* **Note 6**).

7. Swirl the dish to ensure even distribution over the entire plate surface.

8. Incubate the cells at 37 °C with 5% CO_2 for 24 h.

3.2 Cell Preparation and Measurement for Nanoluciferase Complementation Assay

1. Aspirate culture medium and wash the cells with 5 mL of PBS.

2. Add 500 μL of trypsin to the cells dish and incubate at 37 °C for 5 min.

3. Use 10 mL of culture medium to detach the cells by pipetting up and down. Transfer the cells to a 15-mL tube.

4. Centrifuge the cells for 5 min at $400 \times g$ at room temperature.

5. Discard the supernatant and resuspend the cells in 10 mL of PBS.

6. Centrifuge the cells for 5 min at $400 \times g$ at room temperature.

7. Discard the supernatant and resuspend the cells in 13 mL of HBSS supplemented with 1.3 mL of Nano-Glo® Live Cell Reagent (see **Note 7**).

8. Add 100 µL of the cell suspension per well (~50.000 cells/ well) of a white flat-bottom 96-well plate.

9. Incubate the cells at 37 °C for 30 min (see **Note 8**).

10. Add the compounds manually and measure directly the luminescence with a highly sensitive plate reader for several minutes (0.5 s/well) depending on each experiment (see **Note 9**).

4 Notes

1. Opti-MEM is an example of serum-free medium that can be used to dilute X-tremeGENE 9 DNA transfection reagent.

2. It is recommended to use white flat-bottom 96-well plates to record luminescence signal in order to maximize the light output signal.

3. Other location in the receptor may be tested. The NP can be included inside (not at the end) the C-terminus or at any intracellular loop. In case of nonresponsive or suboptimal system, putting the NP elsewhere is one of the options to improve the constructs.

4. In our hands, the best locations for the $G\alpha$ protein insertion were in the loop connecting helices A and B (corresponding to Leu 91 in $G\alpha_{i/o}$, Fig. 3), after helix C (Arg142/143 for $G\alpha_{i/o}$) and at the N-term [9]. However, in case of poor outcomes, other locations, especially in the α-helical domain of the G protein, can be envisaged [14]. Deleting parts of the G protein is also a valid strategy to improve the detection system [9–15].

5. We generally transfected HEK293 cells with a 1:1 ratio of GPCR and $G\alpha$ subunit plasmid solutions diluted 1:10 with empty vector (2 µg of expression construct diluted 1:10 with empty vector). Dilute 5 µL of plasmid coding for a GPCR or $G\alpha$ subunit construct with 45 µL of empty vector (e.g., pcDNA3.1) to obtain a plasmid solution diluted 1:10 with empty vector. These should be seen as starting conditions since the GPCR:$G\alpha$ subunit ratio and the amount of expression constructs used for transfection should be optimized for each GPCR and G protein tested.

6. There is no need to replace with fresh medium before or after the addition of the transfection complex.

7. The use of this reagent is indicative and could be optimized to improve the overall performance.

8. The incubation time should be adapted according to the GPCR and the G protein construct tested.

9. We used the Centro XS^3 LB960 plate reader (Berthold Technologies). There are in principle no minimal requirement, but some luminometer have lower sensitivity that may impair the detection. The measurement parameters should be determined and optimized for each GPCR-G protein pair and can be different on other readers.

Acknowledgements

This work was supported by the Fonds pour la Recherche Scientifique (F.R.S.-FNRS) "Projet de Recherche" Grant T.0111.19, University of Liège (Fonds Spéciaux), Céline Laschet is a FRIA fellow and Julien Hanson a F.R.S.-FNRS research associates.

References

1. Sriram K, Insel PA (2018) G protein-coupled receptors as targets for approved drugs: how many targets and how many drugs? Mol Pharmacol 93:251–258

2. Hauser AS, Attwood MM, Rask-Andersen M, Schiöth HB, Gloriam DE (2017) Trends in GPCR drug discovery: new agents, targets and indications. Nat Rev Drug Discov 16:829–842

3. Kenakin TP (2009) Cellular assays as portals to seven-transmembrane receptor-based drug discovery. Nat Rev Drug Discov 8:617–626

4. Latorraca NR, Venkatakrishnan AJ, Dror RO (2017) GPCR dynamics: structures in motion. Chem Rev 117:139–155

5. Kenakin T (2019) Biased receptor signaling in drug discovery. Pharmacol Rev 71:267–315

6. Okashah N, Wan Q, Ghosh S, Sandhu M, Inoue A, Vaidehi N, Lambert NA (2019) Variable G protein determinants of GPCR coupling selectivity. Proc Natl Acad Sci U S A 116:12054–12059

7. Inoue A, Raimondi F, Kadji FMN, Singh G, Kishi T, Uwamizu A, Ono Y, Shinjo Y, Ishida S, Arang N, Kawakami K, Gutkind JS, Aoki J, Russell RB (2019) Illuminating G-protein-coupling selectivity of GPCRs. Cell 177:1933–1947.e25

8. Zhang R, Xie X (2012) Tools for GPCR drug discovery. Acta Pharmacol Sin 33:372–384

9. Laschet C, Dupuis N, Hanson J (2019) A dynamic and screening-compatible nanoluciferase-based complementation assay enables profiling of individual GPCR–G protein interactions. J Biol Chem 294:4079–4090

10. Hall MP, Unch J, Binkowski BF, Valley MP, Butler BL, Wood MG, Otto P, Zimmerman K, Vidugiris G, MacHleidt T, Robers MB, Benink HA, Eggers CT, Slater MR, Meisenheimer PL, Klaubert DH, Fan F, Encell LP, Wood KV (2012) Engineered luciferase reporter from a deep sea shrimp utilizing a novel imidazopyrazinone substrate. ACS Chem Biol 7:1848–1857

11. Dixon AS, Schwinn MK, Hall MP, Zimmerman K, Otto P, Lubben TH, Butler BL, Binkowski BF, Machleidt T, Kirkland TA, Wood MG, Eggers CT, Encell LP, Wood KV (2016) NanoLuc complementation reporter optimized for accurate measurement of protein interactions in cells. ACS Chem Biol 11:400–408

12. Guan XM, Kobilka TS, Kobilka BK (1992) Enhancement of membrane insertion and function in a type IIIb membrane protein following introduction of a cleavable signal peptide. J Biol Chem 267:21995–21998

13. Hopp TP, Prickett KS, Price VL, Libby RT, March CJ, Pat Cerretti D, Urdal DL, Conlon PJ (1988) A short polypeptide marker sequence useful for recombinant protein identification and purification. Bio/Technology 6:1204–1210

14. Galés C, Van Durm JJJ, Schaak S, Pontier S, Percherancier Y, Audet M, Paris H, Bouvier M (2006) Probing the activation-promoted structural rearrangements in preassembled receptor–G protein complexes. Nat Struct Mol Biol 13:778–786

15. Wan Q, Okashah N, Inoue A, Nehmé R, Carpenter B, Tate CG, Lambert NA (2018) Mini G protein probes for active G protein–coupled receptors (GPCRs) in live cells. J Biol Chem 293:7466–7473

Chapter 11

Imaging of Genetically Encoded FRET-Based Biosensors to Detect GPCR Activity

Luca Bordes, Sergei Chavez-Abiega, and Joachim Goedhart

Abstract

A wealth of assays for screening GPCR activity have been developed. Biosensors that employ Förster Resonance Energy transfer (FRET) are specific and enable dynamic measurements. Moreover, FRET biosensors are ideally suited for the analysis of single living cells. The FRET biosensors described in this manuscript are entirely genetically encoded by plasmids. Here, protocols for employing FRET-based biosensors to detect G protein activity upon GPCR activation are reported. The protocols include details on the isolation of plasmids, transfection, generation of stable cell lines with the FRET biosensors, FRET ratio imaging, and data analysis.

Key words FRET, Biosensor, GPCR, Ratio-imaging, Image analysis

1 Introduction

1.1 GPCR Signaling

G protein-coupled receptors (GPCRs) are proteins that are located in membranes. These receptor proteins are capable of perceiving a wide range of molecules and also photons [1, 2]. The binding of the extracellular components triggers activation of the GPCR. Activation of a GPCR results in a conformational change, which activates so-called "downstream" components. Together, the downstream signaling events compose the information that the cells use to decode (changes in) the extracellular environment [3].

Heterotrimeric G proteins are membrane-associated protein complexes that are directly activated by GPCRs. Heterotrimeric G proteins are composed of α, β, and γ subunits. These protein complexes are often classified according to their α subunit as Gi, Gs, Gq, or G12/13 [4]. This classification reflects the downstream signaling networks that are activated. The Gi and Gs complex, respectively, inhibit and activate cAMP production. The Gq complex is classically related to calcium signaling and PKC activation, and

Sofia Aires M. Martins and Duarte Miguel F. Prazeres (eds.), *G Protein-Coupled Receptor Screening Assays: Methods and Protocols*, Methods in Molecular Biology, vol. 2268, https://doi.org/10.1007/978-1-0716-1221-7_11,
© Springer Science+Business Media, LLC, part of Springer Nature 2021

more recently connected with Rho GTPase activation. Finally, G12/13 is connected to Rho GTPase signaling.

In addition to the activation of heterotrimeric G proteins, the recruitment of G protein-coupled Receptor Kinases and beta-arrestins is a hallmark of GPCR activation [5]. Together, the afore-mentioned "early" signaling events eventually result in the translocation of transcription factors from the cytoplasm into the nucleus. The translocation causes a change in transcriptional activity. Any of the receptor activated molecular events can in principle be used to measure the activity of GPCRs. One way that will be detailed in this chapter is to measure these events with FRET-based biosensors [6]. The working principle of these type of sensors is outlined below.

1.2 Genetically Encoded FRET Biosensors

Fluorescent biosensors enable the real-time imaging of intracellular processes with spatial resolution [7]. Genetically encoded fluorescent biosensors use fluorescent proteins as the fluorophore. In case of a FRET-based biosensor, two spectrally different fluorescent proteins are employed [8]. To date, most FRET biosensors use a cyan fluorescent protein (CFP) and a yellow fluorescent protein (YFP). FRET, which stands for Förster Resonance Energy Transfer, is the radiation-less transfer of energy from the CFP (donor) to the YFP (acceptor). In case of FRET, the CFP emission is reduced and the YFP emission increased relative to the non-FRET situation. Therefore, FRET changes the YFP/CFP emission ratio, which can be detected by fluorescence microscopy [9].

The sensing unit of a FRET biosensor defines what is detected and with what affinity [10]. This part usually comprises one or more naturally occurring proteins or protein domains. A FRET-based biosensor is designed in such a way that a change in the sensing unit results in a change of the YFP/CFP ratio. This change in the optical signal can be detected with a microscope [11]. Below, several FRET-based biosensors that are relevant for measuring GPCR activity will be described. We only discuss the sensors that we have successfully used before [12], but many other biosensors that may be used for detecting GPCR activity are available [13].

1.3 FRET Sensors for GPCR Activity

In Fig. 1, several of the components that are activated downstream of GPCRs are depicted. Any of these events can be used to read out GPCR activity. The events for which we have successfully used FRET-based biosensors are shown in cyan and yellow.

1.3.1 Heterotrimeric G Proteins

FRET sensors for heterotrimeric G proteins are among the most direct read-outs of GPCR activity [6]. The GPCR is actually a guanine exchange factor that catalyzes the exchange of GDP for GTP that is bound by Gα subunit. As such, these biosensors are suitable to measure the kinetics of GPCR activation and their G protein selectivity [12].

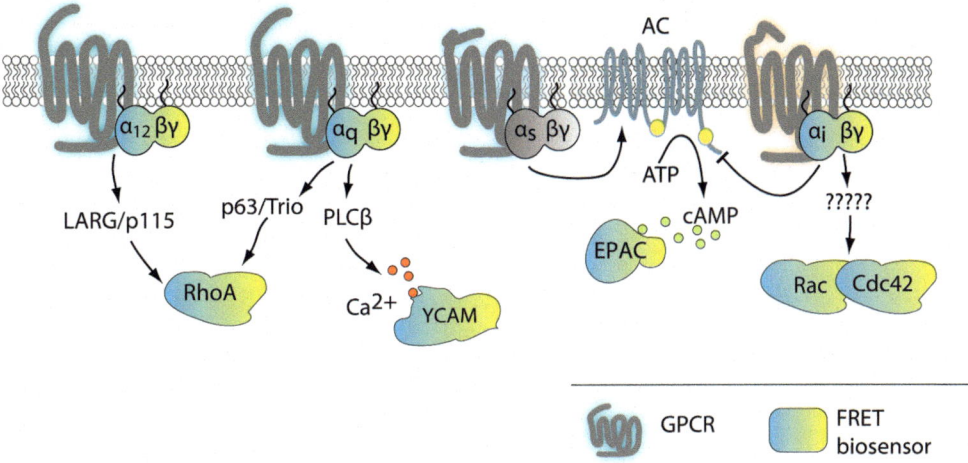

Fig. 1 Cartoon that shows an overview of FRET biosensors in cyan-yellow that can be used to detect the activity of G protein-coupled receptors (*AC* adenylyl cyclase, *ATP* adenosine-triphosphate, *Ca²⁺* free calcium ions, *cAMP* cyclic adenosine monophosphate, *EPAC* EPAC-based cAMP sensor, *GPCR* G protein-coupled receptor, *LARG/p115* guanine exchange factors activated by G12, *p63/Trio* guanine exchange factors activated by Gq, *PLCβ* phospholipase-C beta, *RhoA/Rac/Cdc42* Rho GTPases, *YCam* yellow cameleon calcium sensor)

In general, these sensors use a Gα subunit tagged with the cyan donor and a Gβ or Gγ subunit tagged with a yellow acceptor [14, 15]. The activation of a GPCR results in "activation" of the Gα subunit, which results in a loss of FRET due to a change in distance or orientation between the Gα and Gβγ subunits.

1.3.2 Second Messengers

The levels of two important and intensely studied second messengers, calcium and cAMP, are increased by GPCR activity. Due to signal amplification, relative high concentrations of calcium and cAMP are generated by GPCRs. The detection of these second messengers is a popular method for screening for GPCR activators and inhibitors. The detection of these second messengers can be sufficiently sensitive to measure endogenous GPCR activity.

The FRET biosensors that detect cAMP [16] and calcium [17] employ binding domains from naturally occurring proteins. These are sandwiched between a CFP and YFP. The conformation change induced by binding the second messenger results in a FRET change. This can be either an increase in FRET (in case of the calcium sensor yellow cameleon) or a loss of FRET (in case of EPAC cAMP sensor).

1.3.3 Rho GTPases

The G12/13 family is best known for its activation of RhoA. More recently the molecular details of RhoA activation by Gq have been unraveled [18]. Consequently, RhoA activation is an important outcome of GPCR activation. Next to this, activation of Rac1 and

Cdc42 downstream of GPCRs has been observed, although the molecular connections are less well established. Nevertheless, detection of Rho GTPase activity may provide useful information about GPCR activity.

FRET biosensors for three main classes of Rho GTPases are available [19–21] as well as for the RhoB and RhoC isoforms [22]. We have used a series of sensors that use a Rho GTPase, a CFP-YFP FRET pair and a domain that binds activated, i.e., GTP loaded, Rho GTPases. The exchange of GTP for GDP results in a conformation change that results in a FRET increase.

1.4 Limitations

FRET-based biosensors offer a unique window into GPCR activation in single living cells [23]. These tools are, however, not without limitations.

First, ectopic expression of the probe may change the natural signaling output. The signals may be buffered or amplified [24], which may change the signaling outcome. Ectopic expression also may change the selectivity of the signaling pathway. For instance, when a FRET biosensor for Gi is introduced, the level of this heterotrimeric G protein is increased. It may therefore compete for natural effectors. It is recommended to keep the levels of the biosensors low and close to endogenous levels whenever possible.

Second, several of the FRET sensors work well when the GPCR is overexpressed but show little or no response when endogenous receptors are activated. A good example are the heterotrimeric G protein biosensors, which usually require overexpression of the receptor to detect GPCR activity, although we have successfully detected endogenous signaling [20, 25].

Third, the single cell analysis with FRET-based biosensors is usually low throughput due to the number of cells that can be measured. The throughput can be increased when stable cell lines are used and when the microscopy and addition of compounds is automated.

Finally, the use and detection of FRET biosensors can be challenging and requires dedication and effort. The signals are often low, and the dynamic range of sensors can be low. Therefore, the imaging system should be optimally equipped for these experiments. Next to this, image analysis is an important step to get useful data. For a novice user it is recommended to start with an easy-to-use FRET sensor that can be expressed at moderate levels and shows a good dynamic range. In our hands, the biosensors yellow cameleon 3.60 (YCam) [26] and EPAC [27] are good candidates to start with.

The protocols below focus on imaging of biosensors for heterotrimeric G proteins, but these can be used for the other biosensors as well.

2   Materials

2.1   Plasmids and Storage

1. FRET biosensors. These can be obtained from Addgene or from the authors that reported the plasmid. For example:
   Gq biosensor, addgene plasmid #137810: https://www.addgene.org/137810/
   Gi1 biosensor, addgene plasmid #69623: https://www.addgene.org/69623/
   G13 biosensor, addgene plasmid #112930: https://www.addgene.org/112930/
   YCam biosensor, addgene plasmid #67899: https://www.addgene.org/67899/

2. EPAC: see ref. [27], or contact Kees Jalink (Dutch Cancer Institute, The Netherlands).

3. Rho GTPases: see ref. [19, 21], or contact Yi Wu (UConn Health, USA).

4. mTurquoise2 plasmid.

5. mVenus plasmid.

6. A plasmid with a GPCR: These can be purchased from www.cdna.org or www.addgene.org. Several suitable histamine receptors are available: https://www.addgene.org/browse/article/22426/.

7. DH5α electrocompetent bacteria.

8. Electroporation cuvettes.

9. Electroporation equipment Electroporator 2510.

10. Luria-Bertani medium (LB): 10 g Bacto Tryptone, 5 g Bacto Yeast extract, 10 g NaCl and ddH$_2$O, total volume of 1 L. Autoclaved.

11. Agar plates: 1 L of LB medium supplemented with 15 g agar and 100 µg/mL ampicillin or 50 µg/mL kanamycin. Prepare the plates by pouring 10 mL of LB/Agar per petri dish (diameter 90 mm).

12. Horizontal shaker for growing bacterial suspensions (37 °C and 200 RPM).

13. Glycerol.

14. GeneJET Plasmid Miniprep Kit.

15. NanoDrop™ 2000 Spectrophotometer.

2.2   Insertion of FRET Sensors in piggyBac Plasmids

1. PiggyBac with puromycin resistance (see ref. [28]) and transposase plasmid (see ref. [29]).

2. G12/13 FRET sensor.

3. PCR primers for cloning G12/13 FRET sensor in piggyBac vector. Forward primer with SpeI site (uppercase):

5′-tataACTAGTccccgtcagatccgctagc-3′. Reverse primer with EcoRI site (uppercase): 5′-tataAATTCatgtggtatggctgattatg atc-3′.

4. Pfu DNA polymerase (includes 10× Pfu buffer with $MgSO_4$).

5. dNTPs.

6. PCR apparatus.

7. SpeI enzyme.

8. EcoRI enzyme.

9. Tango buffer (10×).

10. GeneRuler™ DNA Ladder Mix.

11. Orange G DNA loading buffer:100 mg of Orange G, 30% (w/v) glycerol and ddH_2O, total volume 50 mL.

12. 1% (w/v) agarose gel.

13. TAE running buffer: 40 mM Tris, 1 mM ethylenediaminete-traacetic acid (pH 8.0), 4.4% (v/v) glacial acetic acid in ddH_2O mL and 300 nM ethidium bromide.

14. T4 DNA ligase.

15. T4 DNA ligase buffer (10×).

16. DNA electrophoresis equipment (cell, power source, UV visu-alization setup).

17. GeneJET Gel Extraction Kit.

2.3 Cells

1. HEK-293 (ATCC CRL-3216).

2. HeLa (ATCC CCL-2).

3. U2OS (ATCC HTB-96).

2.4 Generation of FRET Sensor Stable Cell Lines

1. Dulbecco's Modified Eagle's Medium-GlutaMAX™.

2. Fetal bovine serum (FBS).

3. Transposase plasmid DNA.

4. piggyBac plasmid DNA with insert, e.g., addgene plasmid #137809: https://www.addgene.org/137809/.

5. Opti-MEM™.

6. Polyethylenimine (PEI).

7. Puromycin.

8. Hank's buffered saline solution (HBSS).

9. Trypsin EDTA solution: 0.05% Trypsin.

2.5 FACS Sorting

1. HEK-293 G12/13 FRET sensor stable cell line.

2. Hank's buffered saline solution (HBSS).

3. Trypsin EDTA solution: 0.05% Trypsin.

4. Dulbecco's Modified Eagle's Medium-GlutaMAX™.

5. Fetal bovine serum (FBS).

6. HF buffer: 2% FBS in HBSS without Mg^{2+}, Ca^{2+} or phenolred, filter sterilized.

7. 5 mg/mL DAPI.

8. HEPES in ddH_2O, pH 7.4.

9. Penicillin.

10. Streptomycin.

2.6 Expression of G Protein-Coupled Receptors

1. 24 mm Glass coverslips.

2. 6-Well culturing plate.

3. Fibronectin solution: 0.014 mg/mL in PBS (*see* **Note 1**).

4. Hank's buffered saline solution (HBSS).

5. Dulbecco's Modified Eagle's Medium-GlutaMAX™.

6. Fetal bovine serum (FBS).

7. GPCR plasmid DNA.

8. mVenus plasmid DNA.

9. mTurquoise2 plasmid DNA.

10. Opti-MEM™.

11. Polyethylenimine (PEI).

12. Microscopy medium: 137 mM NaCl, 5.4 mM KCl, 0.8 mM $MgCl_2$, 1.8 mM $CaCl_2$, 20 mM glucose, 20 mM HEPES pH 7.4 (*see* **Note 2**).

13. Attofluor® cell chamber.

2.7 Transient Expression of FRET Biosensor and GPCR

1. 24 mm Glass coverslips.

2. 6-Well culturing plate.

3. Fibronectin solution: 0.014 mg/mL in PBS (*see* **Note 1**).

4. Hank's buffered saline solution (HBSS).

5. Dulbecco's Modified Eagle's Medium-GlutaMAX™.

6. Fetal bovine serum (FBS).

7. Neon® Transfection system 100 μL kit (includes buffer E2, buffer R and gold tips).

8. Neon pipette, system, and pipette station.

9. GPCR plasmid DNA.

10. G13 biosensor (Addgene Plasmid # 112930).

11. Microscopy medium:137 mM NaCl, 5.4 mM KCl, 0.8 mM $MgCl_2$, 1.8 mM $CaCl_2$, 20 mM glucose, 20 mM HEPES, pH 7.4 (*see* **Note 2**).

12. Attofluor™ cell chamber.

2.8 Imaging FRET Response to Stimuli	1. Widefield microscope with multidimensional acquisition function.
	2. 420/15-excitation filter.
	3. Long pass 455 nm dichroic filter.
	4. 470/30-emission filter.
	5. 535/30-emission filter.
	6. 490/10-excitation filter.
	7. Long pass 515 nm dichroic filter.
	8. Arc lamp.
	9. Solution for eliciting a cellular response, e.g., Calcium chloride: 1 M in ddH$_2$O; Histamine: 100 mM in ddH$_2$O; carbachol: 100 mM in ddH$_2$O; forskolin: 25 mM in DMSO or FBS.
2.9 Ratio-FRET Data Analysis	1. FIJI (https://imagej.net/Fiji) [30].
	2. Microsoft Excel.
	3. PlotTwist [31].

3 Methods

3.1 Plasmids and Storage

To simplify transfections and optimize FRET responses the α, β, and γ subunits are expressed by the same plasmid with an IRES sequence separating the β and γ subunits from the α subunit [32]. The FRET acceptor, cpVenus, is linked to the Gγ or Gβ subunit, and the FRET donor, mTurquoise(2), is inserted in the Gα subunit (*see* **Note 3**). Here we will describe how to handle the plasmids and purify DNA for transfections

1. Obtain the plasmids with FRET-based biosensors and GPCRs from Addgene or another source (*see* **Note 4**).

2. Defrost electro competent DH5α bacteria on ice and add 150 ng of plasmid DNA to 50 μL of bacteria.

3. Transfer the mixture to an electroporation cuvette and set the voltage to 2500 V (pulse length should be 4–6 ms).

4. Resuspend the mixture in 1 mL of LB-medium and incubate in a shaking incubator at 37 °C for 1 h.

5. Discard 500 μL and spin down the bacteria at 9000 × *g*. Discard 400 μL of the supernatant, resuspend the pellet, and plate on a LB-agar plate containing the antibiotic matching the resistance marker of the DNA plasmid.

6. Incubate overnight at 37 °C.

7. Inoculate a bacterial colony from the plate in 5 mL LB-medium containing the antibiotic matching the resistance marker of the DNA plasmid (50 μg/mL kanamycin or 100 μg/mL ampicillin).

8. Incubate the bacterial culture overnight in a shaking incubator at 37 °C.

9. For long-term storage, take 1 mL of bacterial culture and mix with 400 µL of 80% glycerol in a cryovial.

10. Snap freeze the cryovial in liquid nitrogen and store at −80 °C. This frozen suspension can be used to inoculate LB-medium for DNA isolation at a later time.

11. The remaining 4 mL of bacterial culture is used to prepare DNA for transfection. Transfer 2 mL of the bacterial culture to an Eppendorf tube and spin down for 1 min at 16,000 × g.

12. Remove the supernatant and repeat with the residual 2 mL of bacterial culture.

13. Suspend the pellet in 250 µL resuspension buffer from the GeneJET Plasmid Miniprep Kit.

14. Add 250 µL of lysis buffer and mix by inverting.

15. Add 350 µL of neutralization buffer from the GeneJET Plasmid Miniprep Kit and mix by inverting. The suspension should have white precipitation now.

16. Spin the solution down for 5 min at 16,000 × g.

17. Transfer the supernatant to the GeneJET spin column and spin down 1 min at 16,000 × g Discard the flow through and repeat.

18. Wash twice by adding 500 µL of washing buffer from the GeneJET Plasmid Miniprep Kit and centrifuging 1 min at 16,000 × g.

19. Dry the column by spinning down a third time for 1 min at 16,000 × g.

20. Transfer the column to a sterile 1.5 mL Eppendorf tube and add 30 µL of elution buffer from the GeneJET Plasmid Miniprep Kit. Incubate for 3 min at room temperature and centrifuge for 1 min at 16,000 × g.

21. Measure the concentration of your DNA sample with the Thermo Scientific NanoDrop™ and store at −20 °C.

3.2 Inserting FRET Sensors in piggyBac Plasmids

Expression of the FRET biosensor should be optimized to get robust and high-quality data. If the expression is very low, the signal intensities can be too low for detection and analysis. When the expression level is too high, the dynamic range of the FRET change is decreased due to buffering (in case of Rho GTPases and EPAC) or bystander FRET (in case of heterotrimeric G proteins). To control the expression of the FRET sensor, we can generate stable cell lines and use FACS to sort the cells based on fluorescence intensity. Alternatively, FRET sensors can be transiently transfected

(Subheading 3.5). Here we will use the G12/13 FRET sensor and HEK-293 cells to demonstrate how such a stable cell line can be made.

1. Obtain plasmids for the G12/13 FRET sensor, piggyBac vector, and matching transposase (*see* **Note 5**).

2. Prepare constructs for long-term storage according to protocol 3.1.

3. Mix the following reagents (Table 1) carefully by pipetting and centrifuge briefly.

4. Run the PCR with the following program (Table 2):

5. Digest the piggyBac and the PCR product with the following digestion mixture (Table 3).

6. Incubate the digestion mixture for 2 h at 37 °C.

7. Add 5.5 μL of Tango-buffer to the digestion mixture and incubate for another 2 h at 37 °C.

8. Mix with Orange G DNA loading buffer and load the sample in an agarose gel next to a DNA ladder. Run the gel until the insert (PCR amplicon) and the digested piggyBac vector are well separated from the rest of the bands and can be cut out of the gel.

Table 1
Composition of PCR reaction mixture

Reagent	Volume per 50 μL	Final concentration
Pfu polymerase (2.5 U/μL)[a]	1 μL	2.5 U
10× Pfu buffer (w MgSO$_4$)	5 μL	1×
2.5 mM dNTPs	4 μL	0.2 mM
100 ng/μL forward primer[a]	2 μL	4 ng/μL
100 ng/μL reverse primer[a]	2 μL	4 ng/μL
G12/13 FRET sensor plasmid	1 μL	0.04 pg/μL
ddH$_2$O	35 μL	–

[a]For the control experiment replace the Pfu polymerase or primers with ddH$_2$O

Table 2
PCR amplification protocol

Cycle number	Denature	Anneal	Extend
1	5 min at 95 °C		
2–35	1 min at 95 °C	1 min at 49 °C	10 min at 72 °C
36			10 min at 72 °C

Table 3
Composition of the enzyme restriction reaction mixture

Reagent	Volume per 50 µL	Final concentration
Plasmid DNA[a]	25 µL[a]	500 ng[a]
SpeI 10 U/µL	1 µL	0.2 U/µL
EcoRI 10 U/µL	1 µL	0.2 U/µL
Tango-buffer 10×[b]	5 µL	1×
ddH$_2$O	18 µL	–

[a]Regardless of the concentration add 25 µL of the PCR product and add 500 ng of the empty piggyBac plasmid DNA
[b]For optimal digestion SpeI needs to be in 1× Tango and EcoRI in 2× Tango

9. Use the GeneJET Gel Extraction Kit to purify the DNA from the gel.

10. Measure the concentrations of the vector and insert DNAs, and mix in a 1:3 molar ratio to reach a total of 150 ng. To start the ligation process, add 1 µL T4 ligase and 1.5 µL of 10× T4 ligase buffer. Adjust the volume to 15 µL with ddH$_2$O (*see* **Note 6**).

11. Prepare a second ligation mixture without the insert DNA as a negative control.

12. Incubate the ligation mix for 2 h at room temperature.

13. Transform bacteria according to the instructions in Subheading 3.1 starting from **step 2** (*see* **Note 7**).

14. Sequence the isolated plasmids to identify the correct construct.

3.3 Making FRET Sensor Stable Cell Lines

The piggyBac transposase system is a virus-free method to generate stable cell lines by introducing a gene of interest at random TTAA sites in the host genome. A transposase plasmid must also be transfected with your gene of interest (transposon) to guide the piggyBac vector to the TTAA sites and induce integration. The ratio of transposase to transposon can be adjusted to control the number of integrations (*see* **Note 8**).

1. Seed HEK-293 cells one day prior to transfection at 15% confluency in a 10 cm petri dish with DMEM-GlutaMAX™ supplemented with 10% FBS.

2. Mix 500 ng of transposase with 850 ng of piggyBac G12/13 FRET sensor DNA.

3. Dilute the DNA in 100 µL of Opti-MEM®.

4. Add 6 µL of PEI and mix carefully by tapping the tube several times. Incubate for 15–20 min at room temperature.

5. Add the transfection mixture to 30% confluent HEK-293 cells and expand until 70–80% confluent (*see* **Note 9**).

6. Add the antibiotic at the concentration stated by the company. For puromycin, a concentration of 1 μg/mL is advisable when using the HEK-293 cell line.

7. Refresh antibiotics every day until only the puromycin-resistant cells remain alive, but no longer than 5 days.

8. Expand your cells and, if necessary, do another round of selection to get rid of untransfected cells.

9. Keep the HEK-293 G12/13 FRET sensor stable cell line in culture at 37 °C, 5% CO_2 suspended in DMEM-GlutaMAX™ with 10% FBS. Cells should be split before they reach confluency by washing with Mg^{2+}- and Ca^{2+}-free HBSS and incubating for a few minutes with 0.05% trypsin EDTA solution.

10. Resuspend cells in DMEM-Glutamax with 10% FBS and expand until almost confluent before passaging them again.

3.4 FACS Sorting

Pooling cells by Fluorescence Activated Cell Sorting (FACS) is employed to get rid of untransfected cells and to obtain homogeneous levels of sensor expression.

1. Harvest five million HEK-293 G12/13 FRET sensor cells by removing the medium, washing with HBSS without magnesium or calcium and incubating for a few minutes with 0.05% trypsin EDTA solution. Also take wild-type cells along to establish the level of autofluorescence.

2. Stop the trypsinization with a few mL of DMEM-GlutaMAX™ with 10% FBS and spin down at $200 \times g$ for 3 min.

3. Remove supernatant and flick the pellet.

4. Resuspend in 2 mL HF buffer with 1 μg/mL DAPI and spin down at $200 \times g$ for 3 min.

5. Remove supernatant and resuspend in 2 mL of HF buffer.

6. Store on ice and keep in the dark until use.

7. Prepare 15 mL collection tubes with 10 mL DMEM-GlutaMAX™ with 10% FBS and 25 mM HEPES.

8. Use DAPI-stained HEK293 cells to set the gates and identify single living cells, as well as to account for autofluorescence. Load your sample, set the gates for different intensity levels of mTurquoise2, and sort the cells into the tubes.

9. After sorting, spin the collection tubes down at $200 \times g$ for 3 min, remove supernatant, and resuspend in fresh DMEM-GlutaMAX™ with 10% FBS.

10. Add the cell suspension to an appropriate cell culture flask and add 1% penicillin and streptomycin for 1 week.

11. Expand cells and check FRET response (please see Subheading 3.7).

3.5 Transient Expression of G Protein-Coupled Receptors

There are many GPCR transfected stable cell lines commercially available and ready to use. If this is not an option, another possibility is to generate your own stable cell line by transfecting cells with your GPCR of interest. Alternatively, GPCRs can be expressed by transient transfection.

1. Transfer 24 mm glass coverslips into a 6-well plate and coat with 1 mL of fibronectin solution by incubating at 37 °C for 1 h.

2. Wash the coverslips with 1 mL of HBSS.

3. Transfer 2 mL of recently harvested HEK-293 G12/13 FRET sensor stable cell line to the 6-well plate at a density of 75,000 cells/mL. Let the cells attach and grow overnight with DMEM-GlutaMAX™ with 10% FBS at 37 °C.

4. Add 500 ng of GPCR plasmid DNA to 100 μL Opti-MEM® and mix carefully by tapping the tube.

5. Do the same for the control samples with mTurquoise or CFP and cpVenus or YFP (*see* **Note 10**).

6. Add 2.3 μL PEI, mix again, and incubate for 20 min at room temperature.

7. Add the transfection mixtures to the seeded cells.

8. Incubate cells for 24–48 h with transfection mixture at 37 °C.

9. When ready to image transfer the coverslip with transfected cells to a 35 mm Attofluor™ cell chamber and incubate for 20 min with microscopy medium (*see* **Note 11**).

3.6 Transient Expression of FRET Biosensor and GPCR

For pilot experiments it can be preferable to transiently transfect a FRET biosensor and GPCR instead of generating stable cell lines. When transfecting multiple plasmids electroporation often achieves higher transfection efficiency than PEI-based transfections. Control experiments can be followed according to protocol 3.5 for the mTurquoise and mVenus plasmid DNA.

1. Transfer 24 mm glass coverslips into a 6-well plate and coat with 1 mL of fibronectin solution by incubating at 37 °C for 1 h.

2. Wash the coverslips with 1 mL of HBSS, and add 1.9 mL of DMEM-Glutamax with 10% FBS to each well.

3. Add 3 mL of buffer E2 from Neon® Transfection system to the microporation tube and push it in place until you hear a "click."

4. For HEK-293 cells set voltage at 1100 V, pulse width at 20 ms and pulse amount at 2.

5. Harvest five million HEK-293 G12/13 FRET sensor cells by removing the medium, washing with HBSS without magnesium or calcium and incubating for a few minutes with 0.05% trypsin EDTA solution.

6. Add a 5 mL of DMEM-GlutaMAX™ with 10% FBS and spin down at $200 \times g$ for 3 min.

7. Remove the supernatant and suspend in 3 mL of buffer R.

8. Dilute 200,000 cells in 120 μL of buffer R.

9. Mix ~1 μg G protein FRET sensor plasmid with ~1 μg GPCR plasmid DNA to a maximum volume of 12 μL of DNA solution and add to the harvested HEK-293.

10. Place a 100 μL microporation tip on the microporation pipette and take up 100 μL of cell suspension. Aspirate the sample using the microporation pipette and be sure to remove any air bubbles from within the tip. Air bubbles cause an electric disconnection and may generate an arc during the microporation leading to lower transfection efficiency and cell viability.

11. Insert the pipette vertically until you hear a "click."

12. Deliver the electric pulse by pressing the start button.

13. Add the cell suspension to the fibronectin-coated wells.

14. Incubate cells for 24–48 h with transfection mixture at 37 °C.

15. When ready to image transfer the coverslip with transfected cells to a 35 mm Attofluor™ cell chamber and incubate for 20 min with microscopy medium (*see* **Note 11**).

3.7 Imaging FRET Response to Stimuli

The dynamic range of the FRET sensor with your GPCR of choice has to be established by adding a known activating compound. Here we use the histamine 1 receptor and calcium sensing receptor as examples.

1. Use a wide field microscope with a sequential imaging option and a mercury lamp for excitation light. Insert a 420/15-excitation filter cube and a long pass 455 nm dichroic filter. Set up a FRET CFP channel with a band pass 470/30-emission filter and a FRET YFP channel with a band pass 535/30-emission filter. Also, set up a YFP-only channel by inserting a 490/10-excitation filter cube and a long pass 515 dichroic filter in combination with the same 535/30-emission filter for direct YFP imaging. Thus, for every time point, three images will be acquired (*see* **Note 12**).

2. Find cells with the transmission light and adjust your focal plane.

3. Switch off the transmission light and turn on the mercury lamp to find fluorescent cells.

4. Start live mode with the FRET CFP or FRET YFP channel and refocus. Get as many healthy-looking fluorescent cells in a field of view as possible. Check the focus again and stop live mode.

5. Set up a time-lapse to acquire the FRET CFP, FRET YFP, and YFP-only channel sequentially every 2 s for approximately 8 min.

6. Start the time-lapse and obtain a baseline of at least 10 images before adding 100 μM histamine. For the calcium sensing receptor incubate the cells with low calcium microscopy medium (0.5 mM $CaCl_2$) for 20 min before adding 10 μM $CaCl_2$ (*see* **Note 13**).

7. Save the images and discard the coverslip. Rinse the cell chamber with demi-water and dry.

3.8 Ratio-FRET Data Analysis

To calculate true sensitized emission, we need to correct for several factors.

The largest factors are the bleed-through from the donor fluorescent protein (CFP/mTurquoise2) into the YFP channel and the direct excitation of the acceptor. These values are dependent on the excitation source, filters, and exposure time. For pilot experiments these bleed-through values can be ignored and the YFP/CFP ratio can be calculated directly for a rough FRET estimate.

1. Open the mTurquoise/CFP only images in ImageJ and draw regions of interest (ROI) to select cells and background (i.e., area without cells). Measure the counts in the FRET CFP and FRET YFP channels and subtract the background to each cell. Then divide the FRET YFP channel through the FRET CFP channel. This will be the correction value gamma. In our system this is approximately 0.55.

2. Open the YFP only images in ImageJ, and draw ROIs to select cells and background. Measure the counts in the FRET YFP and the YFP-only channels and subtract the background to each cell. Then divide the FRET YFP channel through the YFP-only channel. This will be the correction value beta. In our system this is around 0.075.

3. Open the data from the FRET biosensor experiment in ImageJ and draw ROIs to select cells and background. Measure the intensity of the ROIs in the three channels and export the data to Microsoft Excel in separate tabs (*see* **Note 14**).

4. Subtract the background from the data and place the result in a new tab.

5. Correct the FRET YFP channel for mTurquoise/CFP bleed-through and direct excitation of YFP with this formula:

Corrected FRET YFP = (FRET YFP − background)
                  − (FRET CFP − background) ∗ gamma
                  − (YFP − only − background) ∗ beta

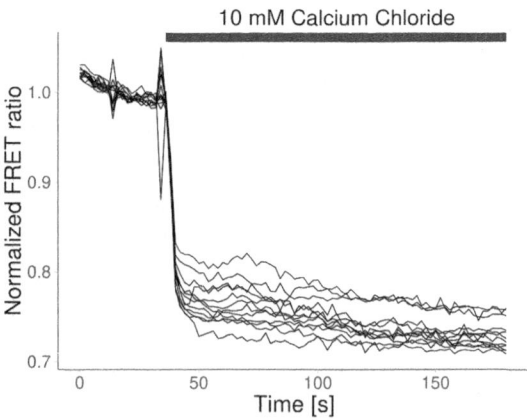

Fig. 2 Example outcome of a FRET ratio time-lapse imaging experiment. The lines depict the responses detected in individual cells with a Gi FRET sensor. The cells also overexpressed a Calcium Sensing Receptor which was stimulated with calcium chloride

6. To calculate the sensitized emission, divide the corrected FRET YFP channel by the FRET CFP channel (*see* **Note 15**).

7. A template to facilitate **steps 3–6** is available here, together with instructions and example data: https://github.com/JoachimGoedhart/Quantify-FRET-ratio.

8. Optionally, the resulting data can be visualized with PlotTwist. To this end copy-paste the data or save a CSV file and import in PlotTwist: https://huygens.science.uva.nl/PlotTwist/.

9. The expected outcome of the analysis (Fig. 2) is a 0.1–0.2 average drop in normalized ratio FRET upon stimulation with histamine or calcium (*see* **Notes 16** and **17**).

4   Notes

1. HEK-293 cells detach easily from the substrate during washing steps and display morphological changes upon stimulation of a RhoA activating receptor. The glass coverslips, although great for high-resolution imaging, are untreated and will be more slippery than regular culturing disposables. To image the activation of the calcium sensing receptor it is necessary to deplete the medium of calcium, to maximize the observed response. However, this also increases detachment and cell movement. To minimize detachment of HEK293 cells, coating the glass surface is essential for FRET imaging. We found fibronectin does not affect the FRET response or normal cell signaling, while decreasing the detaching and crawling of cells. Note that coating is not always necessary and, in some cases, can affect GPCR signaling.

2. To reduce the calcium in the microscopy medium, simply reduce the amount of CaCl$_2$ to 0.5 mM.

3. The insertion position and orientation of the fluorescent proteins are crucial in the design of a FRET sensor. Theoretically, the highest FRET response can be achieved if the dipole moments of the donor and acceptor are parallel to each other and the distance in the activated state is smaller than 10 nm.

4. Plasmids from Addgene are supplied as agar stabs and can be stored for up to 2 weeks at 4 °C. Agar stabs contain transformed bacteria; therefore **steps 2–6** can be skipped when these are used. Plasmids from alternative source are often supplied as plasmid DNA, which needs to be transformed into *E. coli* bacteria.

5. A piggyBac transfection kit is commercially available but can also be obtained from the Sanger institute: https://www. sanger.ac.uk/science/tools/piggybac-transposase-resources. We found the puromycin resistant piggyBac plasmids to be the most efficient in selection, although the plasmids with blasticidin and hygromycin are also effective. The concentration and selection time differ between antibiotics, but for puromycin 2–3 days of 1 µg/mL is sufficient for selection of HEK-293 cells. It is advisable to check first the recommendations of the fabricant for your cell line of interest, and then perform a kill curve to adjust the conditions. Also, check if the piggyBac transfection kit is suitable for use with DH5α cells, as there are some that cannot be cloned into DH5α cells.

6. This ratio is highly dependent on the size of the receptor and piggyBac vector. Most piggyBac plasmids are 4000–5000 base pairs long and most FRET sensor plasmids are 2500–4000 base pairs long. Adjust the concentration of either the insert or vector when your FRET sensor or piggyBac is a different size.

7. There should be fewer colonies on the control plate than on the ligated plate. Inoculate at least three colonies if the control plate is almost empty, and inoculate more according to the relative number of colonies between the plates.

8. Although transfection with PEI is very cheap, there are other transfection methods available that can be more efficient. We found that the Neon transfection kit from Invitrogen is able to reach high transfection efficiency even with multiple piggyBac plasmids simultaneously [33]. In addition, the tips that are used for electroporation can be reused to reduce the costs [34].

9. To make sure sufficient cells survive the selection it is advisable to start from a 30% confluent 100 mm cell culture dish. The cells will have a day to express the genes in the piggyBac plasmids without getting overcrowded. The transfected cells

need to be in a proliferative state for the antibiotic selection to work. This approach should yield enough transfected and selected cells in 1 week.

10. These FRET controls have to be measured the first time to establish the system-specific bleed-through from CFP/mTurquoise2 in the YFP-channel and the direct activation of mVenus/YFP by the FRET excitation.

11. Because most of the culturing media contain growth factors, hormones, and other GPCR activating compounds, we recommend to incubate the cells in minimal medium to ensure the receptor activity is as low as possible prior to stimulation. In the case of the calcium sensing receptor the cells require an additional starvation step of 20 min with microscopy medium that contains 0.5 mM Ca^{2+}.

12. These filter settings are used as a guideline. www.fpbase.org contains detailed information about the excitation and emission spectra of CFP/mTurquoise2 and YFP/cpVenus and should be consulted to adjust the settings of your microscope. Take notice to keep at least 10 nm between the excitation light and the detected emission wavelengths.

13. Because the stimuli need to be added quickly and accurately start by pipetting 10 μL with a P-20, take the tip off and transfer it to a P-200 while keeping the plunger pressed down. Add the stimulus by releasing the plunger with the tip in the microscopy medium in the ring to dilute the stimulus and pipette up and down to homogeneously distribute the stimulus in the medium. Do this while imaging, preferably when the shutter is closed.

14. Make sure that the ROIs encompass the selected cell during entire stack and that there is no floating debris, as this could dramatically affect the outcome of your traces.

15. To compare the data from different cells, it is recommended to normalize the FRET ratio to the baseline values and display the normalized YFP/CFP ratio. Normalization can be achieved within the Excel sheet or using PlotTwist.

16. It could be useful to create a FRET ratio image using ImageJ. This can be done by following the same formula for calculating sensitized emission and applying this to the image calculator function. However, subpixel alignment of the channels is crucial.

17. Figure 2 can be recreated by using this link or this URL:
    https://huygens.science.uva.nl:/PlotTwist/?data=5;;;
    fold;1,5;&vis = dataasline;1;;;1;;&layout=;TRUE;TRUE;
    0,180;;;;6;X;480;600&color = none&label=;;TRUE;Time [s];
    Normalized FRET ratio;TRUE;24;24;18;8;;;&stim = TRUE;

bar;36,180;10 mM Calcium Chloride;&url=https://raw. githubusercontent.com/JoachimGoedhart/Quantify-FRET-ratio/master/Example-data_processed/FRET-ratio_Galphai. csv.

Acknowledgments

S.C.-A. was funded by Marie Sklodowska-Curie Actions of the European Union's Horizon 2020 program as part of the CaSR Biomedicine Network.

References

1. Dorsam RT, Gutkind JS (2007) G-protein-coupled receptors and cancer. Nat Rev Cancer 7:79–94. https://doi.org/10.1038/nrc2069

2. Lohse MJ, Maiellaro I, Calebiro D (2014) Kinetics and mechanism of G protein-coupled receptor activation. Curr Opin Cell Biol 27:87–93

3. Hilger D, Masureel M, Kobilka BK (2018) Structure and dynamics of GPCR signaling complexes. Nat Struct Mol Biol 25:4–12. https://doi.org/10.1038/s41594-017-0011-7

4. Wettschureck N, Offermanns S (2005) Mammalian G proteins and their cell type specific functions. Physiol Rev 85:1159–1204. https://doi.org/10.1152/physrev.00003.2005

5. Peterson YK, Luttrell LM (2017) The diverse roles of arrestin scaffolds in G protein-coupled receptor signaling. Pharmacol Rev 69:256–297. https://doi.org/10.1124/pr.116.013367

6. Lohse MJ, Nuber S, Hoffmann C (2012) Fluorescence/bioluminescence resonance energy transfer techniques to study G-protein-coupled receptor activation and signaling. Pharmacol Rev 64:299–336. https://doi.org/10.1124/pr.110.004309

7. Chavez-Abiega S, Goedhart J, Bruggeman FJ (2019) Physical biology of GPCR signalling dynamics inferred from fluorescence spectroscopy and imaging. Curr Opin Struct Biol 55:204–211. https://doi.org/10.1016/j.sbi.2019.05.007

8. Miyawaki A, Niino Y (2015) Molecular spies for bioimaging — fluorescent protein-based probes. Mol Cell 58:632–643. https://doi.org/10.1016/j.molcel.2015.03.002

9. Goedhart J, Hink MA, Jalink K (2014) An introduction to fluorescence imaging techniques geared towards biosensor applications. Methods Mol Biol 1071:17–28. https://doi.org/10.1007/978-1-62703-622-1_2

10. Okumoto S, Jones A, Frommer WB (2012) Quantitative imaging with fluorescent biosensors. Annu Rev Plant Biol 63:663–706

11. Pietraszewska-Bogiel A, Gadella TWJ (2011) FRET microscopy: from principle to routine technology in cell biology. J Microsc 241:111–118. https://doi.org/10.1111/j.1365-2818.2010.03437.x

12. Van Unen J, Rashidfarrokhi A, Hoogendoorn E, Postma M (2016) Quantitative single cell analysis of signaling pathways activated immediately downstream of histamine receptor isoforms. Mol Pharmacol 90(3):162–176. https://doi.org/10.1124/mol.116.104505

13. Greenwald EC, Mehta S, Zhang J (2018) Genetically encoded fluorescent biosensors illuminate the spatiotemporal regulation of signaling networks. Chem Rev 118:11707–11794. https://doi.org/10.1021/acs.chemrev.8b00333

14. Janetopoulos C, Jin T, Devreotes P (2001) Receptor-mediated activation of heterotrimeric G-proteins in living cells. Science 291:2408–2411. https://doi.org/10.1126/science.1055835291/5512/2408

15. Adjobo-Hermans MJW, Goedhart J, van Weeren L et al (2011) Real-time visualization of heterotrimeric G protein Gq activation in living cells. BMC Biol 9:32. https://doi.org/10.1186/1741-7007-9-32

16. Ponsioen B, Zhao J, Riedl J et al (2004) Detecting cAMP-induced Epac activation by fluorescence resonance energy transfer: Epac as a novel cAMP indicator. EMBO Rep 5:1176–1180. https://doi.org/10.1038/sj.embor.7400290

17. Miyawaki A, Griesbeck O, Heim R, Tsien RY (1999) Dynamic and quantitative Ca^{2+} measurements using improved cameleons. Proc Natl Acad Sci U S A 96:2135–2140. https://doi.org/10.1073/pnas.96.5.2135

18. Rojas RJ, Yohe ME, Gershburg S et al (2007) Galphaq directly activates p63RhoGEF and Trio via a conserved extension of the Dbl homology-associated pleckstrin homology domain. J Biol Chem 282:29201–29210. https://doi.org/10.1074/jbc.M703458200

19. van Unen J, Reinhard NR, Yin T et al (2015) Plasma membrane restricted RhoGEF activity is sufficient for RhoA-mediated actin polymerization. Sci Rep 5:14693. https://doi.org/10.1038/srep14693

20. Reinhard NR, Mastop M, Yin T et al (2017) The balance between Gαi-Cdc42/Rac and Gα12/13-RhoA pathways determines endothelial barrier regulation by sphingosine-1-phosphate. Mol Biol Cell 28(23):3371–3382. https://doi.org/10.1091/mbc.E17-03-0136

21. Kedziora KM, Leyton-Puig D, Argenzio E et al (2016) Rapid remodeling of invadosomes by Gi-coupled receptors. J Biol Chem 291:4323–4333. https://doi.org/10.1074/jbc.M115.695940

22. Reinhard NR, van Helden SF, Anthony EC et al (2016) Spatiotemporal analysis of RhoA/B/C activation in primary human endothelial cells. Sci Rep 6:25502

23. van Unen J, Woolard J, Rinken A et al (2015) A perspective on studying G-protein-coupled receptor signaling with resonance energy transfer biosensors in living organisms. Mol Pharmacol 88:589–595. https://doi.org/10.1124/mol.115.098897

24. Haugh JM (2012) Live-cell fluorescence microscopy with molecular biosensors: what are we really measuring? Biophys J 102:2003–2011. https://doi.org/10.1016/j.bpj.2012.03.055

25. van Unen J, Stumpf AD, Schmid B et al (2016) A new generation of FRET sensors for robust measurement of Gαi1, Gαi2 and Gαi3 activation kinetics in single cells. PLoS One 11:e0146789. https://doi.org/10.1371/journal.pone.0146789

26. Nagai T, Yamada S, Tominaga T et al (2004) Expanded dynamic range of fluorescent indicators for Ca^{2+} by circularly permuted yellow fluorescent proteins. Proc Natl Acad Sci 101 (29):10554–10559

27. Klarenbeek J, Goedhart J, van Batenburg A et al (2015) Fourth-generation epac-based FRET sensors for cAMP feature exceptional brightness, photostability and dynamic range: characterization of dedicated sensors for FLIM, for ratiometry and with high affinity. PLoS One 10:e0122513. https://doi.org/10.1371/journal.pone.0122513

28. Balasubramanian S, Matasci M, Kadlecova Z et al (2015) Rapid recombinant protein production from piggyBac transposon-mediated stable CHO cell pools. J Biotechnol 200:61–69. https://doi.org/10.1016/J.JBIOTEC.2015.03.001

29. Yusa K, Zhou L, Li MA et al (2011) A hyperactive piggyBac transposase for mammalian applications. Proc Natl Acad Sci U S A 108:1531–1536. https://doi.org/10.1073/pnas.1008322108

30. Schindelin J, Arganda-Carreras I, Frise E et al (2012) Fiji: an open-source platform for biological-image analysis. Nat Methods 9:676–682. https://doi.org/10.1038/nmeth.2019

31. Goedhart J (2020) PlotTwist: a web app for plotting and annotating continuous data. PLoS Biol 18:e3000581

32. Goedhart J, van Weeren L, Adjobo-Hermans MJW et al (2011) Quantitative co-expression of proteins at the single cell level—application to a multimeric FRET sensor. PLoS One 6:1–8. https://doi.org/10.1371/journal.pone.0027321

33. Behringer R, Gertsenstein M, Nagy KV, Nagy A (2017) Integrating piggyBac transposon transgenes into mouse fibroblasts by electroporation. Cold Spring Harb Protoc. 2017:pdb.prot092601. https://doi.org/10.1101/pdb.prot092601

34. Brees C, Fransen M (2014) A cost-effective approach to microporate mammalian cells with the Neon transfection system. Anal Biochem 466:49–50. https://doi.org/10.1016/j.ab.2014.08.017

Chapter 12

cAMP Biosensor Assay Using BacMam Expression System: Studying the Downstream Signaling of LH/hCG Receptor Activation

Darja Lavogina, Tõnis Laasfeld, Maris-Johanna Tahk, Olga Kukk, Anni Allikalt, Sergei Kopanchuk, and Ago Rinken

Abstract

Cyclic adenosine monophosphate (cAMP) serves as a second messenger for numerous G-protein-coupled receptors. Changes in cellular cAMP levels reflect the biological activity of various GPCR-specific agents, including protein hormones. cAMP biosensors based on detection of Förster-type resonance energy transfer (FRET) offer unique advantages including the ratiometric nature of measurement, adjustable affinity toward detected molecule, capability of monitoring kinetics of cAMP release, and compatibility with the multi-well format and fluorescence plate reader platforms. In this chapter, we introduce the optimized version of the previously reported method to achieve sufficient and reproducible level of cAMP biosensor protein expression with the means of BacMam transduction system. As a practical challenge, we address the applicability of the designed assay for screening of biological activity of human hormones, including human chorionic gonadotropin (hCG) bearing different posttranslational modifications.

Key words cAMP assay, BacMam, Baculovirus, Epac-S^{H188}, FRET, Protein hormone activity

1 Introduction

From a variety of cellular cAMP assays [1], Förster resonance energy transfer (FRET)-based biosensors allow real-time detection of changes in cellular cAMP levels. The assay can be performed by either Fluorescent Lifetime Imaging (FLIM), or detection of Sensitized Emission (SE). The SE measurement relies on the cAMP-responsive change in relative fluorescence of donor and acceptor fluorophores genetically fused to a cAMP-binding protein moiety. The sensor proteins are freely distributed in the cell cytosol; binding of cAMP leads to a conformational change associated with an increase in the distance between the fluorophores, which in turn results in a decrease in acceptor fluorescence and an increase in

Sofia Aires M. Martins and Duarte Miguel F. Prazeres (eds.), *G Protein-Coupled Receptor Screening Assays: Methods and Protocols*, Methods in Molecular Biology, vol. 2268, https://doi.org/10.1007/978-1-0716-1221-7_12,

donor fluorescence. The calculated acceptor/donor fluorescence emission intensity ratio depends on the intracellular cAMP concentration. In our laboratory, we have successfully used Epac2-camps biosensor [2] and Epac-S^{H74} biosensor [3], and have now moved on to a next generation Epac-S^{H188} biosensor [4] . The new generation sensor has somewhat higher affinity [4] but still reflects physiological changes in intracellular cAMP from sub- to high micromolar concentrations, and so is well suited for GPCR ligand screening, enabling the study of activation kinetics as well as drug potencies. In our laboratory, the pharmacological validity of the cAMP biosensor expressed using the BacMam system (baculovirus for gene delivery to mammalian cells) was successfully confirmed by studies of various biological targets in different cell lines: melanocortin MC1 receptor in B16F10 murine melanoma cell line [5], recombinant melanocortin MC4 receptor in CHO-K1 cells [6], recombinant dopamine receptor subtypes in HEK293 cells [7], recombinant LH/hCG receptor in COS-7 or MDCK cells [8–10], endogenous FSH receptor in KGN cells [8], etc. While the assay conditions for the particular receptor of interest may require additional optimization, the BacMam technology is compatible with a broad range of cells, including primary and stem cells [11]. Using BacMam provides advantages such as adjustable protein expression levels, low cytotoxicity to host cells, ease of use, and safety in production and handling (Biosafety Level 1 or 2, dependent on the requirements for the mammalian cells). Since the constant need for transfection reagents is eliminated, the relatively low cost of BacMam system is also a considerable advantage.

In 2014, Klarenbeek and Jalink described a protocol on how to use the $^{T}Epac^{VV}$ cAMP biosensor for microscopy on live cells [12]. In 2015, we reported a general protocol for the generation of the BacMam virus and for setting up the subsequent assay in a fluorescence microplate reader with simultaneous dual emission mode of SE (emission ~480 nm and ~ 530 nm, excitation ~430 nm) [13]. Here, we introduce a modernized version of this assay, including a recently developed technique for titration of baculoviruses. We will provide a detailed protocol for sufficient and adjustable expression of the biosensor proteins in most mammalian cell lines, focusing more closely on the application of the sensor for the determination of biological activity of hormone hCG (according to our previous studies reported in [9]).

We describe the sequential steps required to generate and harvest the BacMam viruses with the cAMP biosensor gene cloned into the baculovirus genome under the control of Cytomegalovirus (CMV) promoter. We will cover the expression of the biosensor protein in mammalian cells and give a protocol for a simple, ratiometric cAMP assay to determine the potencies of GPCR ligands and the kinetics of cAMP generation in real time. The selection of the best cAMP sensor construct is vital, and we suggest consulting further materials for detailed information and the best choice for your application [1, 4, 12]. Our protocol is applicable for the generation of the BacMam expression system with any desired biosensor DNA construct.

2 Materials

All solutions for molecular cloning and the assay are made with Milli-Q water unless stated otherwise. Diligently follow all waste disposal regulations when disposing waste materials and/or biological materials.

2.1 Cell Culture

1. Sf9 isolated from pupal ovarian tissue of the fall armyworm *Spodoptera frugiperda* (*see* **Note 1**).

2. Madin-Darby Canine Kidney (MDCK) cells stably expressing human luteinizing hormone/chorionic gonadotropin receptor (LHCGR)—a kind gift from Dr. Prema Narayan (Southern Illinois University); the initial MDCK cell line had been obtained from ATCC (catalog number CRL 2935) (*see* **Note 2**).

3. Sf9 growth medium: EX-CELL 420™.

4. LHCGR-MDCK growth medium: Dulbecco's Modified Eagle's Medium and Ham's F-12 nutrient mixture (DMEM/Ham's F12) supplemented with 10% fetal bovine serum (FBS), 100 U/mL penicillin, 100 μg/mL streptomycin, and 0.25 μg/mL amphotericin B.

5. Polylysine-coated plastic dishes.

6. Humidified CO_2 incubator.

7. Laminar flow cabinets (*see* **Notes 3** and **4**).

2.2 Plasmids and Generation of BacMam Virus

1. Epac-S^{H188} gene in pcDNA3.1 expression vector: mTurq2Δ_Epac(CD,ΔDEP)_td cp^{173}Ven kindly provided by Dr. Kees Jalink from the Netherlands Cancer Institute.

2. *E. coli* cells: DH5α for amplification of plasmid DNA and DH10BAC for the production of bacmid DNA.

3. Baculovirus vector: pFastBac1 vector.

4. Restriction enzymes FastDigest RruI, FastDigest EcoRI, Eco105I (SnaBI), and EcoRI.

5. Silica bead DNA gel extraction kit.

6. Ligase.

7. Plasmid DNA purification kit.

8. Spectrophotometer NanoDrop 1000.

9. Agar-LB plates: 2% agar-LB plates with 100 μg/mL X-Gal (5-bromo-4-chloro-3-indolyl-beta-D-galacto-pyranoside), 50 μg/mL IPTG (isopropyl-beta-D-thiogalactopyranoside), 50 μg/mL kanamycin, 10 μg/mL tetracycline, 7 μg/mL gentamycin.

10. 2% Agar-LB plates with 50 μg/mL ampicillin.

11. Transfection reagent: polyethylene imide-based reagent such as ExGen500.

12. Cytation 5™ cell imaging multi-mode reader.

2.3 cAMP Biosensor Protein Expression

1. Recombinant baculovirus.

2. 70–80% confluent mammalian cells on 10 mm Petri dish(es).

3. DPBS and trypsin with a suitable concentration (here, 0.25%) for detachment of cells.

4. Serum-free DMEM/Ham's F12 for transduction with the recombinant baculovirus.

5. Sodium butyrate at the final concentration of 5–10 mM for enhancing protein expression after the transduction with the recombinant baculovirus.

6. TC10™ Automated Cell Counter.

7. Polylysine-coated 96-well plate with flat transparent bottom.

2.4 Fluorescence Microplate Reader for the cAMP Assay

1. Fluorescence from Epac-S^{H188} biosensor is measured using a microplate reader (*see* **Notes 5–8**).

2. Excitation filter for CFP-analogous proteins (*see* **Note 9**).

3. Simultaneous dual emission filter set for CFP- and YFP-analogous proteins (*see* **Note 10**).

4. DPBS supplemented with Ca^{2+} (1.2 mM) and Mg^{2+} (0.5 mM) as the assay buffer.

2.5 Equipment for Dilution Series of Ligands

1. 96-Well plate, a set of pipettes, and suitable pipette tips for preparation of dilution series.

2. 0.5 mg/mL solution of BSA (Fraction V) in DPBS supplemented with Ca^{2+} (1.2 mM) and Mg^{2+} (0.5 mM) as the dilution buffer.

3. Stock solutions of ligands and a positive control (we used Forskolin) (*see* **Note 11**).

2.6 Software

1. MATLAB version 2014a or higher.

2. Aparecium 2.0 or higher or ICSETools (available at www.gpcr.ut.ee/software.html).

3. GraphPad Prism 5 or higher.

3   Methods

3.1   Cloning, Generation, and Collection of BacMam Virus

1. Clone the Epac-S^{H188} construct under the control of the cytomegalovirus (CMV) promoter into the pFastBac1 vector using the restriction enzymes EcoRI together with Eco105I (SnaBI) for pFastBac1, and FastDigest RruI together with FastDigest EcoRI for pcDNA3.1(+), respectively (see **Notes 12** and **13**).

2. Isolate the desired donor and recipient fragments using agarose gel electrophoresis and purify the fragment from the agarose gel with a silica bead kit. Next, ligate the fragments, transform the ligation mixture into *E. coli* DH5α cells (in our case, using heat shock method), and plate the cells onto 2% agar-LB plates containing 50 µg/mL ampicillin.

3. Amplify the selected bacterial colony.

4. Purify the pFastBacMam plasmid with the biosensor insert using any DNA purification kit. In brief: lyse the bacterial pellet, neutralize the lysate, and wash the DNA with ice-cold isopropanol and 70% ethanol (some kits provide DNA binding columns for easy washing procedure).

5. Remove and dry the residual alcohol and elute the plasmid DNA in Milli-Q water. Determine the purity and the concentration of the DNA with UV-Vis spectrophotometer (in our case, NanoDrop 1000).

6. Transform pFastBacMam plasmid into *E. coli* DH10BAC competent cells and plate them on agar-LB plates supplemented with 100 µg/mL X-Gal, 50 µg/mL IPTG, 50 µg/mL kanamycin, 10 µg/mL tetracycline, and 7 µg/mL gentamycin.

7. Amplify the selected white bacterial colony (see **Note 14**).

8. Purify the recombinant bacmid DNA using the same lysis, neutralization, and alcohol solutions as in plasmid purification. Please note that the use of DNA-binding columns should be tested in a pilot experiment because bacmid DNA is a large molecule (~140 kbp) and might clog the column.

9. Transfect the purified bacmid DNA into Sf9 insect cells using the polyethylene imide-based transfection reagent (in our case, ExGen500; see **Note 15**) to prepare BacMam virus stocks according to the Invitrogen Life Technologies Bac-to-Bac expression system manual [14].

10. Incubate the cells for 3–5 days (until the viability is below 50%; see **Note 16**) and transfer the supernatant fraction that contains recombinant baculoviruses (P0) to 10 mL Sf9 cell suspension at a density of 1.5–1.8×10^6 cells/mL.

11. Harvest the P1 virus by centrifugation at $1600 \times g$ for 10–15 min (collect the supernatant fraction). Determine the titer of the stock and store it at +4 °C or aliquot it and store at −90 °C as an initial transfection virus stock (see **Note 17**).

12. Amplify P1 virus by infecting the desired volume of Sf9 cells at the density of $1.5–2 \times 10^6$ cells/mL at the multiplicity of infection (MOI) 0.01–0.1, or if the titer of the virus stock is unknown, add up to 0.1–1% of the total end volume (virus + cells).

13. Harvest P2 virus after 3–5 days (whenever the Sf9 viability is below 50%; *see* **Note 18**) by centrifugation of cells for 10–15 min at $1600 \times g$. Transfer the supernatant to sterilized (*e.g.*, autoclaved, UV-light- or plasma-cleaner-treated) tubes and determine the titer. Store the aliquots at +4 °C until the day of the experiment (*see* **Notes 17–20**). For longer storage, the virus can be aliquoted and kept at −90 °C.

3.2 Determination of Virus Titer with Cell Size Change-Based Assay

In our laboratory, viral titers are determined using a cell size change-based assay [15].

1. Seed Sf9 cells onto 24-well cell culture plates at 2×10^5 cells/well in 250 μL of EX-CELL 420™ cell culture medium and allow for the cells to adhere for 20–60 min (*see* **Note 21**). Check proper adhesion using a microscope.

2. Make threefold serial dilutions of harvested virus supernatant in EX-CELL 420™ medium and add to each well a 250 μL aliquot of diluted virus (in duplicates).

3. Incubate the cells in the presence of virus for 24 h and thereafter image the microplate using Cytation 5 (bright-field, 4x objective) and the correct protocol, which can be downloaded from the following link: http://gpcr.ut.ee/icse%2D%2D-baculovirus-quantification-assay.html (*see* **Note 22**).

4. Analyze the images using Aparecium software according to the software manual (http://gpcr.ut.ee/uploads/ApareciumHelp/Aparecium.html?Welcome1.html).

5. Plot the average cell diameter *versus* log (virus dilution) on a graph and calculate the virus concentration in infectious viral particles per mL (ivp/mL) from a sigmoidal dose-response curve using eq. (1):

$$\text{Virus concentration (ivp/mL)} = \frac{\frac{1}{ED_{50}} \times 50\% \text{of infected cells}}{V} \quad (1)$$

where *V*—sample volume in wells in milliliters (here, 0.5 mL); ED_{50}—50% effective virus dilution corresponding to dilution at which the average cell diameter has changed by 50%; 50% of infected cells—50% of the cells in wells at the time of infection (here, 1×10^5 cells) given that the number of cells is roughly equal to the number of infective viral particles and the proportion of secondary infection is minimal [16].

3.3 Epac-S^{H188} Biosensor Protein Expression

1. Seed mammalian cells (expressing the desired recombinant GPCR) on Petri dishes 1–2 days before transduction at about $(0.8–1.5) \times 10^6$ cells/dish in complete cell culture medium (*see* **Note 23**).

2. On the day of transduction, aspirate the medium from cells and add 1.5–2 mL of serum-free medium (*see* **Note 24**) and 1–1.5 mL of the concentrated virus stocks of BacMam Epac-S^{H188} baculovirus to achieve MOI of approximately 10–400 (*see* **Note 25**).

3. After 2–3 h of incubation in serum-free medium, remove the virus solution and trypsinize the cells according to the usual protocol. Collect the obtained suspension and dilute into the routine culture medium supplemented with sodium butyrate at 5–10 mM final concentration (*see* **Notes 26** and **27**). Count and seed the cells onto a 96-well plate at a density of 40,000 cells per well (*see* **Note 28**). Before the assay, the cells are further incubated for 20–40 h to allow the expression of Epac-S^{H188} protein (*see* **Notes 29** and **30**).

3.4 cAMP Assay

1. On the day of the assay, replace growth medium with 90 μL/well (in case of activation assay) or 80 μL/well (in case of competition assay, see below) of DPBS supplemented with Ca^{2+} (1.2 mM) and Mg^{2+} (0.5 mM) 1 h before the start of measurement (*see* **Notes 31** and **32**).

2. Prepare dilution series of ligands of interest and control compounds in 0.5 mg/mL solution of BSA in DPBS (*see* **Note 33**).

3. Measure the fluorescence using a fluorescence plate reader at 30 °C or 37 °C, dependent on the conditions of plate incubation prior to the measurement (*see* **Note 29**).

4. First, detect the initial fluorescence intensities for the non-stimulated cells by exciting at a wavelength suitable for mTurquoise excitation (in our case, 427(20) nm) and measuring simultaneous dual emission of the excited fluorophores at wavelengths corresponding to mTurquoise emission (in our case, 480(20) nm) and Venus emission (in our case, 530 (20) nm).

5. For the receptor agonist dose-response assays (activation assays), add the ligand solutions (at 10× concentrations) to wells (at 10 μL/well), achieving the final volume of 100 μL.

6. Measure the responses over at least 60 min, taking measurements every 30 s–5 min.

7. If you would like to measure the ability of antagonists to block the agonist response (competition assay), preincubate the cells with an antagonist (at 10 μL/well) for 10–60 min to achieve equilibrium before the addition of agonist (at 10 μL/well). Measure the responses, as mentioned in the previous point.

3.5 Data Analysis

1. Calculate the change in FRET (ΔFRET) using Eq. (2):

$$\Delta\text{FRET} = \frac{\frac{I^{530 \text{ nm}}_{t=0}}{I^{480 \text{ nm}}_{t=0}} - \frac{I^{530 \text{ nm}}}{I^{480 \text{ nm}}}}{\frac{I^{530 \text{ nm}}_{t=0}}{I^{480 \text{ nm}}_{t=0}}}. \tag{2}$$

as described by Mazina et al. [5].

2. For equilibrium state analysis, plot the data as ΔFRET *versus* time and choose the time-point where the signals have reached the upper plateau (in the case of hCG, we used 60 min; Fig. 1a, b). Analyze pharmacological data by means of nonlinear least squares regression analysis using the commercial program GraphPad Prism™ (GraphPad Software Inc., CA) version 5 or higher. Plot the data and perform fitting of sigmoidal dose-response curves (Fig. 1b) to calculate corresponding pEC_{50} values.

3. To establish the kinetic parameters for activation of receptor pathways by agonists, plot the data as ΔFRET *versus* time (Fig. 1a). Choose one to two ΔFRET *vs.* time curves per individual measurement (*see* **Note 34**). Using GraphPad Prism™, fit each curve to a one-phase association equation with Y_0 value was fixed at 0. To calculate the value of the apparent association rate constant k_{on}^{app}, divide the rate constant k value obtained for each curve by the corresponding concentration of ligand (*see* **Note 35**).

3.6 Time Frame

The entire protocol takes approximately 3–4 weeks to complete. In case the P2 viral stocks have been prepared in advance, the subsequent workflow is limited to procedures 6–8, which take 2–3 days to complete.

1. Cloning the cAMP biosensor gene into pFastBac vector and purification of plasmid DNA of pFastBacMam
    (restriction + agarose-gel electrophoresis to isolate and purify donor and recipient DNA: 1 day)
    (ligation + transformation of the recombinant pFastBac-Mam plasmid: 1 day + purification of DNA: 1 h).

2. Transforming plasmid DNA of pFastBacMam into DH10BAC cells and purification of bacmid DNA
    (transformation: 4 h + color-selection of colonies: 2 days + purification of bacmid: 2 h).

3. Transfection of Sf9 cells with bacmid DNA
    (transfection: 30 min + incubation: 3–5 days).

4. Harvesting of P1 BacMam virus, determination on virus titer
    (harvesting by centrifugation: 10 min + preparing 24-well plates with Sf9 for titration experiment: 1 h + incubation with virus: 24 h + titration experiment: 10 min).

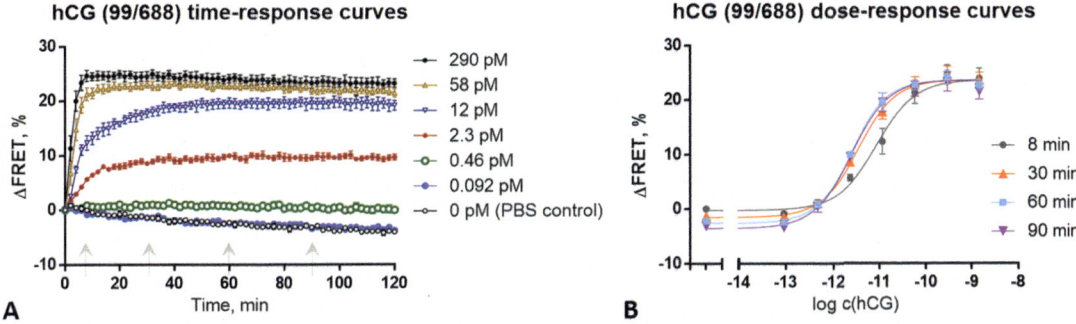

Fig. 1 hCG-triggered activation of cAMP pathway in LHCGR-overexpressing MDCK cells measured with Epac-S^{H188} biosensors. LHCGR-MDCK cells were transduced with BacMam virus for 3 h and further incubated for 24 h at 30 °C in complete growth medium supplemented with 10 mM sodium butyrate. Subsequently, growth medium was replaced with PBS, and after 1 h preincubation three measurements were taken before addition of ligands (the fluorescence intensities in each channel were averaged to obtain signal at time-point 0). Next, cells were treated with different concentrations of hCG NIBSC standard 99/688, and the measurement was immediately started. (**a**) Time-response curves (the curve for 1400 pM hCG is not shown for clarity). The light gray arrows mark time-points which have been used in panel **b**. (**b**) Dose-response curves for signal measured at different times post agonist addition. The graphs show averaged triplicate data from a single representative experiment

5. Amplification of BacMam virus, harvesting of P2 virus, determination on virus titer.

   (transduction with P1 virus: 5 min + incubation with virus 3–5 days + harvesting by centrifugation: 10 min + titration experiment: 24 h)

6. Transduction of mammalian cells with P2 BacMam virus

   (transduction: 5 min + incubation with virus: 2–3 h + counting and seeding the cells on the assay plate: 20 min + incubation with complete medium and 5–10 mM sodium butyrate: 20–40 h).

7. Performing cAMP assay

   (substitution of growth medium with DPBS: 5 min + preparation of dilution series of ligand 30 min + duration of the measurement: 10 min–2 h).

8. Data analysis

   (quick analysis: 30 min, thorough analysis depends on the assay design and research questions posed).

4 Notes

1. Sf9 insect cells are cultured in suspension with EX-CELL 420™ growth medium in a 27 °C incubator in a non-humidified environment.

2. Mammalian cells used for GPCR studies are grown as an adherent monolayer on polylysine-coated plastic dishes and

maintained at 37 °C and 5% CO_2 in a humidified incubator in the complete growth medium. In the case of cells transiently transfected with the receptor, or semi-stable cell lines, antibiotics such as geneticin (G418, 400 µg/mL) should be added to ensure stable expression of recombinant GPCR of interest.

3. For the work with cells and baculoviruses, use aseptic technique in laminar flow cabinets (sterile conditions).

4. Baculoviruses and Sf9 cell culture require biosafety level 1 (BSL 1). Some common mammalian cell lines (*e.g.*, HEK293 cells) may require BSL 2 laboratory.

5. In our case, we used the microplate readers PHERAstar, or Synergy NEO.

6. The gain is adjusted to enable an adequate measurement window (we used 1500 for both emission channels in case of PHERAstar, and 80 for both emission channels in case of NEO).

7. The focal height is adjusted according to the plate type and apparatus (we used 4.5 mm in case of PHERAstar, and 3.5 mm in case of NEO for Thermo Scientific™ BioLite cell culture-treated 96-well plates).

8. The cycle duration and the number of cycles are adjusted according to the need for monitoring the speed of cAMP production (we used 30 s interval and a total number of 250 cycles in case of PHERAstar, and 2 min interval and a total measurement time of 2 h in case of NEO).

9. We used 427(20) nm in case of PHERAstar, and 420(50) nm in case of NEO.

10. We used 480(20) nm and 530(20) nm in case of PHERAstar, and 485(20) nm and 540(25) nm in case of NEO.

11. It is recommended to optimize the storage conditions of protein hormone stock solutions. In the case of hCG, the solution is stable at 4 °C for several weeks. For the proteins stored at −90 °C, repeated freeze-thaw cycles should be avoided. It is also advisable to run a pilot experiment with a fresh protein hormone standard (*e.g.*, from National Institute for Biological Standards and Control) and a protein hormone standard stored at the same conditions as the sample of interest, to assess the possible partial loss of biological activity. In the case of lyophilized powders, we recommend reconstitution in 0.5 mg/mL BSA solution in DPBS to avoid loss of active compound due to nonspecific binding of protein hormone to the plastic. In case ligand or control compound stocks must be prepared in DMSO, we recommend using cell culture-grade DMSO and preparing stock solution with such a concentration that the final total volume percentage of DMSO upon addition to the cells will be ≤0.1%.

12. If you use another construct or another expression vector, you will need to apply the suitable restriction enzymes.

13. The Polyhedrin promoter is removed from the pFastBac1 vector during cloning to ensure low promoter interference during virus amplification.

14. DH10BAC cells contain lacZ gene cassette that is disrupted during in vivo insertion of the recombinant DNA between the transposon sequences. If the lacZ gene is not intact, the enzyme β-galactosidase is not expressed in the cells, and its added substrate X-gal cannot be cleaved. The colony selection is based on the fact that the characteristic blue dye product is not present in the colonies containing the inserted recombinant gene of interest.

15. You can also use other types of transfection reagent or procedures for the initial P1 virus production—*e.g.*, Lipofectamine, Cellfectin, calcium phosphate precipitation, DNA microinjection, and electroporation.

16. The growth and viability of transfected cells should be monitored and compared to the non-transfected control. Once there is a clear difference (decrease in cell number and viability in transfected cells), the P1 virus can be collected.

17. Baculovirus stocks are frequently filtered through a 0.22–0.45 μm hydrophilic syringe filter (*e.g.*, polytetrafluoroethylene or surfactant-free cellulose acetate membrane) to ensure the sterile conditions in the stock solution. Be aware that this results in loss of virus (especially in case of 0.22 μm pore size) since the capsid of the budded baculovirus is 200–400 nm in length and 40–50 nm in diameter.

18. Cells will stop dividing a couple of days after infection with MOI 0.01–1, the viability usually drops on days 3–5.

19. You can add a small amount of FBS or DMSO to stored viral stocks to freeze them for a longer time at −90 °C. Depending on the virus, this addition may not be necessary (we have seen that baculoviruses in EX-CELL 420™ medium stay active for years when stored at −90 °C). It is advisable to determine the titer of the stock solution after one freeze-thaw cycle; the obtained titer value can be subsequently used for MOI calculation in future experiments.

20. Another amplification may be required to obtain high-titer baculovirus stocks of 10^8–10^9 ivp/mL.

21. The assay has been successfully used also with Sf9 cell-derived MIMIC™ Sf9 cells (ThermoFisher Scientific).

22. You can use any other equipment that allows measuring of cell diameter.

23. You can also perform the transduction directly on the assay plate using a multichannel pipette, but exchanging the media is tedious, and we have thus opted for transduction on Petri dishes. Transfection of cells on Petri dish also ensures subsequently lower well-to-well variability, compared to the transfection on the assay plate.

24. The volume at the time of transduction should be as low as possible because apart from MOI, the higher concentration of the required infectious viral particles also improves the probability of virus-cell interaction. The use of serum-free culture medium is suggested, because FBS increases the viscosity of the solution and reduces the mobility of virus particles, hence lowering the probability of virus-cell interaction. However, if the viability of the chosen cell line strongly depends on the presence of FBS, you can nonetheless use FBS-supplemented serum and increase MOI to achieve sufficient expression.

25. You can adjust the expression level of cAMP biosensor protein in a virus dose-dependent manner. The pilot experiments should be started with the suggested MOI, and if the sensor signal in cells is too high or too low, use lower or higher MOI, respectively, in the subsequent experiments.

26. 10 mM sodium butyrate (an histone deacetylase inhibitor) is generally well tolerated by most cell lines; however, you can optimize its concentration to enhance protein expression and avoid any cytotoxic effects. You can also use valproate instead of butyrate for the same effect.

27. If the majority of cells are detached from the dish after 2–3 h incubation with BacMam virus, removing the virus is optional. The complete medium can simply be added to the cells. The three- to fourfold increase in volume dilutes the virus concentration, and the FBS increases the viscosity of the solution; hence the subsequent transduction is extensively reduced. Also, baculoviruses are insect viruses and cause very low cytopathic effects in mammalian cells compared to adenovirus or lentivirus, which in their wild-type form infect mammalian cells. Thus, it is not problematic if the cells are incubated in the presence of baculovirus for 24 h.

28. The optimal number of cells per well might vary, depending on the type of cells. A preliminary optimization of the cell density is advisable, as the latter can affect the measurement window.

29. In some mammalian cell lines, protein expression is more efficient at 30 °C. Consider testing the optimal cAMP sensor expression temperature for your cells.

30. The assay can be performed 20–48 h after transduction.

31. The use of culture media is not recommended for several reasons: elevated background fluorescence (due to the relatively low excitation wavelength of the biosensor) and issues with cell viability in case of bicarbonate-buffered media (if the plate reader is not connected to the CO_2 controller).

32. Buffer composition may influence binding of ligand to your GPCR of interest—*e.g.*, potencies of melanocortin receptor ligands have been shown to depend on divalent cation concentration [6, 17].

33. For ligands with high nonspecific binding, the use of low-retention pipette tips is necessary. In the case of protein hormones with extremely high affinity (low picomolar EC_{50} values), we recommend changing pipette tips after every dilution. The range of dilution series must be optimized for every ligand individually.

34. To ensure maximum quality of data (i.e., representing not too fast kinetics yet a sufficient measurement window), the chosen curve usually features data measured at agonist concentrations above its EC_{50} value, but not the agonist concentrations giving maximal activation.

35. As the concentration of the receptor in the used biological system is usually not known, an assumption can be made that it is the same in all independent measurements, thus avoiding the need to compensate the calculated k_{on}^{app} values for the receptor concentration.

Acknowledgments

We thank Dr. Prema Narayan (Southern Illinois University) for LHCGR-MDCK cells, and Professor Kees Jalink group (The Netherlands Cancer Institute) for cAMP sensor plasmids. The work has been financed by the Estonian Ministry of Education and Science (PRG454, PSG230) and by Enterprise Estonia (grant EU48695).

References

1. Hill SJ, Williams C, May LT (2010) Insights into GPCR pharmacology from the measurement of changes in intracellular cyclic AMP; advantages and pitfalls of differing methodologies. Br J Pharmacol 161:1266–1275

2. Nikolaev VO, Bünemann M, Hein L et al (2004) Novel single chain cAMP sensors for receptor-induced signal propagation. J Biol Chem 279:37215–37218

3. Klarenbeek JB, Goedhart J, Hink MA et al (2011) A mTurquoise-based cAMP sensor for both FLIM and ratiometric read-out has improved dynamic range. PLoS One 6:e19170

4. Klarenbeek J, Goedhart J, van BA et al (2015) Fourth-generation Epac-based FRET sensors for cAMP feature exceptional brightness, photostability and dynamic range: characterization of dedicated sensors for FLIM, for ratiometry and with high affinity. PLoS One 10: e0122513

5. Mazina O, Reinart-Okugbeni R, Kopanchuk S et al (2012) BacMam system for FRET-based cAMP sensor expression in studies of melanocortin MC1 receptor activation. J Biomol Screen 17:1096–1101

6. Link R, Veiksina S, Tahk M-J et al (2020) The constitutive activity of melanocortin-4 receptors in cAMP pathway is allosterically modulated by zinc and copper ions. J Neurochem. 153:346–361

7. Mazina O, Tõntson L, Veiksina S et al (2013) Application of baculovirus technology for studies of G protein-coupled receptor signaling. In: Fesenko O, Yatsenko L, Brodin M (eds) Nanomaterials imaging techniques, surface studies, and applications. Springer, New York, NY, pp 339–348

8. Mazina O, Allikalt A, Tapanainen JS et al (2017) Determination of biological activity of gonadotropins hCG and FSH by Förster resonance energy transfer based biosensors. Sci Rep 7:42219

9. Koistinen H, Koel M, Peters M et al (2019) Hyperglycosylated hCG activates LH/hCG-receptor with lower activity than hCG. Mol Cell Endocrinol 479:103–109

10. Mazina O, Luik T, Kopanchuk S et al (2015) Characterization of the biological activities of human luteinizing hormone and chorionic gonadotropin by a Förster resonance energy transfer-based biosensor assay. Anal Lett 48:2799–2809

11. Kost TA, Condreay JP (2002) Recombinant baculoviruses as mammalian cell gene-delivery vectors. Trends Biotechnol 20:173–180

12. Klarenbeek J, Jalink K (2014) Detecting cAMP with an Epac-based FRET sensor in single living cells. In: Zhang J, Ni Q, Newman RH (eds) Fluorescent protein-based biosensors: methods and protocols. Humana Press, Totowa, NJ, pp 49–58

13. Mazina O, Allikalt A, Heinloo A et al (2015) cAMP assay for GPCR ligand characterization: application of BacMam expression system. In: Prazeres DMF, Martins SAM (eds) G Protein-coupled receptor screening assays: methods and protocols. Springer, New York, NY, pp 65–77

14. Life Technologies Corporation Bac-to-Bac® Baculovirus Expression System, http://tools.thermofisher.com/content/sfs/manuals/bactobac_man.pdf

15. Laasfeld T, Kopanchuk S, Rinken A (2017) Image-based cell-size estimation for baculovirus quantification. BioTechniques 63:161–168

16. O'Reilly DR, Miller LK, Luckow VA (1994) Baculovirus expression vectors: a laboratory manual. Oxford University Press, New York

17. Kopanchuk S, Veiksina S, Petrovska R et al (2005) Co-operative regulation of ligand binding to melanocortin receptor subtypes: evidence for interacting binding sites. Eur J Pharmacol 512:85–95

Chapter 13

FLIPR Calcium Mobilization Assays in GPCR Drug Discovery

Grzegorz Woszczek, Elisabeth Fuerst, and Thomas J. A. Maguire

Abstract

Intracellular calcium mobilization can be measured using several methods varying in indicator dyes and devices used. In this chapter, we describe the fluorescence-based method (FLIPR Calcium 4 Assay) developed by Molecular Devices for a FlexStation and routinely used in our laboratory for detecting intracellular calcium changes. The assay is designed to study calcium mobilization induced by majority of GPCRs and calcium channels and allows for simultaneous concentration-dependent analysis of several receptor agonists and antagonists, useful in receptor characterization and drug discovery projects.

Key words G-protein-coupled receptors, Calcium mobilization, Fluorescence, FlexStation, Airway smooth muscle cells, Mast cells

1   Introduction

As G-protein-coupled receptors (GPCRs) can activate several signaling pathways dependent on G protein subunits they bind to [1], there have been extensive efforts to develop new, sensitive, and simple-to-use assays to measure second messengers as readout for GPCR activation.

In this protocol, we will focus on calcium mobilization assays. Calcium mobilization is the result of activation of GPCRs binding to specific $G\alpha$ subunits G_q and G_i [2]. G_q-coupled receptors activate classical phospholipase C pathways, leading to an increase of inositol tri-phosphate (IP_3) levels. IP_3 binds to IP_3-sensitive calcium channels on the endoplasmic reticulum, resulting in the release of calcium from the endoplasmic reticulum into the cytoplasm [3]. Changes in intracellular calcium reflecting GPCR activation can be measured easily and precisely using fluorescent calcium-sensitive dyes. Activation of GPCRs binding to different $G\alpha$ subunits, e.g., G_s, $G_{12/13}$, or GPCRs not binding to G proteins at all, however, cannot be measured using this assay except cells transfected with a specific G protein, $G\alpha_{16}$. This promiscuous G protein has the ability to couple to almost all GPCRs and force

Sofia Aires M. Martins and Duarte Miguel F. Prazeres (eds.), *G Protein-Coupled Receptor Screening Assays: Methods and Protocols*, Methods in Molecular Biology, vol. 2268, https://doi.org/10.1007/978-1-0716-1221-7_13,
© Springer Science+Business Media, LLC, part of Springer Nature 2021

them to signal through calcium [4]. This is an interesting method since screening of a ligand library is possible without knowledge of specific GPCR signaling pathway under physiological conditions.

The first attempts of calcium measurement were done in the 1920s, with the first reliable measurement performed by Ridgeway and Ashley in 1967 using the calcium-sensitive photoprotein aequorin, derived from the jellyfish, to detect changes in calcium during muscle contraction in barnacle [5]. However, despite many well-described advantages of this method [6, 7], there are disadvantages of using aequorin as calcium indicator in GPCR drug discovery. Firstly, cells have to be transfected with the apoprotein aequorin, and secondly light emission is very low [8]. In the 1980s, Roger Tsien, the Nobel Prize laureate in Chemistry in 2008, and colleagues developed a range of fluorescent calcium indicators including Rhod-2, Fura-2, Indo-1, and Fluo-3 [9]. Fluo-3 and also Fluo-4, an improved version of Fluo-3, do not need UV-light for excitation but can be excited with 485 nm wavelength, making them suitable for use in a standard molecular biology laboratory.

In this chapter, we describe the fluorescence-based method FLIPR Calcium 4 Assay developed by Molecular Devices for a FlexStation and routinely used in our laboratory for detecting intracellular calcium changes. The principle behind the FLIPR calcium mobilization assay is the direct fluorescent measurement of intracellular calcium using a calcium sensitive dye. During incubation of cells with the loading dye, the calcium-sensitive dye (acetoxymethyl (AM) ester) enters the cell where it is processed by esterases (AM groups are cleaved) in the cytoplasm, retaining the dye within the cell. In contrast, the masking dye (included in the loading dye) does not pass the cell membrane and reduces extracellular background fluorescence, which is often generated by remaining extracellular calcium-sensitive dye or other extracellular factors. After GPCR stimulation, calcium is released from the endoplasmic reticulum into the cytoplasm. The intracellular calcium-sensitive dye binds to cytoplasmic calcium, thereby causing a fluorescent signal. Using a FlexStation device, fluorescence, as indicator of calcium release, can be measured over time. There are several advantages of the FLIPR calcium 4 assay over conventional calcium mobilization assays. The first one is the removal of several washing steps from the assay protocol; the loading dye is not removed from the cells before measuring calcium mobilization and there is no additional washing of cells required. This not only prevents loss or detachment of cells during washing steps, especially weakly adherent cells, but also reduces mechanical stimulation of cells due to addition of washing buffers and the risk of remaining dye after incomplete or inconsistent washing. In addition, having fewer assay steps saves time and allows high throughput. Other advantages are low background due to the special quenching (masking) technology, reduced well-to-well variation, reproducibility, high-quality data, high sensitivity of the assay, and ease of use. However, it is to note that the content of the assay kit is not fully disclosed, especially with regard to the masking and calcium sensing dyes, making it difficult to optimize the experiments if required. There are two plate formats available (96-well and 384-well) for FlexStation and assay protocols should be adjusted accordingly. Lower throughput devices such as the 96-well FlexStation can only pipette/measure 8 wells simultaneously (one column of 96-well plate); thus, the total time of measuring of the whole plate needs to be considered while planning experiments.

The assay is designed to study calcium mobilization induced by majority of GPCRs and calcium channels and allows for simultaneous concentration-dependent analysis of several receptor agonists and antagonists, useful in receptor characterization and drug discovery projects. We have successfully applied this protocol for calcium mobilization experiments with various adherent cells (endothelial cells [10], airway epithelial cells [11], airway smooth muscle cells) and non-adherent cells (peripheral blood mononuclear cells [12], monocytes [13], lymphocytes [14], dendritic cells [15], platelets [16], and mast cells [17]). The same protocol was also used for calcium measurement in cells overexpressing specific GPCRs [16–18].

As an example, we describe here protocols for calcium mobilization experiments in human airway smooth muscle cells (adherent cells) stimulated by sphingosine-1-phosphate (S1P) and in human mast cells (LAD2) (non-adherent cells) stimulated by leukotriene D4 (LTD$_4$).

2 Materials

2.1 Cell Culture

1. Primary human airway smooth muscle cells were grown from bronchial biopsies by explant culture [19, 20].

2. Primary human airway smooth muscle cells (ASM) can also be obtained from commercial providers (e.g., Lonza).

3. Human mast cells (LAD2) were provided by Dr. Arnold Kirshenbaum, NIAID, NIH, USA [21].

4. Phosphate-buffered saline (PBS) without calcium and magnesium, pH 7.4: sodium chloride (NaCl), 9 g/L; potassium phosphate monobasic (KH$_2$PO$_4$), 144 mg/L; sodium phosphate dibasic (Na$_2$HPO$_4$-7H$_2$O), 795 mg/L.

5. Trypsinization buffer: 0.05% Trypsin-EDTA in PBS.

6. Culture medium: Dulbecco's Modified Eagle's Medium (DMEM) supplemented with 10% heat-inactivated Fetal Bovine Serum (FBS), L-glutamine (2 mM), sodium pyruvate

(1 mM), 1× nonessential amino acids, gentamicin (50 µg/mL), and fungizone (2 µg/mL) or appropriate cell culture medium (*see* **Note 1**).

7. Culture medium (LAD2): StemPro34 medium with additional StemPro34 nutrient supplement and penicillin/streptomycin (50 units/mL, 50 µg/mL); L-glutamine (200 mmol/L); stem cell factor (SCF), 100 ng/mL.

8. HEPES buffer stock solution: 1 M HEPES.

9. Serum-free medium: DMEM culture medium without FBS or appropriate serum-free cell culture medium.

10. RPMI 1640 medium supplemented with 25 mM HEPES and L-glutamine.

11. 96-Well plates, black wall, clear bottom, sterile.

2.2 Calcium Mobilization Assay

1. Hank's Balanced Salt Solution (HBSS) with calcium, magnesium, and without phenol red: calcium chloride ($CaCl_2$) anhydrous, 140 mg/L; magnesium chloride ($MgCl_2$-$6H_2O$), 100 mg/L; magnesium sulfate ($MgSO_4$-$7H_2O$), 100 mg/L; potassium chloride (KCl), 400 mg/L; potassium phosphate monobasic (KH_2PO_4), 60 mg/L; sodium bicarbonate ($NaHCO_3$), 350 mg/L; sodium chloride (NaCl), 8 g/L; sodium phosphate dibasic (Na_2HPO_4) anhydrous, 48 mg/L; D-glucose (Dextrose), 1 g/L.

2. HBSS/HEPES solution: Add 2 mL 1 M HEPES solution to 98 mL HBSS to obtain HBSS supplemented with 20 mM HEPES.

3. FLIPR Calcium 4 Assay Kit Bulk (Molecular Devices) (*see* **Note 2**).

4. RPMI medium 1640 supplemented with 25 mM HEPES and L-glutamine.

5. 96-Well polypropylene plates.

6. Multipette dispensing 100 µL volumes.

7. Combitips for 10 mL volume.

8. Leukotriene D_4 receptor (Cayman Chemicals).

9. Sphingosine-1 phosphate (Enzo Life Sciences).

10. Histamine.

11. Bradykinin.

12. Compound library.

2.3 Running the Assay

1. Flex Station 3 microplate reader.

2. Flex Station Pipet Tips, 96-well tips, 200 µL capacity.

3. Soft Max Pro software.

3   Methods

Cell type and experimental conditions are important factors determining cell preparation for calcium mobilization assay and should be tested for each new experiment. It is recommended to have 90–100% cell confluence at the time of adding the loading dye. For some cells, use of serum-free conditions prior to an assay is advised to ensure cell cycle synchronization and removal of serum effect (*see* **Note 3**). Adherent cells need to be seeded into black wall 96-well plates at least 24 h before the assay to ensure attachment of cells. Weakly adherent cells such as HEK293 and non-adherent cells (LAD2) need to be seeded on poly-D-lysine pre-coated 96-well plates to avoid alterations of fluorescence caused by detached cells (*see* **Note 4**).

3.1   Preparation of Cells and Loading Dye (Adherent Cells)

1. Grow human airway smooth muscle cells to 80–90% confluence in tissue culture flasks using DMEM culture medium.

2. Remove medium, wash with PBS, and harvest cells with trypsin-EDTA solution. Resuspend cells in DMEM culture medium.

3. Count and seed 10,000 cells/100 µL in black wall 96-well plate with 100 µL of cell suspension per well using a dispenser pipette. Let grow for 24 h at 37 °C in a humidified incubator with a 5% CO_2 atmosphere. After 24 h, cells should be attached to the microwell plastic.

4. Remove DMEM cell culture medium and replace with serum-free DMEM medium (100 µL/well). Incubate for 18 h at 37 °C in a humidified incubator with a 5% CO_2 atmosphere.

5. When using FLIPR4 bulk kit, add 100 mL of HBSS/HEPES solution (Subheading 2.2, **item 3**) to 1 vial of Component A (loading dye powder) from the FLIPR Calcium 4 Assay Kit and vortex rigorously to ensure that content is completely dissolved. Aliquot appropriately and store aliquots at −20 °C (*see* **Note 5**).

3.2   Preparation of Cells and Loading Dye (Non-Adherent Cells)

1. Pre-coat a 96-well black wall plate with poly-D-lysine to enhance attachment of cells to a plate at least a day before experiments (*see* **Note 4**).

2. Grow LAD2 cells to around 500,000 cells/mL in tissue culture flasks using serum-free StemPro34 medium supplemented with StemPro-34 supplement and Stem Cell Factor.

3. Count, spin down, and resuspend required number of LAD2 cells (100,000 cells/well) in warm RPMI 1640 medium supplemented with 25 mM HEPES and L-glutamine.

4. Seed LAD2 cells at 100,000 cells/100 µL in black wall 96-well plate with 100 µL cell suspension per well using a dispenser pipette.

5. Incubate for 4 h at 37 °C in a humidified incubator with a 5% CO_2 atmosphere.

6. Centrifuge plate at $100 \times g$ (brake off) for 5 min to ensure cells attach to the plate bottom before reading assay.

7. When using FLIPR4 bulk kit, add 100 mL of HBSS/HEPES solution to 1 vial of Component A (loading dye powder) from the FLIPR Calcium 4 Assay Kit and vortex rigorously to ensure that content is completely dissolved. Aliquot appropriately and store aliquots at −20 °C (*see* **Note 5**).

3.3 Loading Cells

1. After 18 h serum-free conditions, remove medium and replace with pre-warmed RPMI 1640 (100 µL/well) (Subheading 3.1 for adherent cells). This step is not required for non-adherent cells (see Subheading 3.2).

2. Add 100 µL of prepared and pre-warmed (37 °C) loading dye buffer (Component A) per well using dispenser pipette (*see* **Notes 2, 5,** and **6**).

3. Incubate plate for 1 h at 37 °C in dark incubator (*see* **Note 7**).

4. If antagonists are tested, add antagonists directly to wells for time required and at concentration calculated for the total volume of experiment.

5. For some cell types, an inhibitor of the anion-exchange protein (Probenecid) may be required to increase dye retention intra-cellularly, but we have not observed an advantage of using it with the cells and conditions tested in our lab.

3.4 Preparation of Compound Plate

1. Dilute compounds of interest in RPMI 1640 medium to relevant concentration. Please note that the typical volume added to the well with cells is 25 µL. As described above, the volume of loading buffer + RPMI 1640 in the cell well is 200 µL, after addition of 25 µL of compound it is 225 µL. Therefore, it is important to adjust the concentration in the compound plate by a factor of 9 to obtain the concentration of interest in the final readout plate. Vehicle controls have to be added to plates since some vehicles (DMSO, ethanol) can affect calcium measurements (*see* **Note 8**). In addition, the use of positive and negative controls is recommended (*see* **Note 9**).

2. Pipette compounds in a 96-well polypropylene plate. 50–75 µL excess volume has to be added for adequate transfer in a Flex Station device (this volume depends on the type of plate used; it should be tested before running experiments).

3.5 Reading Assay Plate

1. Turn on the FlexStation device and set up the following parameters for fluorescence measurements using Soft Max Pro: Excitation wavelength—485 nm, emission wavelength—525 nm, emission cutoff—515 nm; PMT sensitivity—high; Pipette height—180 μL; Transfer volume—25 μL; Addition speed (rate)—1; Set temperature to 37 °C (*see* **Note 10**).

2. Set up parameters of the experiment: reading time, e.g., 120 s, reading intervals, e.g., 4 s, compound addition time, e.g., 17 s.

3. Make sure the bottom of the plate is clean and has not been touched prior to reading the assay plate. It can be cleaned with tissue to avoid interferences with fluorescent measurement (*see* **Notes 11** and **12**).

4. Place black wall 96-well cell plate, compound plate, and tips in appropriate compartments of the device and wait approximately 10 min before starting the protocol (*see* **Note 13**).

3.6 Data Analysis

Soft Max Pro software is used to analyze and visualize data.

1. Obtain a fluorescent baseline for each well prior to compound addition (depending on the protocol setup, in our case at 17 s).

2. Baseline should be steady and similar between wells on the plate (assuming that the same cells have been used) (*see* **Note 14**). If this is the case, delta values (peak value - baseline value) can be calculated and used for analysis.

3. Results are either displayed in 96-well format in arbitrary units of peak fluorescence values or shown as response curves (fluorescence traces) (Figs. 1 and 2). Fluorescence readings from each time point and area under the curve can also be used for data analysis.

4. After activation of GPCRs, fluorescence readings rapidly increase within seconds (seen by a steep curve) until a peak value is noted. After that peak, fluorescence usually decreases slowly but steadily and within 3 min after compound addition fluorescent baseline is usually reached. For testing unknown receptor agonists/antagonists, a range of concentrations should be used for screening (Figs. 3 and 4), allowing for creation of concentration response curves and calculation of EC$_{50}$ or IC$_{50}$ values (Fig. 2).

5. Variation between plates can occur due to incubation time, cell number variation, and receptor expression, just to mention a few factors. To compare values across assay plates, a positive control should be added to each plate and peak values should be expressed as percentage of peak response of positive control (*see* **Note 9**) (Fig. 2).

A

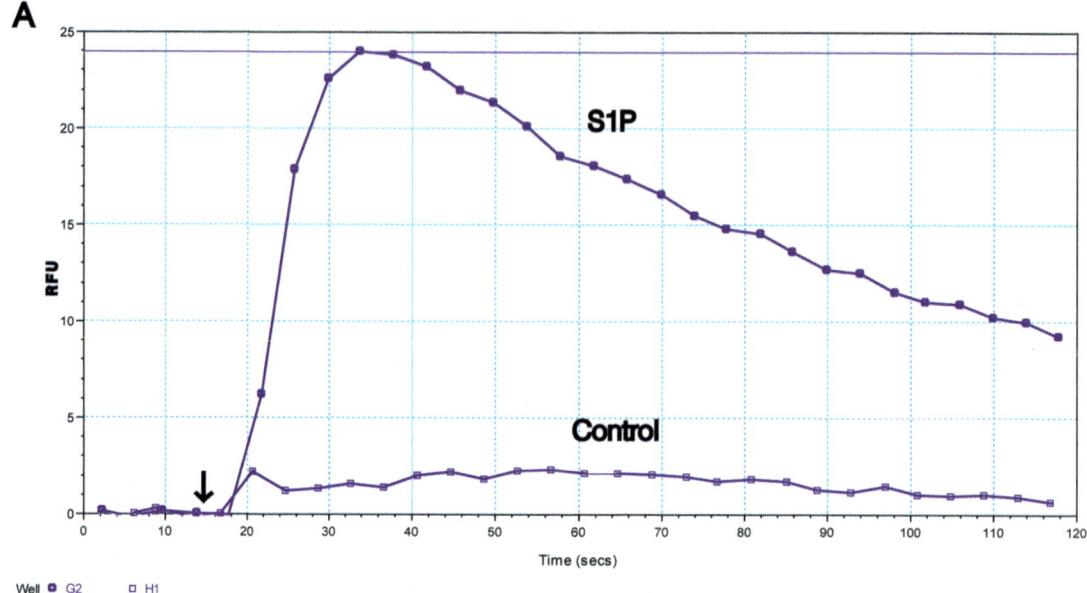

Well ● G2   □ H1
Peak   23.975   2.293

B

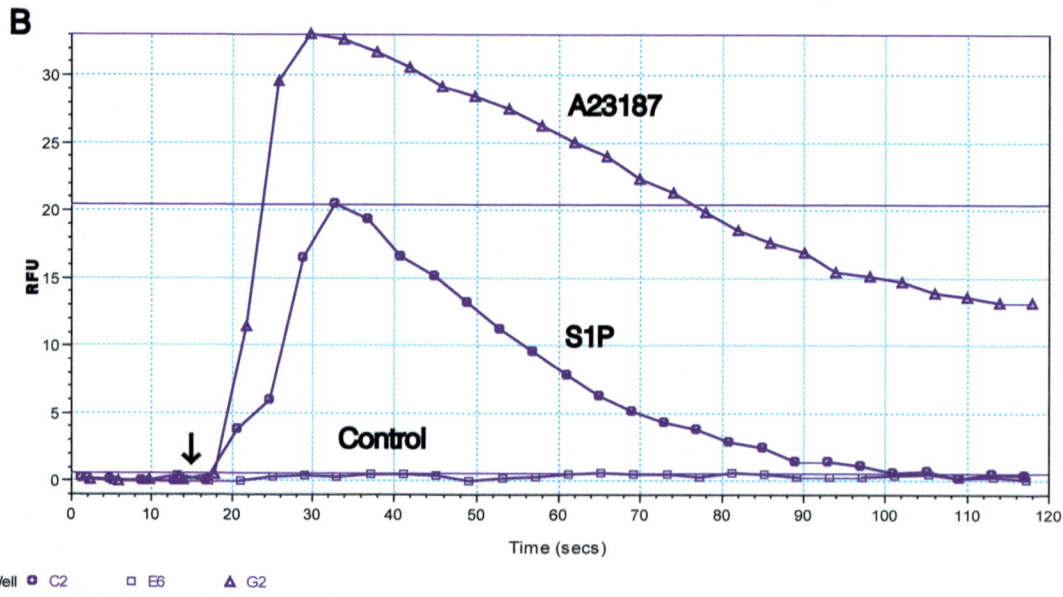

Well ● C2   □ E6   ▲ G2
Peak   20.415   0.597   32.958

Fig. 1 Calcium mobilization assay. Traces of intracellular calcium changes in human airway smooth muscle cells in response to (**a**) sphingosine-1-phosphate (S1P) (300 nmol/L) and vehicle control, and (**b**) calcium ionophore A23187 (1 μmol/L), S1P (100 nmol/L) and vehicle control. Data shown as relative fluorescence units (RFU) from a single experiment. Arrows represent compound addition time (17 s)

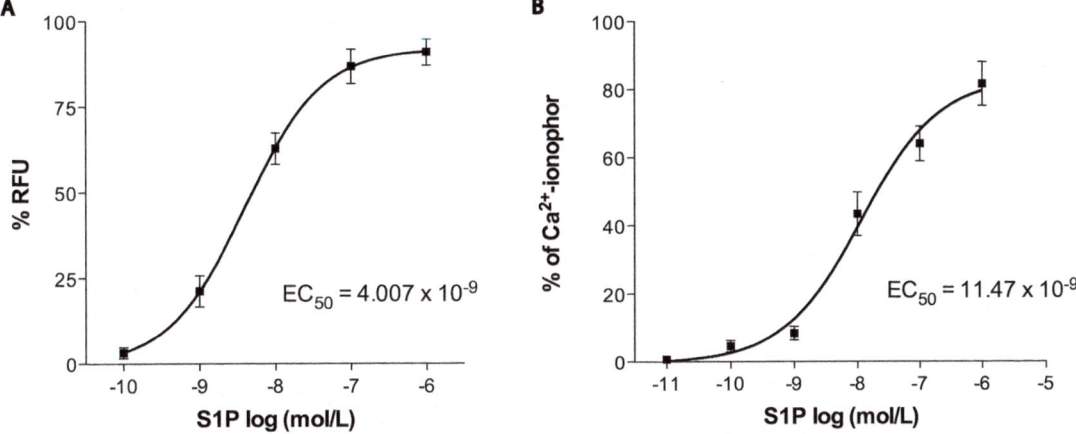

Fig. 2 Calcium mobilization assay. Concentration-response curves for sphingosine-1-phosphate (S1P) in human airway smooth muscle cells. Data presented as (**a**) % of maximum response to S1P (1 μmol/L) and (**b**) % of response to calcium ionophore (A23187, 1 μmol/L). Data shown as mean ± SEM of baseline corrected peak calcium fluxes from 3 separate experiments run in duplicate. *RLU* relative fluorescence units, *EC$_{50}$* half maximal effective concentration calculated using GraphPad Prism

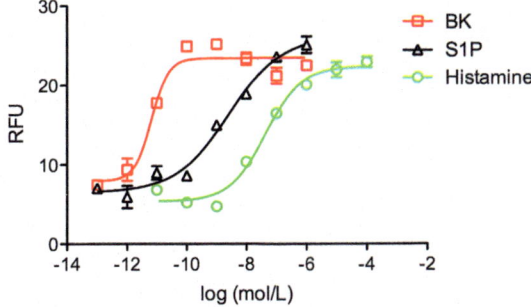

Fig. 3 Calcium mobilization in human airway smooth muscle cells in response to different GPCR agonists. Cells were stimulated with different concentrations of bradykinin (BK), sphingosine-1-phosphate (S1P) and histamine (Histamine) and calcium mobilization measured as described in the protocol. Data shown as mean ± SE of baseline corrected peak calcium fluxes from a single experiment run in triplicate. *RFU* relative fluorescence units

4   Notes

1. Some GPCRs interact with factors in serum, e.g., it is known that sphingosine-1-phosphate receptors are activated and rapidly desensitized (downregulated) by sphingosine-1-phosphate, a lipid mediator found in serum. It is advised to change to serum-free conditions at least 18 h before starting the calcium mobilization assay.

2. The FLIPR calcium 4 assay kit from Molecular Devices comes in different sizes. The Explorer Kit contains 10 vials of loading dye powder and Component B Buffer (HBSS /HEPES

A

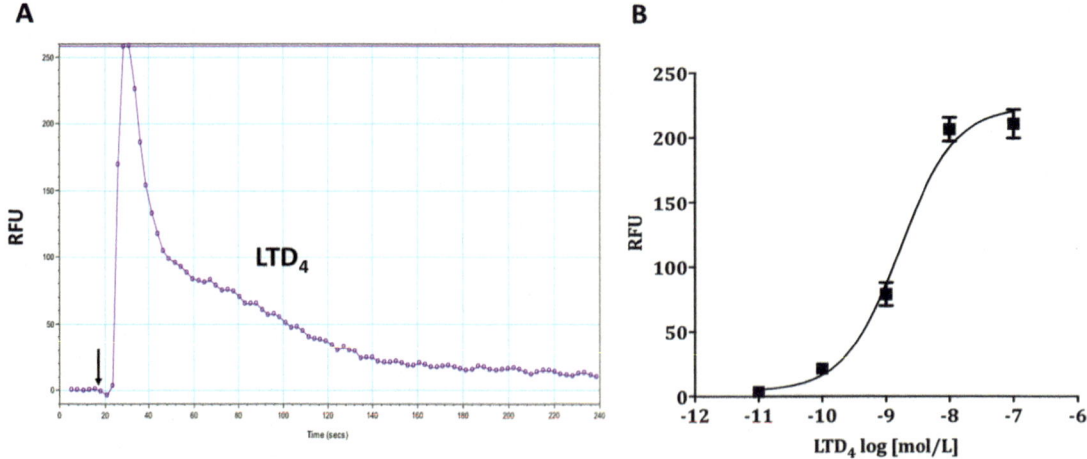

B

Fig. 4 Calcium mobilization assay (non-adherent cells). (**a**) Traces of intracellular calcium changes in human mast cells (LAD2) in response to leukotriene D_4 (LTD$_4$) (10 nmol/L). (**b**) Concentration-response curves for leukotriene D_4 in LAD2 cells. Data shown as mean $\pm$ SE of baseline corrected peak calcium fluxes from a single experiment run in triplicate. *RFU* relative fluorescence units

solution) used for dilution. In larger kits, e.g., bulk kit, Component B Buffer is not included and HBSS/HEPES buffer has to be prepared.

3. Experimental conditions may differ depending on cell type and GPCR studied. Weakly adherent (e.g., HEK293) or non-adherent (e.g., peripheral blood cells) cells need to be plated in poly-D-lysine coated plates as described in **Note 4**. Adherent cells need to be plated at least 1 day before the experiment having enough time to attach properly. Airway smooth muscle cells are plated in full culture medium before medium is changed to RPMI 1640 and loading buffer is added. When serum starvation is required, cells are first incubated in serum-deprived medium for at least 18 h before medium is changed to RPMI 1640 and loading buffer is added. For plasmid/siRNA transfection experiments the best time point for calcium measurement should be tested. In our experience, 48 h post-transfection was the best time for analysis.

4. Poly-D-lysine solutions can be obtained from Sigma-Aldrich. Coat plate according to a protocol the day before the assay; wash at least twice with PBS. Leave plates overnight or at least for a couple of hours to ensure that wells are dry before plating cells. Seed cells in RPMI 1640 medium at least 1 h before adding loading dye. To make sure cells stick to bottom of plate, centrifuge plate at $100 \times g$ (brake off) for 5 min before measuring fluorescence. Cells that are not attached or are easily dislodged will interfere with fluorescent measurement.

5. The Bulk Kit consists of 10 vials of loading dye powder, where one vial is sufficient for 10 96-well plates. Dissolve one vial of Compound A in 100 mL HBSS + 20 mM HEPES and make sure that the whole content is fully dissolved. Once dissolved in HBSS/HEPES, aliquots can be prepared (10 mL aliquot is sufficient for one 96-well plate) and frozen at -20 °C. Since the loading dye is light sensitive, it is recommended to wrap aliquots in aluminum foil. For thawing aliquots, place in a 37 °C water bath to pre-warm before adding to cells. It is not recommended to re-freeze loading buffer.

6. The advantage of using a dispenser pipette is fast pipetting and small excess volume required. However, the dispense speed of these pipettes is relatively high, and cells, especially weakly adherent or non-adherent, might be dislodged. It is therefore important to pipette as gently as possible and to carefully pipet down the side of the well when working with non-adherent cells.

7. The loading dye is light sensitive. Try to avoid light, pipette fast, always incubate in dark incubators, and wrap plate in aluminum foil.

8. DMSO required for solubilization of some compounds can interfere with fluorescence measurements. Always use a vehicle control with the same concentration of DMSO as used for compound stimulations.

9. Calcium ionophore A23187 (Sigma-Aldrich) can be used as a positive control for experiments. Alternatively, verified GPCR agonists can be used as positive controls, e.g., ATP or ADP when the corresponding receptors are expressed in studied cells. Negative control is either medium alone or medium with respective vehicle.

10. In our experience, the best results are obtained when temperature is set to 37 °C allowing cells to function at physiological temperature. Much lower responses were observed in experiments run at room temperature.

11. Flex Station devices read from the bottom of the plate; therefore it is important that this part has not been touched and is not dirty to avoid interference with the fluorescent measurement.

12. When non-adherent cells are used, centrifuge cell plate before starting measurement as described in **Note 4**.

13. Wait for 10 min after placing plates in a FlexStation to warm plates (cells/media). This step allows cells to adjust to temperature, especially if they have been centrifuged or transported at room temperature.

14. If baseline is not steady, this might be due to dislodged cells. Use poly-D-lysine coated plates as described in **Note 4** and centrifuge cell plate right before the measurement. In addition, the speed of compound addition can be reduced in the settings of a FlexStation device avoiding possible cell detachment.

References

1. Marinissen MJ, Gutkind JS (2001) G-protein-coupled receptors and signaling networks: emerging paradigms. Trends Pharmacol Sci 22:368–376

2. Berridge MJ (1993) Inositol trisphosphate and calcium signalling. Nature 361:315–325

3. Clapham DE (1995) Calcium signaling. Cell 80:259–268

4. Offermanns S, Simon MI (1995) G alpha 15 and G alpha 16 couple a wide variety of receptors to phospholipase C. J Biol Chem 270:15175–15180

5. Ridgway EB, Ashley CC (1967) Calcium transients in single muscle fibers. Biochem Biophys Res Commun 29:229–234

6. Stables J, Green A, Marshall F, Fraser N, Knight E, Sautel M, Milligan G, Lee M, Rees S (1997) A bioluminescent assay for agonist activity at potentially any G-protein-coupled receptor. Anal Biochem 252:115–126

7. Stables J, Mattheakis LC, Chang R, Rees S (2000) Recombinant aequorin as reporter of changes in intracellular calcium in mammalian cells. Methods Enzymol 327:456–471

8. Brini M, Pinton P, Pozzan T, Rizzuto R (1999) Targeted recombinant aequorins: tools for monitoring $[Ca^{2+}]$ in the various compartments of a living cell. Microsc Res Tech 46:380–389

9. Minta A, Kao JP, Tsien RY (1989) Fluorescent indicators for cytosolic calcium based on rhodamine and fluorescein chromophores. J Biol Chem 264:8171–8178

10. Woszczek G, Chen LY, Nagineni S, Alsaaty S, Harry A, Logun C, Pawliczak R, Shelhamer JH (2007) IFN-gamma induces cysteinyl leukotriene receptor 2 expression and enhances the responsiveness of human endothelial cells to cysteinyl leukotrienes. J Immunol 178:5262–5270

11. Chen LY, Woszczek G, Nagineni S, Logun C, Shelhamer JH (2008) Cytosolic phospholipase A2alpha activation induced by S1P is mediated by the S1P3 receptor in lung epithelial cells. Am J Physiol Lung Cell Mol Physiol 295: L326–L335

12. Wilson GA, Butcher LM, Foster HR, Feber A, Roos C, Walter L, Woszczek G, Beck S, Bell CG (2014) Human-specific epigenetic variation in the immunological Leukotriene B4 Receptor (LTB4R/BLT1) implicated in common inflammatory diseases. Genome Med 6:19

13. Chen LY, Eberlein M, Alsaaty S, Martinez-Anton A, Barb J, Munson PJ, Danner RL, Liu Y, Logun C, Shelhamer JH, Woszczek G (2011) Cooperative and redundant signaling of leukotriene B4 and leukotriene D4 in human monocytes. Allergy 66:1304–1311

14. Parmentier CN, Fuerst E, McDonald J, Bowen H, Lee TH, Pease JE, Woszczek G, Cousins DJ (2012) Human T(H)2 cells respond to cysteinyl leukotrienes through selective expression of cysteinyl leukotriene receptor 1. J Allergy Clin Immunol 129:1136–1142

15. Woszczek G, Chen LY, Nagineni S, Shelhamer JH (2008) IL-10 inhibits cysteinyl leukotriene-induced activation of human monocytes and monocyte-derived dendritic cells. J Immunol 180:7597–7603

16. Foster HR, Fuerst E, Lee TH, Cousins DJ, Woszczek G (2013) Characterisation of P2Y (12) receptor responsiveness to cysteinyl leukotrienes. PLoS One 8:e58305

17. Foster HR, Fuerst E, Branchett W, Lee TH, Cousins DJ, Woszczek G (2016) Leukotriene E4 is a full functional agonist for human cysteinyl leukotriene type 1 receptor-dependent gene expression. Sci Rep 6:20461

18. Woszczek G, Chen LY, Nagineni S, Kern S, Barb J, Munson PJ, Logun C, Danner RL, Shelhamer JH (2008) Leukotriene D (4) induces gene expression in human monocytes through cysteinyl leukotriene type I receptor. J Allergy Clin Immunol 121:215–221

19. Chan V, Burgess JK, Ratoff JC, O'Connor BJ, Greenough A, Lee TH, Hirst SJ (2006) Extracellular matrix regulates enhanced eotaxin expression in asthmatic airway smooth muscle cells. Am J Respir Crit Care Med 174:379–385

20. Fuerst E, Foster HR, Ward JP, Corrigan CJ, Cousins DJ, Woszczek G (2014) Sphingosine-1-phosphate induces pro-remodelling response in airway smooth muscle cells. Allergy 69:1531–1539

21. Kirshenbaum AS, Akin C, Wu Y, Rottem M, Goff JP, Beaven MA, Rao VK, Metcalfe DD (2003) Characterization of novel stem cell factor responsive human mast cell lines LAD 1 and 2 established from a patient with mast cell sarcoma/leukemia; activation following aggregation of FcepsilonRI or FcgammaRI. Leuk Res 27:677–682

Chapter 14

Live Cell Imaging and Optogenetics-Based Assays for GPCR Activity

Xenia Meshik and Narasimhan Gautam

Abstract

GPCRs are responsible for activation of numerous downstream effectors. Live cell imaging of these effectors therefore provides a real-time readout of GPCR activity and allows for better understanding of temporal dynamics of GPCR-mediated signaling. Opsins, or optically activatable GPCRs, allow for these signaling pathways to be activated in a spatiotemporally precise and reversible manner. Here, we describe optogenetic methods for activating Gi, Gq, and Gs signaling pathways. Additionally, we present assays for detecting activation of these pathways in real time through live cell imaging of Gβγ translocation, PIP_3 increase, PIP_2 hydrolysis, cAMP production, and cell migration. These assays can be utilized for GPCR-targeted drug development, as well as for studies of a wide range of GPCR-mediated physiological processes.

Key words Optogenetics, Live cell imaging, GPCRs, Opsins, G proteins, Gβγ translocation, Second messengers, Cell migration

1 Introduction

G protein-coupled receptors (GPCRs) are a large family of transmembrane proteins, which initiate numerous downstream signaling pathways. These pathways are often involved in pathological processes, making GPCRs a frequent target in drug development [1]. Therefore, reliable methods for controlling and detecting the activity of GPCRs and their downstream signaling pathways are in high demand. Our lab has utilized live cell imaging to detect GPCR activation through downstream processes such as: (a) G protein activation, as demonstrated by translocation of Gβγ subunits [2–4] and fluorescence resonance energy transfer (FRET) between α and β subunits [5], (b) response of second messengers such as PIP_3 [3, 6], PIP_2, cAMP [3], and Ca^{2+} [7], and (c) cell behaviors such as migration [3, 6] and neurite extension [3]. Additionally, we have developed methods to control GPCR-mediated pathways through

Sofia Aires M. Martins and Duarte Miguel F. Prazeres (eds.), *G Protein-Coupled Receptor Screening Assays: Methods and Protocols*, Methods in Molecular Biology, vol. 2268, https://doi.org/10.1007/978-1-0716-1221-7_14,
© Springer Science+Business Media, LLC, part of Springer Nature 2021

the use of optically activatable GPCRs, known as opsins. These optogenetic methods allow for activation of native signaling pathways with fine spatial and temporal precision and subcellular control not achievable with traditionally used agonists and antagonists [8].

Here, we describe assays developed in our lab which combine optogenetics and live cell imaging to initiate and detect GPCR signaling. These assays involve transient transfections of epithelial or macrophage cells with opsins, which are then optically activated and the resulting cellular response monitored through live cell imaging. To induce Gi-mediated signaling, cells are transfected with parapinopsin, the activity of which can be detected through real-time imaging of Gβγ translocation, PIP$_3$ response, or cell migration. To induce Gq-mediated signaling, cells are transfected with melanopsin, the activity of which can be detected through real-time imaging of Gβγ translocation or PIP$_2$ hydrolysis. To induce Gs-mediated signaling, cells are transfected with CrBlue, a chimeric Gs-coupled opsin, the activity of which can be detected through real-time imaging of Gβγ translocation or cAMP production. The ability to activate these native signaling cascades with fine spatiotemporal precision has great potential for use in drug development and other studies of GPCR-mediated processes.

2   Materials

2.1   Cell Culture

1. HeLa cells (*see* **Note 1**).

2. HeLa culture media (*see* **Note 2**): 1× Minimum Essential Medium (MEM) containing Earle's salts and L-glutamine, supplemented with 10% dialyzed fetal bovine serum (DFBS) (*see* **Note 3**) and 1% of penicillin/streptomycin (10,000 U/mL penicillin and 10,000 mg/mL streptomycin) (*see* **Note 4**) and stored at 4 °C (*see* **Note 5**).

3. RAW 264.7 cells (*see* **Note 6**).

4. RAW culture media (*see* **Note 2**): Dulbecco's modified Eagle's medium (DMEM) containing 4500 mg/L glucose, L-glutamine, sodium pyruvate, and sodium bicarbonate, supplemented with 10% dialyzed fetal bovine serum (DFBS) (*see* **Note 3**) and 1% of penicillin/streptomycin (10,000 U/mL penicillin and 10,000 mg/mL streptomycin) (*see* **Note 4**) and stored at 4 °C (*see* **Note 7**).

5. 0.02% Ethylenediaminetetraacetic acid (EDTA): 0.2 g/L EDTA in Dulbecco's phosphate-buffered saline (DPBS) without calcium or magnesium.

6. T75 tissue culture flasks.

7. Serological pipettor.

8. 10 mL serological pipettes.

9. 15 mL centrifuge tubes.

10. Hemocytometer.

11. 1.5 mL vials.

12. P1000, P100, P10, and P2 micropipettes and tips.

13. Sterile incubator maintained at 37 °C and 5% CO_2.

14. Water bath maintained at 37 °C.

2.2 Transfections

1. Lonza Nucleofector 2b electroporator (*see* **Note 8**).

2. Lonza Cell Line Nucleofector Kit V: Nucleofector Cell Line Solution V, Supplement 1, electroporation cuvettes, plastic pipettes (*see* **Note 9**). Combine Nucleofector Cell Line Solution V and Supplement 1 and store at 4 °C.

3. Cellvis 35-mm diameter cell culture dishes with a 10-mm glass-bottom well (*see* **Notes 10** and **11**).

4. Cellvis 29-mm diameter cell culture dishes with a 20-mm glass-bottom well (*see* **Note 12**).

2.3 DNA Constructs: Gβγ Translocation Assay for Gi-Coupled Receptor Activity

1. 1 μg Parapinopsin-Venus cDNA (*see* **Note 13**).

2. 0.1 μg mCh-γ9 cDNA (*see* **Note 14**).

2.4 DNA Constructs: Gβγ Translocation Assay for Gq-Coupled Receptor Activity

1. 1 μg Melanopsin cDNA (*see* **Note 15**).

2. 0.1 μg mCh-γ9 cDNA.

2.5 DNA Constructs: Gβγ Translocation Assay for Gs-Coupled Receptor Activity

1. 1 μg CrBlue cDNA (*see* **Note 16**).

2. 0.1 μg mCh-γ9 cDNA.

2.6 DNA Constructs: PIP₃ Assay for Gi-Coupled Receptor Activity

1. 1 μg Parapinopsin-Venus cDNA.

2. 1 μg PH-AKT-mCh cDNA (*see* **Note 17**).

2.7 DNA Constructs: PIP₂ Hydrolysis Assay for Gq-Coupled Receptor Activity

1. 1 μg Melanopsin cDNA.

2. 1 μg PH-PLCδ-mCh cDNA (*see* **Note 18**).

2.8 DNA Constructs: cAMP Assay for Gs-Coupled Receptor Activity

1. 1 μg CrBlue cDNA.
2. 1 μg GFP-ΔEPAC-mRFP cDNA (*see* **Note 19**).

2.9 DNA Constructs: Cell Migration Assay for Gi-Coupled Receptor Activity

1. 1 μg Parapinopsin-Venus cDNA.

2.10 Imaging

1. Inverted fluorescence microscope with camera and compatible software (*see* **Note 20**).
2. 63× oil immersion lens.
3. 445-nm and 488-nm lasers or LEDs for photoactivation (*see* **Note 21**).
4. 488-nm, 515-nm, and 594-nm lasers for fluorescence imaging (*see* **Note 22**).
5. Software-controlled localized photoactivation unit (*see* **Note 23**).

2.11 Reagents

1. Hank's Balanced Salt Solution (HBSS) supplemented with 1 g/L glucose.
2. 50 mM Stock solution of 11-*cis*-retinal dissolved in ethanol (*see* **Note 24**).

3 Methods

3.1 Cell Culture

1. Culture cells in a T75 flask with the appropriate media in a cell culture incubator maintained at 37 °C and 5% CO_2.
2. Once the cells in a flask are close to confluent (*see* **Note 25**), aspirate the media, add 5–6 mL EDTA, and place the flask in the incubator for 10 min (*see* **Note 26**).
3. Using a serological pipettor and 10 mL pipette, detach the cells from the bottom of the flask by repeatedly triturating (*see* **Note 27**).
4. Transfer the cells in EDTA to a 15 mL centrifuge tube and spin for 90 s at around $210 \times g$ (1173 rpm).
5. Aspirate the EDTA with a vacuum aspirator without disturbing the cell pellet.
6. Add 3 mL of warm media with a 1 mL micropipette and resuspend the cells via gentle trituration.
7. Place 10 μL of cell suspension on a hemocytometer and count to obtain the cell density.

3.2 Transfections (See Note 28)

1. Add 100 μL of Nucleofector V solution containing Supplement 1 to an empty 1.5 mL vial.

2. Add the appropriate amounts of cDNA to the vial.

3. Add 500 μL of warm media to an empty 1.5 mL vial and leave the vial uncapped in the incubator.

4. Calculate the volume of cell suspension corresponding to three million cells and transfer this volume to a 1.5 mL vial (*see* **Note 29**).

5. Centrifuge the vial of cells for 10 min at $100 \times g$.

6. Once centrifugation is complete, aspirate the media carefully so as to not disturb the cell pellet.

7. Using a 1 mL pipette, add the Nucleofector-cDNA mix to the cells and triturate gently to resuspend the cells (*see* **Note 30**).

8. Transfer the cells to a Lonza electroporation cuvette and electroporate cells using the appropriate setting (*see* **Note 31**).

9. Remove the uncapped vial containing 500 μL of media from the incubator, add the media to the electroporation cuvette, and immediately transfer the contents of the electroporation cuvette back into the vial (*see* **Notes 32** and **33**).

10. Place the uncapped vial into the incubator for 10 min.

11. Plate the cells in glass-bottom dishes (*see* **Note 34**).

12. Prior to imaging, aspirate media from the dish and replace with 1 mL of warm HBSS.

13. Working under a red light lamp, add 10 μL of 1 mM 11-*cis*-retinal to the dish and incubate for at least 20 min (*see* **Notes 24** and **35**).

3.3 Imaging: Gβγ Translocation Assay for Gi-Coupled Receptor Activity

For general imaging comments, please *see* **Note 36.**

1. Using the 63× objective, locate a cell expressing mCh-γ9 and Parapinopsin-Venus at the plasma membrane (*see* **Note 37**) and acquire an image of mCh-γ9 using the 594-nm laser.

2. Select a rectangular photoactivation region around all or part of the cell and photoactivate parapinopsin in the selected region with 365–460 nm light (*see* **Notes 21** and **38**).

3. Acquire an image of mCh-γ9 using the 594-nm laser and compare the intracellular mCh intensity with that in the pre-photoactivation image (Fig. 1). An increase in intracellular mCh-γ9 intensity is indicative of parapinopsin and Gi protein activation (*see* **Note 39**).

3.4 Imaging: Gβγ Translocation Assay for Gq-Coupled Receptor Activity

1. Using the 63× objective, locate a cell expressing mCh-γ9 at the plasma membrane and acquire an image using the 594-nm laser.

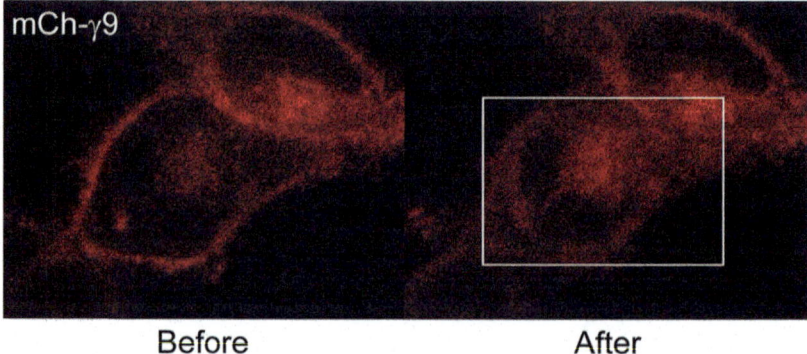

Fig. 1 Gβγ translocation in response to parapinopsin photoactivation with 445-nm laser light in a HeLa cell expressing Parapinopsin-Venus and mCh-γ9. White rectangle represents the photoactivated region

2. Select a rectangular photoactivation region around all or part of the cell and photoactivate melanopsin inside the selected region with 488-nm light (*see* **Note 40**).

3. Acquire an image of mCh-γ9 using the 594-nm laser and compare the intracellular mCh intensity with that in the pre-photoactivation image (*see* **Note 41**). An increase in intracellular mCh-γ9 intensity is indicative of melanopsin and Gq protein activation.

3.5   Imaging: Gβγ Translocation Assay for Gs-Coupled Receptor Activity

1. Using the 63× objective, locate a cell expressing mCh-γ9 at the plasma membrane and acquire an image using the 594-nm laser.

2. Select a rectangular photoactivation region around all or part of the cell and photoactivate CrBlue inside the selected region with 445-nm light (*see* **Note 42**).

3. Acquire an image of mCh-γ9 using the 594-nm laser and compare the intracellular mCh intensity with that in the pre-photoactivation image. An increase in intracellular mCh-γ9 intensity is indicative of CrBlue and Gs protein activation.

3.6   Imaging: PIP₃ Assay for Gi-Coupled Receptor Activity

1. Using the 63× objective, locate a cell expressing Parapinopsin-Venus at the plasma membrane (*see* **Note 37**) and PH-AKT-mCh uniformly in the cytosol.

2. Select a rectangular photoactivation region around all or part of the cell and initiate a protocol in which every 5 s parapinopsin in the selected region is photoactivated with 365–460 nm light and an image is acquired using the 594-nm laser (*see* **Notes 21** and **38**).

3. Compare the distribution of PH-AKT-mCh over time. An increase in PH-AKT-mCh at the plasma membrane in the photoactivated region is indicative of Gi-mediated PIP₃ generation induced by parapinopsin activation.

3.7 Imaging PIP₂ Hydrolysis Assay for Gq-Coupled Receptor Activity

1. Using the 63× objective, locate a cell expressing PH-PLC-δ-mCh at the plasma membrane.

2. Select a rectangular photoactivation region around all or part of the cell and initiate a protocol in which every 5 s melanopsin in the selected region is photoactivated with the 488-nm laser and an image is acquired using the 594-nm laser.

3. Compare the distribution of PH-PLCδ-mCh over time. An increase in cytosolic PH-PLCδ-mCh in the photoactivated region is indicative of Gq-mediated PIP₂ hydrolysis induced by melanopsin activation (*see* **Note 43**).

3.8 Imaging: cAMP Assay for Gs-Coupled Receptor Activity

1. Using the 63× objective, locate a cell expressing cAMP sensor GFP-ΔEPAC-mRFP.

2. Select a rectangular photoactivation region around a cell and initiate an imaging protocol in which every 5 s CrBlue in the selected region is photoactivated with 445-nm laser and images are acquired of the FRET donor and acceptor channels (*see* **Note 44**).

3. Measure the FRET signal over time (*see* **Note 45**). A decrease in FRET is indicative of Gs-mediated cAMP increase induced by CrBlue activation (*see* **Notes 46** and **47**).

3.9 Imaging: Cell Migration Assay for Gi-Coupled Receptor Activity

1. Using the 63× objective, locate a cell expressing Parapinopsin-Venus at the plasma membrane (*see* **Note 37**).

2. Select a rectangular photoactivation region approximately 5 × 15 μm in size and place it at one edge of the cell.

3. Initiate an imaging protocol in which every 5 s parapinopsin in the selected region is photoactivated with 365–460 nm light and fluorescence or brightfield images of the cell are acquired (*see* **Notes 21** and **38**).

4. Lamellipodia extension and movement of the cell toward the photoactivated region is indicative of polarized parapinopsin activation and Gi signaling (*see* **Note 48**). Continuously move the photoactivation region such that it remains in front of the leading edge (Fig. 2) (*see* **Note 49**).

4 Notes

1. We use HeLa cells for the Gβγ translocation, cAMP, and PIP₂ hydrolysis assays because these cells' large flat morphology makes it easy to measure fluorescence intensity changes in the cytosol. Other cells with similar morphology, such as fibroblasts, would also be suitable.

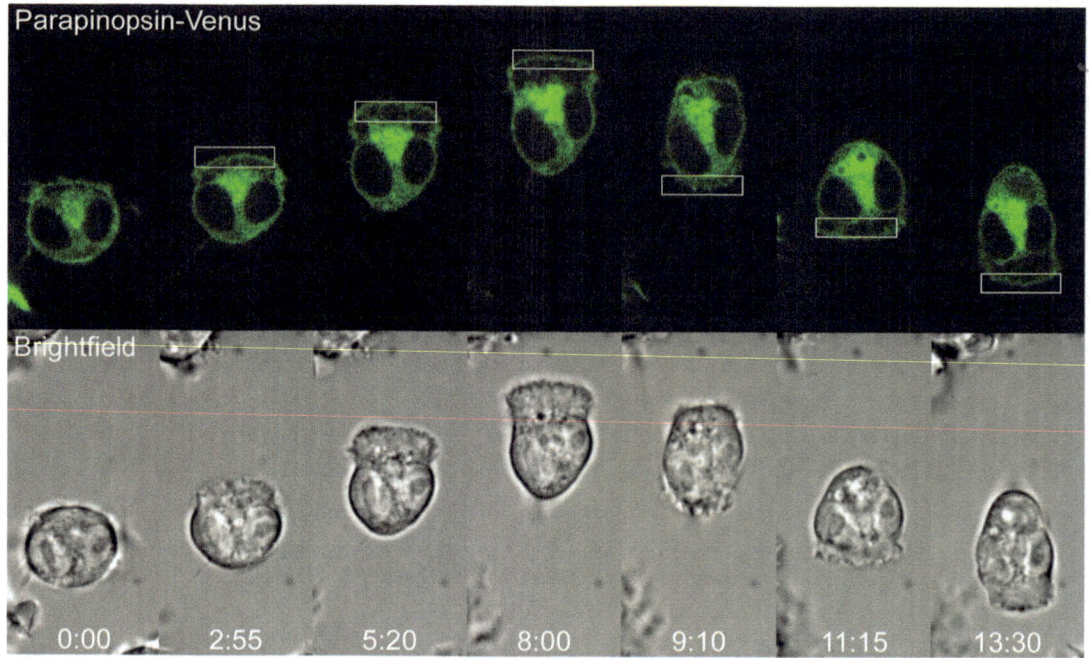

Fig. 2 Directional migration in response to localized parapinopsin activation with 445-nm laser light in a RAW cell expressing Parapinopsin-Venus. White rectangle represents the photoactivated region. Time is in min:sec

2. If using a different cell type, use the recommended media for that cell type.

3. We keep DFBS at −80 °C and thaw it in a 37 °C water bath before use. Once the DFBS is completely thawed, we put it in a 56 °C water bath for 30 min to heat inactivate the complement proteins, followed by 5 min on ice. We then aliquot the DFBS into 50 mL tubes and store at −20 °C.

4. We keep penicillin/streptomycin at −20 °C and thaw it in a 37 °C water bath before use. We then make 5 mL aliquots and store them at −20 °C.

5. To make HeLa media, we combine 445 mL of MEM, 50 mL of DFBS, and 5 mL of penicillin/streptomycin for a total of 500 mL. We aliquot the media into 50 mL tubes and keep at 4 °C until use.

6. We use RAW macrophage cells for cell migration assays because, unlike HeLa cells, macrophages are motile. We also use them for the PIP$_3$ assay because immune cells are known to exhibit an increase in PIP$_3$ in response to activation of Gi-coupled GPCRs [9]. RAW cells also exhibit low levels of basal movement, making it possible to use the migration assay to assess GPCR activity. However, other motile cell types can also be used.

7. To make RAW media, we combine 445 mL of DMEM, 50 mL of DFBS, and 5 mL of penicillin/streptomycin for a total of 500 mL. We aliquot the media into 50 mL tubes and keep at 4 °C until use.

8. We use electroporation for the majority of our transfections. Electroporation is preferred for RAW cells because it results in higher expression levels compared to lipofectamine-based methods. For HeLa cells, electroporation and lipofectamine-based transfections result in similar expression levels.

9. The composition of these reagents is proprietary, so they must be purchased from Lonza. If using a different electroporator, use the reagents recommended for that electroporator. If using a different cell line with the Lonza electroporator, use the recommended Nucleofector solution for that cell line as suggested on the Lonza website.

10. We use dishes with a 0.13–0.16 mm thick glass bottom, which allows us to image with high magnification and short working distance objectives.

11. We use the 10-mm well dishes for RAW cells because RAW cells are typically healthy only during the first 12 h after electroporation. We therefore plate them at a higher density for faster recovery and ease of finding cells to assay.

12. We use the 20-mm well dishes for HeLa cells because HeLa cells are healthy for up to 48 h after transfection (both for electroporation and lipofectamine-based transfections). We often assay the cells for 1–2 days after transfection, and the 20-mm well provides the cells with more area as they divide.

13. Previously, we have used the Gβγ translocation assay with blue opsin, green opsin, and red opsin, the absorption peaks of which are 414 nm, 540 nm, and 560 nm, respectively. These are all Gi-coupled opsins. However, we found that imaging any fluorescent protein in the 445–595 nm excitation range causes activation of green and red opsins. This makes it impossible to have temporal control over GPCR activation while imaging. Blue opsin, on the other hand, is spectrally separate enough from fluorescent proteins in the yellow-red range such that imaging these proteins does not photoactivate it. This makes blue opsin more suitable for imaging-based assays [3]. Parapinopsin is another Gi-coupled opsin that is excited by light in the UV-blue range. It has an added advantage of being bistable, meaning it can be optically switched on and off by different wavelengths of light [10]. For these reasons we prefer to use parapinopsin in our assays. Using Venus-tagged parapinopsin allows us to visually confirm its expression in a cell.

14. Upon GPCR activation, the associated Gα subunit assumes the GTP-bound form and dissociates from the Gβγ subunits. The Gβγ subunits then translocate from the plasma membrane to intracellular membranes. Tagging the γ subunit with mCherry allows us to visualize this translocation in real time through imaging.

15. Melanopsin is a Gq-coupled opsin with an absorption peak around 480 nm [11].

16. CrBlue is a chimeric GPCR, which has the extracellular chromophore-binding domains of blue opsin and the intracellular Gs-coupling domains of jellyfish opsin. Jellyfish opsin's absorption peak is around 500 nm, making it incompatible with live cell imaging assays involving most fluorescent proteins. CrBlue, on the other hand, has the spectral properties of blue opsin while still being capable of activating Gs-dependent signaling pathways [3].

17. Pleckstrin homology (PH) domain of the protein AKT is known to bind to phosphatidylinositol (3,4,5)-trisphosphate (PIP$_3$) and can therefore be used as a sensor for Gi-mediated PIP$_3$ increase [12].

18. Hydrolysis of phosphatidylinositol 4,5-biphosphate (PIP$_2$) to form diacyl glycerol (DAG) and inositol 1,4,5-triphosphate (IP$_3$) is a Gq-mediated process. The PH domain of phospholipase C δ (PLCδ) binds to PIP$_2$ at the plasma membrane. As PIP$_2$ is hydrolyzed, PH-PLCδ remains bound to IP$_3$ and translocates to the cytosol. PH-PLCδ-mCh increase in the cytosol can therefore be used as a sensor for Gq-mediated PIP$_2$ hydrolysis [13].

19. When Gs family proteins are activated by GPCRs, the Gαs subunit activates adenylyl cyclase which catalyzes the conversion of ATP to cAMP. cAMP, in turn, mediates downstream signaling by binding to Exchange Protein Activated by cAMP (EPAC). The conformational change induced in EPAC by cAMP binding allows it to function as a FRET sensor for cAMP. Here, the N-terminal-fused GFP acts as the donor and the C-terminal fused mRFP acts as an acceptor. Upon cAMP binding, EPAC unfolds which leads to an increased distance between GFP and mRFP and, consequently, a decreased FRET signal [14]. This construct therefore functions as a sensor for Gs-mediated cAMP production.

20. For most of the experiments described here, we have used a Leica DMI6000 inverted fluorescence microscope combined with a Yokogawa spinning disc confocal unit. The system is controlled through Andor iQ software, and images are acquired with an Andor iXon EMCCD camera.

21. We use a 445-nm solid-state laser for photoactivation of para-pinopsin and CrBlue and a 488-nm solid-state laser for photo-activation of melanopsin. We have also found that LEDs in the 365–460 nm range can activate parapinopsin with similar effi-ciency and spatiotemporal precision as the laser.

22. Most of our imaging constructs have fluorescent proteins which are excited with light in the yellow-red range, since they can be imaged with 515 and 594 nm lasers without exciting parapinopsin and CrBlue. When working with mela-nopsin, we image with the 594 nm laser only. The 488-nm laser is needed for imaging the cAMP FRET sensor.

23. For most of the experiments described here, we have used an Andor Revolution FRAPPA unit for localized photoactivation. We have also used LEDs coupled to an Andor Mosaic DMD array. Both of these are controlled through Andor iQ software and allow us to photoactivate opsins in select cells or specific regions of a single cell. Subcellular control over GPCR activity is necessary in applications which require polarized signaling, such as cell migration.

24. 11-*cis*-retinal is a chromophore required for opsin activation. It isomerizes to all-*trans* form when exposed to blue light, so great care must be taken to avoid light exposure when working with retinal. We work in a dark room with electronics displays covered and use a red-light lamp for illumination. We found that 9-*cis*-retinal works equally well for opsin-based assays and is commercially available, so we often use it instead of 11-*cis*-retinal.

25. For RAW cells, we find that a T75 flask with 8–10 million cells, three million cells, or one million cells reaches confluency in 1 day, 2 days, or 3 days, respectively. For transfections we typically use RAW cells plated 1 or 2 days prior. We also avoid transfecting RAW cells that have been passaged more than 15 times. For HeLa cells, we find that a T75 flask with six million cells, two million cells, or one million cells reaches confluency in 1 day, 2 days, or 3 days, respectively.

26. For HeLa cells, 5 min of incubation with EDTA is sufficient to detach the cells.

27. Alternatively, a cell scraper can be used to detach cells.

28. The transfection method described here is electroporation-based, as it works well for both RAW and HeLa cells. Occa-sionally, we use a lipofectamine-based method to transfect HeLa cells. We plate 0.1 million cells in 29-mm glass-bottom dishes with 20-mm diameter wells and transfect the cells the next day using standard lipofectamine protocols. The cells can then be imaged the following day. We find that for RAW cells, lipofectamine-based transfections do not yield high enough expression of our desired constructs.

29. Lonza protocols recommend using 1–3 million cells per transfection. We find no decrease in cell viability or transfection efficiency when using three million cells compared to lower amounts.

30. We use the 1 mL pipette for this step since the larger opening in the pipette tip could minimize damage to the cells during the resuspension process.

31. We use Lonza electroporator protocol T-20 for RAW cells and I-013 for HeLa cells. The parameters associated with Lonza protocols are proprietary.

32. This step should be performed quickly. We often remove the media-containing vial from the incubator immediately before electroporating the cells to minimize the time required for this step.

33. If a light-colored aggregate of dead cells is floating in the cuvette after the electroporation, avoid transferring it to the media-containing vial.

34. We typically plate RAW cells in 35-mm diameter glass-bottom dishes with 10-mm diameter wells. We pipette 50–100 μL of the cell suspension into the well, leave the dishes in the incubator for 1 h until cells are adherent, and add 1 mL of warm media to the dish. For HeLa cells, we typically use 29-mm glass-bottom dishes with 20-mm wells and pipette 50–100 μL of the cell suspension into the well, immediately followed by 1 mL of media. Since HeLa cells can be imaged for up to 48 h after transfection, the larger diameter wells allow cells more space to divide.

35. We dissolve retinal in ethanol to a stock concentration of 50 mM and keep aliquots in lightproof vials at −80 °C until use. Prior to use, we dilute the retinal in warm HBSS to a concentration of 1 mM.

36. RAW cells can be imaged 4–10 h following electroporation. HeLa cells can be imaged 4–48 h following electroporation.

37. In our experience, cells transfected with multiple constructs usually express all or none of the constructs. Therefore, if using an untagged opsin, it is usually sufficient to find a cell with good expression of mCh-γ9 or other fluorescently tagged proteins and assume that it expresses the opsin.

38. We find that one pulse of 0.5 μW 445-nm laser light (equivalent to 3% laser power with our laser with the 63× objective in place) scanned over the sample at the rate of 0.8 ms/μm^2 is sufficient to activate parapinopsin with subcellular precision. We find that 5 pulses induce maximal activation of parapinopsin while sacrificing some of the subcellular precision.

39. Alternatively, G protein activation can also be detected using FRET-based methods. We have previously used CFP-tagged αo subunits and YFP-tagged β1 subunits to detect activation of Gi/o family proteins by muscarinic 2 (M2) receptor. Activation of M2 with carbachol induces a decrease in CFP-YFP FRET due to dissociation of the α and β subunits, whereas inhibition of M2 with atropine has the opposite effect. This FRET method was also shown to effectively detect activation of Gq proteins by muscarinic 3 (M3) receptor [5].

40. We find that a single pulse of 488-nm laser light is sufficient to activate melanopsin.

41. Melanopsin-induced βγ translocation reverses quite rapidly ($t_{1/2}$ ~5 s), so it is best to take a snapshot immediately after photoactivation.

42. We find that 5–10 pulses of 445-nm laser light are needed to activate CrBlue.

43. Ca^{2+} sensing can also be used to monitor G protein activity. Gq-mediated IP_3 formation is known to increase cytosolic Ca^{2+} by stimulating Ca^{2+} release from the endoplasmic reticulum [15]. Gi-mediated Ca^{2+} influxes have also been reported [16]. We have previously used the Ca^{2+} sensor Fluo-4 to monitor Ca^{2+} oscillations in response to activation of Gi protein-coupled receptor α2AR by norepinephrine [7].

44. To acquire an image of the donor channel, we excite GFP with the 488-nm laser and image GFP emission with the 515/20 filter in place. To acquire an image of the acceptor channel, we excite GFP with the 488-nm laser and image mRFP emission with the 628/20 filter in place.

45. FRET signal can be calculated as (Acceptor Fluorescence)/(Donor Fluorescence).

46. Gs-mediated cAMP increase can also be induced pharmacologically. We found that addition of 25 μM forskolin and 100 μM phosphodiesterase inhibitor, IBMX, induces a significant FRET decrease in the GFP-ΔEPAC-mRFP cAMP sensor. Forskolin stimulates adenylate cyclase which leads to cAMP increase. IBMX inhibits degradation of cAMP by phostphodiesterases [3].

47. Gi-mediated signaling is known to inhibit adenylyl cyclase, which in turn inhibits cAMP [17]. Thus, cAMP sensors can also potentially be used to detect Gi-coupled GPCR activity.

48. Polarized activation of the Gi signaling is known to induce directional migration in macrophages [18]. Activation of Rac and Cdc42 at the leading edge promotes lamellipodia formation and extension, whereas RhoA activity at the trailing edge promotes retraction [19]. Together, these polarized signals cause directional movement of the cell.

49. In neurons, the same polarized activation of parapinopsin results in neurite extension and growth cone formation [3]. This can serve as another cell behavior-based assay for parapinopsin activation and Gi-mediated signaling.

Acknowledgments

This work was funded by the NIH through NIGMS grants GM069027, GM107370, and GM122577. We thank Akihisa Terakita and Mitsumasa Koyanagi for parapinopsin cDNA and Ignacio Provencio for melanopsin cDNA.

References

1. Hauser AS, Attwood MM, Rask-Andersen M, Schioth HB, Gloriam DE (2017) Trends in GPCR drug discovery: new agents, targets and indications. Nat Rev Drug Discov 16:829–842. https://doi.org/10.1038/nrd.2017.178

2. Saini DK, Kalyanaraman V, Chisari M, Gautam N (2007) A family of G protein betagamma subunits translocate reversibly from the plasma membrane to endomembranes on receptor activation. J Biol Chem 282:24099–24108. https://doi.org/10.1074/jbc.M701191200

3. Karunarathne WK, Giri L, Kalyanaraman V, Gautam N (2013) Optically triggering spatiotemporally confined GPCR activity in a cell and programming neurite initiation and extension. Proc Natl Acad Sci U S A 110:E1565–E1574. https://doi.org/10.1073/pnas.1220697110

4. Akgoz M, Kalyanaraman V, Gautam N (2004) Receptor-mediated reversible translocation of the G protein betagamma complex from the plasma membrane to the Golgi complex. J Biol Chem 279:51541–51544. https://doi.org/10.1074/jbc.M410639200

5. Azpiazu I, Gautam N (2004) A fluorescence resonance energy transfer-based sensor indicates that receptor access to a G protein is unrestricted in a living mammalian cell. J Biol Chem 279:27709–27718. https://doi.org/10.1074/jbc.M403712200

6. Karunarathne WK, Giri L, Patel AK, Venkatesh KV, Gautam N (2013) Optical control demonstrates switch-like PIP3 dynamics underlying the initiation of immune cell migration. Proc Natl Acad Sci U S A 110:E1575–E1583. https://doi.org/10.1073/pnas.1220755110

7. Giri L, Patel AK, Karunarathne WK, Kalyanaraman V, Venkatesh KV, Gautam N (2014) A G-protein subunit translocation

embedded network motif underlies GPCR regulation of calcium oscillations. Biophys J 107:242–254. https://doi.org/10.1016/j.bpj.2014.05.020

8. Karunarathne WK, O'Neill PR, Gautam N (2015) Subcellular optogenetics—controlling signaling and single-cell behavior. J Cell Sci 128:15–25. https://doi.org/10.1242/jcs.154435

9. Servant G, Weiner OD, Herzmark P, Balla T, Sedat JW, Bourne HR (2000) Polarization of chemoattractant receptor signaling during neutrophil chemotaxis. Science 287:1037–1040. https://doi.org/10.1126/science.287.5455.1037

10. Kawano-Yamashita E, Koyanagi M, Wada S, Tsukamoto H, Nagata T, Terakita A (2015) Activation of transducin by bistable pigment parapinopsin in the pineal organ of lower vertebrates. PLoS One 10:e0141280. https://doi.org/10.1371/journal.pone.0141280

11. Kumbalasiri T, Provencio I (2005) Melanopsin and other novel mammalian opsins. Exp Eye Res 81:368–375. https://doi.org/10.1016/j.exer.2005.05.004

12. James SR, Downes CP, Gigg R, Grove SJ, Holmes AB, Alessi DR (1996) Specific binding of the Akt-1 protein kinase to phosphatidylinositol 3,4,5-trisphosphate without subsequent activation. Biochem J 315 (Pt 3):709–713. doi:https://doi.org/10.1042/bj3150709

13. Stauffer TP, Ahn S, Meyer T (1998) Receptor-induced transient reduction in plasma membrane PtdIns(4,5)P2 concentration monitored in living cells. Curr Biol 8:343–346. https://doi.org/10.1016/s0960-9822(98)70135-6

14. van der Krogt GN, Ogink J, Ponsioen B, Jalink K (2008) A comparison of donor-acceptor pairs for genetically encoded FRET sensors:

application to the Epac cAMP sensor as an example. PLoS One 3:e1916. https://doi.org/10.1371/journal.pone.0001916

15. Bootman MD (2012) Calcium signaling. Cold Spring Harb Perspect Biol 4:a011171. https://doi.org/10.1101/cshperspect.a011171

16. Dorn GW 2nd, Oswald KJ, McCluskey TS, Kuhel DG, Liggett SB (1997) Alpha 2A-adrenergic receptor stimulated calcium release is transduced by Gi-associated G(beta gamma)-mediated activation of phospholipase C. Biochemistry 36:6415–6423. https://doi.org/10.1021/bi970080s

17. Dessauer CW, Chen-Goodspeed M, Chen J (2002) Mechanism of Galpha i-mediated inhibition of type V adenylyl cyclase. J Biol Chem 277:28823–28829. https://doi.org/10.1074/jbc.M203962200

18. O'Neill PR, Gautam N (2014) Subcellular optogenetic inhibition of G proteins generates signaling gradients and cell migration. Mol Biol Cell 25:2305–2314. https://doi.org/10.1091/mbc.E14-04-0870

19. Ridley AJ (2001) Rho GTPases and cell migration. J Cell Sci 114(Pt 15):2713–2722

Split-Tobacco Etch Virus (Split-TEV) Method in G Protein-Coupled Receptor Interacting Proteins

Marta Alonso-Gardón and Raúl Estévez

Abstract

Split-TEV assay enables the identification of protein–protein interaction in mammalian cells. This method is based on the split of tobacco etch virus (TEV) protease in two fragments, where each fragment is fused to the candidate proteins predicted to interact. If there is indeed an interaction between both proteins, TEV protease reconstitutes its proteolytic activity and this activity is used to induce the expression of some reporter genes. However, some studies have detected unspecific interaction between membrane proteins due to its higher tendency to aggregate. Here we describe a variation of the Split-TEV method developed with the aim to increase the specificity in the study of G protein-coupled receptor (GPCR) interacting proteins. This approach for monitoring interactions between GPCRs is an easy and robust assay and offers good perspectives in drug discovery.

Key words Split-TEV, G protein-coupled receptor, Protein, protein interaction, Membrane proteins

1 Introduction

1.1 The Split-TEV Method

The split-tobacco etch virus method or Split-TEV is a protein complementation technique used to detect interaction between proteins [1]. The method is based on the split of TEV protease in two fragments, where each fragment is fused to the candidate proteins that are going to be analyzed for interactions. If there is interaction between both proteins, the tobacco virus protease reconstitutes its proteolytic activity and this activity is used to transcriptionally activate the expression of different reporter genes [2].

The modification of the original split-TEV method [3] that we implemented in our laboratory specifically for membrane proteins [4] consists on the fusion of one of the studied proteins to an artificial protein containing the N-TEV terminal fragment, the TEV protease recognition site, and the chimeric transcription factor GV. The expression of this protein is controlled under the strong

Sofia Aires M. Martins and Duarte Miguel F. Prazeres (eds.), *G Protein-Coupled Receptor Screening Assays: Methods and Protocols*, Methods in Molecular Biology, vol. 2268, https://doi.org/10.1007/978-1-0716-1221-7_15,

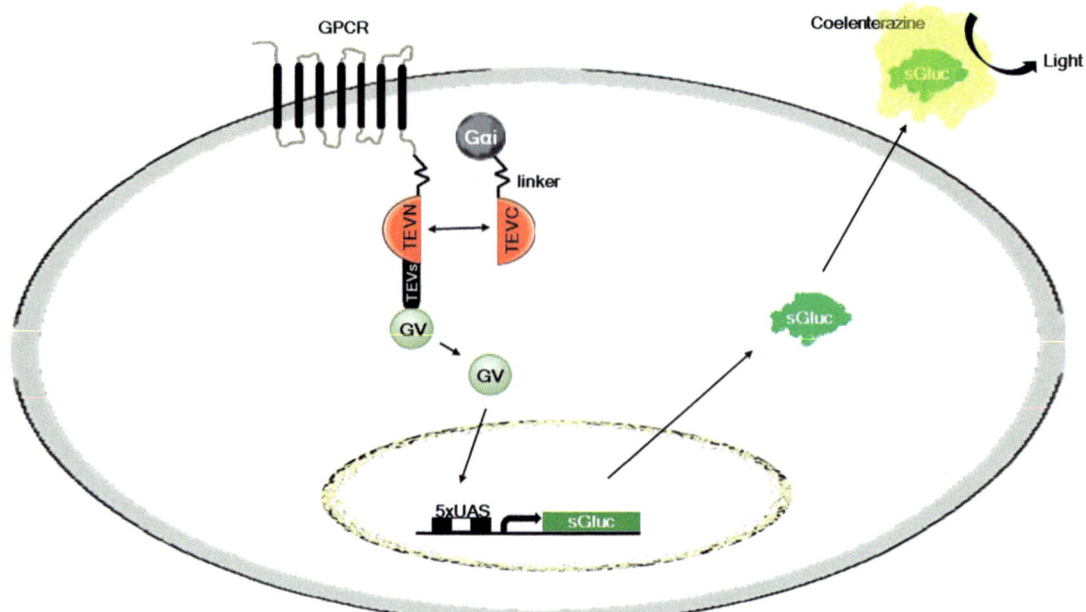

Fig. 1 Schematic representation of GPCR and Gαi subunit interaction by Split-TEV method. The GPCR is fused through a flexible linker to the TEV-N (residues 1–118) terminal fragment, the TEV recognition site, and the chimeric transcription factor GV, under the control of a CMV promoter. The other protein of interest, Gαi, is fused with a flexible linker to the TEV-C (residues 119–242) fragment under the control of a TK promoter. The interaction between the GPCR and Gαi subunit allows the reconstitution of TEV protease activity and the release of the GV transcription factor. Liberated GV enters the nucleus and activates the expression of the gene reporter *Gaussia luciferase* (Gluc) through binding to Gal4 responsive elements (5×UAS). The secretable form of Gluc is released to the culture medium where it reacts with coelenterazine, producing light that is detected with a luminometer

promoter cytomegalovirus (CMV). The other studied protein is fused just to the C-TEV terminal fragment. In this case, the HSV-TK promoter controls the expression of this fusion protein, which provides low levels of expression. This low expression was found to be very important to reduce unspecific interactions between membrane proteins [4]. If both proteins interact with each other, the TEV reconstitutes its activity proteolyzing the TEV protease recognition site and releasing the transcription factor that reaches the cell nucleus. Subsequently, the transcription of the reporter gene is activated generating a measurable activity (Fig. 1).

Although the Split-TEV method has been previously described as a technique that could identify interactions between cytosolic and membrane proteins [3], our group found that when analyzing interactions between membrane proteins using the described method, a high percentage of false interactions arise [4]. It was suggested [3] that playing with different cell lines, transfection efficiencies, and different expression levels could be a strategy to avoid false interactions. Therefore, we performed several modifications on the original described Split-TEV assay in order to increase the specificity of membrane protein interactions in a defined manner and in an easy way.

Thus, and apart from fusing different promoters for each of the fused fragments, in our configuration of the method, we selected different options that were already originally performed [3], or introduced several modifications [4]. These options are detailed below.

1. The transcription factor of choice was GAL4-VP16 (vector pBT3-N; P03230) (GV), which contains the DNA-binding domain of the yeast GAL4 promoter and the Herpes simplex virus VP16 transactivation domain. The reporter gene selected was carried by the pNEBR-X1 Gluc plasmid that contains the cDNA of a secretable form of the highly active Gaussia luciferase (Gluc). This plasmid is composed of a UAS sequence (Upstream Activation Sequence) that has 5 specific elements of response to the yeast GAL4 promoter, ensuring high levels of expression. Furthermore, Gluc presents a high stability and its activity is quite resistant to changes in pH or ionic strength [5]. Furthermore, measurements of interaction are done from the supernatant, which is very convenient and affordable.

2. Considering the TEV protease, we worked with the TEV S219V variant in order to prevent self-proteolysis inhibition of TEV protease. As the catalytic domain of the TEV tends to cut itself between residues 218 and 219 in vitro, this S219V mutant is more stable. Moreover, the K_M of TEV S219V ($K_M = 0.041$) is lower than the K_M of the wild-type variant ($K_{MTEV} = 0.061$), thus having an increased affinity for its substrate [6]. In this assay, the TEV protease is divided in two halves: the N-terminal fragment (N-TEV, amino acids 1–118) and the C-terminal fragment (C-TEV, amino acids 119–242). The complementation of both fragments results in a 40% recovery of catalytic activity in relation to the whole TEV, which is the highest percentage of activity found in TEV fragments [3]. These two fragments are fused to our proteins of interest by means of a linker sequence of glycines and serines (GGGGSGGGGS) that allows greater mobility of the construct, a feature that is needed for TEV fragments reconstitution.

3. Regarding the TEV recognition substrate, we chose to work with a high-affinity enzyme substrate (TEVs) whose sequence is ENLYFQS. TEVs has a high K_{cat}/K_M ratio ($4.51 \, \mathrm{mM}^{-1} \, \mathrm{s}^{-1}$) with the variant S219V TEV compared to other substrates [7].

4. Finally, in order to normalize the transfection efficiency in each experimental condition, we co-transfected cells with the pCMV-βGal plasmid and monitored beta-galactosidase activity using a luminescence kit.

1.2 Interactions
with GPCRs

Interactions between transmembrane proteins are essential for most physiological and cellular functions such as cognition, inflammation, metabolism, cell proliferation, or differentiation [8]. The most abundant proteins of the cell surface that regulate these processes in eukaryotes are the seven-transmembrane receptor family known as the G protein coupled receptor (GPCR) family. The GPCR diversity is dramatically high, which allows cells to react to a variety of stimuli ranging from hormones to neurotransmitters [9]. As a result, these receptors are important targets in the development of many therapeutic drugs [10]. Agonist stimulation allows receptor's activation and transmits extracellular signals across the plasma membrane after coupling with heterotrimeric G proteins (a combination between one of the 16Gα, 5Gβ and 13Gγ described subunits), which initiate a cascade of downstream second messengers pathways [11]. Sustained GPCR activation prompts the phosphorylation of their carboxyl terminus and intracellular loops. This promotes the recruiting of the multifunctional protein β-arrestin, which downregulates the GPCR from the membrane [12].

In this chapter, we detail the application of the Split-TEV approach to study the interaction of the sphingosine-1-phosphate receptor (S1P1R) model GPCR. S1P1R has a high affinity for the ligand sphingose 1-phosphate (S1P) [13], which activates the receptor and stimulates Gi-dependent intracellular signaling cascades [14]. Thus, we describe the study of S1P1R interaction with the Gαi subunit and β-arrestin upon receptor activation as an example of our method for GPCRs. We also include the study of S1P1R homo-oligomerization because different studies have demonstrated that GPCRs are able to homo-oligomerize in many cell types [15].

We used a fusion of S1P1R with N-TEV fragment, TEV protease recognition site, and transcription factor GV (S1P1R-linker-NTEV-TEVs-GV). All the other proteins were fused to the C-TEV fragment (S1P1R-linker-CTEV, Gαi-linker-CTEV, β-linker-arrestin-CTEV). We used the surface antigen 4F2hc as a negative control, since it is a membrane protein that holds no relationship with S1P1R.

2   Materials

2.1 Plasmids

All expression plasmids were constructed by polymerase chain reaction (PCR) using KOD Hot Start polymerase and the Multisite Gateway System. For a scheme of the constructs generated see Fig. 2.

1. To clone the PCR products, we added the attB1, attB2, attB5r, or attB5 recombination sites to our proteins of interest. We then obtained the Entry Clones by a BP reaction (Gateway

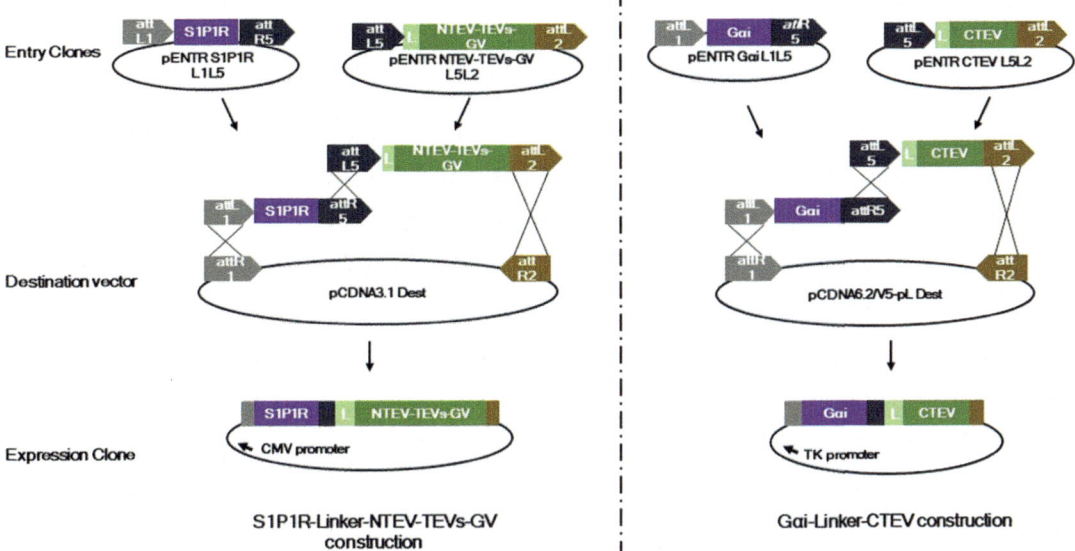

Fig. 2 Schematic representation of the Gateway Multisite System (Invitrogen) cloning two types of fusion proteins used in Split-TEV method. Left: combination of two different entry clones, one containing attL1,R5 sequences (pENTR S1P1R L1R5) and the other one containing attL5,2 sequence (pENTR TEV-N-TEVs-GV L5L2) with both being cloned into a destination vector containing attR1,2 recombination sequences. The destination vector used is pCDNA3.1, which contains a CMV promoter. Right: schematic combination of pENTR Gαi L1R5 and pENTR TEV-C L5L2, cloned into pCDNA6.2/V5-pL destination vector (attR1, 2) with a TK promoter, resulting in low expression of the protein. Similar reactions are performed for S1P1R-Linker-TEV-C, 4F2-Linker-TEV-C, and β-arrestin-Linker-TEV-C proteins

cloning®), which were later recombined with pcDNA3.1 Dest vector or pcDNA6.2/V5-pL Dest vector containing the HSV-TK promoter by a LR reaction.

2. S1P1R-Linker-NTEV-TEVs-GV plasmid. S1P1R is fused to N-TEV fragment, the TEV protease recognition site, and the chimeric transcription factor GV. They were cloned in a pcDNA3.1 Dest vector (*see* **Note 1**).

3. S1P1R-Linker-CTEV, ARRB2-Linker-CTEV, Galphai-Linker-CTEV, and 4F2-Linker-CTEV plasmids. Each protein is fused to the C-TEV fragment and they are cloned in the pcDNA6.2/V5-pL Dest vector (*see* **Note 2**).

4. Reporter gene pNEBr-X1Gluc.

5. pCMV-βGal vector. This plasmid was used to normalize transfection levels. (Vector obtained from β-Galactosidase Detection Kit II).

2.2 Cell Culture

1. We used HeLa cells or HEK293T cells although other heterologous cell lines may be used (*see* **Note 3**).

2. Cell culture medium: Dulbecco's modified Eagle's medium (DMEM), supplemented with 1 mM sodium pyruvate, 2 mM L-glutamine, 100 U/mL streptomycin, 100 mg/mL penicillin, and 5% (v/v) fetal bovine serum (FBS).

3. Phosphate-Buffered Saline (PBS): 137 mM NaCl, 2.7 mM KCl, 10 mM Na_2HPO_4, 2 mM KH_2PO_4, pH 7.4.

4. Cell detachment solution: Trypsin-EDTA 1×.

5. Neubauer chamber.

6. 100 mm Culture plate.

7. 6-Well culture plates.

8. OptiMEM Glutamax medium.

9. Cell transfection reagent: transfectin.

10. Expression plasmids coding for S1P1R-Linker-NTEV-TEVs-GV, S1P1R-Linker-CTEV, Galphai-Linker-CTEV, and Arrestin-Linker-CTEV.

2.3 Split-TEV Assay

1. Native coelenterazine solution: 5 μM.

2. Cell scraper.

3. Lysis solution: 1% Triton X-100, 150 mM NaCl, 1 mM PMSF, 1 mg/L leupeptin/pepstatin, and 2 mg/L aprotinin in PBS.

4. β-Galactosidase Detection Kit II.

2.4 Equipment

1. Cell culture facility including flow hood and incubator set to 5% CO_2 and 37 °C.

2. TD-20/20 Luminometer.

3. GraphPad Prism software.

3 Methods

3.1 HeLa Cell Culture and Transfection

Grow HeLa cells in a 100 mm culture plate with culture medium at 37 °C, 90% relative humidity, and 5% CO_2.

1. On day 1, rinse the cells with sterile PBS and incubate them with 1 mL of a trypsin-EDTA solution for 5 min at 37 °C.

2. Resuspend the cells in 10 mL of culture medium detaching mechanically the cells from the plate using a pipette.

3. Count the cells in a Neubauer chamber and seed 400,000 cells per well in a 6-well plate (*see* **Note 4**). Add culture medium to a final volume of 2 mL per well.

4. Leave the 6-well plate for 24 h at 37 °C and 5% CO_2.

5. On day 2, transiently transfect the cells with 1 μg of two plasmid DNAs containing the proteins of interest fused to N-TEV or C-TEV. Co-transfect cells with 0.3 μg of the pNEBr-X1Gluc vector containing the reporter gene and 0.27 μg of the pCMV-βGal vector. Transfection is performed following the Transfectin™ protocol (*see* **Notes 5** and **6**).

6. Wash the cultured cells with PBS and add 1.5 mL of OptiMEM.

7. Add the DNA-transfectin mixture carefully to each well and incubate for 4 h at 37 °C and 5% CO_2.

8. Remove the OptiMEM medium, wash the cells with PBS, and add 2 mL of culture medium in each well.

9. Incubate the cells for 48 h (*see* **Note 7**) in order to reach optimal expression of the transfected cDNAs.

10. On day 4, recover 20 μL of supernatant per each transfection condition from cell plates and place in an Eppendorf.

11. Add 20 μM native coelenterazine (in PBS) to each supernatant and immediately analyze them in the luminometer.

12. In order to normalize the coelenterazine readings with the transfection levels, solubilize each cell experimental group with lysis solution (100 μL for 6-well plates).

13. Scrap the cells with the cell scraper, resuspend the cell lysate, and transfer to a 1.5 mL Eppendorf.

14. Incubate the cell lysates for 1 h at 4 °C in an orbital.

15. Centrifuge the cell lysates at $18,400 \times g$ for 10 min at 4 °C.

16. Recover the supernatant, which contains the soluble proteins.

17. Mix 30 μL of each solubilized with 200 μL of β-Galactosidase (196 μL Reaction buffer plus 4 μL (Reaction Substrate from the β-Galactosidase Detection Kit II).

18. Incubate the β-Galactosidase reactions at RT for 1 h.

19. Record the light emitted in the luminometer twice (*see* **Note 8**).

20. Correct the coelenterazine luminescence readings obtained in each experimental group obtained with the mean values of β-Galactosidase readings obtained in each group (*see* **Note 9**).

21. Generate a bar graph using GraphPad Prism program to visualize the data and calculate mean, standard deviations, and standards errors of the mean.

22. Determine the statistical significance between groups using one-way ANOVA with Bonferroni multiple-comparison tests. For an example of different measurements see Fig. 3.

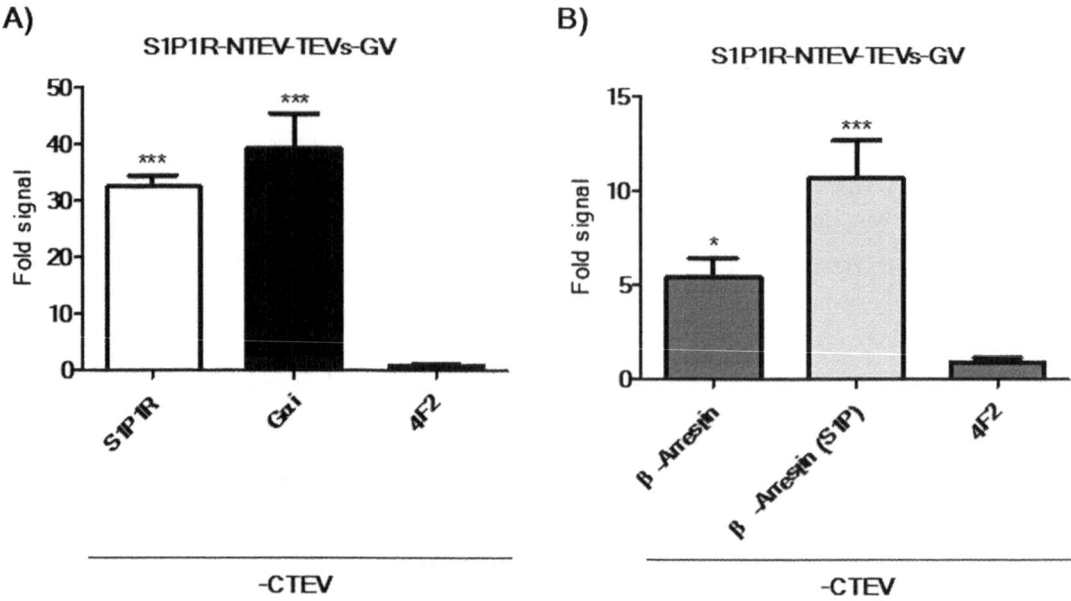

Fig. 3 Split-TEV assay in HeLa cells to detect S1P1R interacting proteins. The title of the graph indicates the GPCR fused to TEV-N fragment, the TEV recognition site, and the transcription factor, and on the bottom is shown the other protein candidate to interact fused to the TEV-C fragment under the control of a low-expression promoter TK. (**a**) S1P1R constitutively interacts with itself (homo-oligomerization) and with the Gαi subunit. (**b**) Interaction of S1P1R with β-arrestin upon receptor activation with Sphingosine 1-Phosphate (S1P). Data correspond to at least three independent experiments. $*P < 0.05$ and $***P < 0.001$ in Bonferroni's multiple comparison test against the negative control 4F2

4   Notes

1. For cloning two different DNAs in one final vector, the first construct is generated in an entry clone with the sequences attB1 and attB5r, whereas the second fragment is introduced in an entry clone with the attB5 and attB2 sequences. The STOP codon of the first cDNA amplified (attB1, attB5r) needs to be removed so that the final recombinant protein can be expressed. The reaction is made with enzyme LR clonase II+.

2. Depending on the protein one may clone the TEV at the N terminus or the C-terminus. Different results may be obtained, so it is important to analyze what is the better combination to perform this technique.

3. It is important to choose properly the heterologous cell line because expression levels could change between different expression systems.

4. For a better transfection efficiency, the cells are seeded at a 35% confluence, in order to reach a 70–80% confluence within the following 24 h, which is the optimal confluence for the transfection procedure.

5. It is important to transfect one negative control group with only the protein fused to N-TEV and the GV transcription factor. The N-TEV-protein alone may present certain degree of auto-proteolysis and could give background signals that have to be subtracted.

6. To transfect cell in each well prepare a total of 2.57 μg cDNA and mix in 250 μL of OptiMEM. In another Eppendorf mix 250 μL of OptiMEM with 2.57 μL of Transfectin lipidic reagent (1:1 DNA-Lipidic reagent solution). Then mix the DNA-OptiMEM and Transfectin-OptiMEM together by gently pipetting and tapping and incubate it for 20 min at RT.

7. For best results, conduct the Split-TEV recordings 24–48 h after transfection.

8. Process from the first sample to the last recording the light emission in the luminometer twice, in the same order. β-Galactosidase activity remains more than 1 h after incubation, and in this way an average of the 2 values obtained can give a more accurate measure.

9. The construct containing the fused protein in the N-TEV fragment + TEV recognition site + the GV transcription factor must be transfected alone to give the background signal to each protein (may be different in each construct). The luminescence value divided by the galactosidase activity value obtained for the N-TEV-construct alone is used as a reference to calculate the induction times of the signal obtained in the experimental groups with respect to this group that is considered the basal signal (induction fold).

References

1. Wehr MC, Galinski S, Rossner MJ (2015) Monitoring G protein-coupled receptor activation using the protein fragment complementation technique split TEV. Methods Mol Biol 1272:107–118. https://doi.org/10.1007/978-1-4939-2336-6_8

2. Djannatian MS, Galinski S, Fischer TM, Rossner MJ (2011) Studying G protein-coupled receptor activation using split-tobacco etch virus assays. Anal Biochem 412:141–152. https://doi.org/10.1016/j.ab.2011.01.042

3. Wehr MC, Laage R, Bolz U et al (2006) Monitoring regulated protein-protein interactions using split TEV. Nat Methods 3:985–993. https://doi.org/10.1038/nmeth967

4. Capdevila-Nortes X, López-Hernández T, Ciruela F, Estévez R (2012) A modification of the split-tobacco etch virus method for monitoring interactions between membrane proteins in mammalian cells. Anal Biochem 423:109–118. https://doi.org/10.1016/j.ab.2012.01.022

5. Rathnayaka T, Tawa M, Sohya S et al (2010) Biophysical characterization of highly active recombinant Gaussia luciferase expressed in Escherichia coli. Biochim Biophys Acta 1804:1902–1907. https://doi.org/10.1016/j.bbapap.2010.04.014

6. Kapust RB, Tözsér J, Fox JD et al (2001) Tobacco etch virus protease: mechanism of autolysis and rational design of stable mutants with wild-type catalytic proficiency. Protein Eng 14:993–1000. https://doi.org/10.1093/protein/14.12.993

7. Tözsér J, Tropea JE, Cherry S et al (2005) Comparison of the substrate specificity of two potyvirus proteases. FEBS J 272:514–523. https://doi.org/10.1111/j.1742-4658.2004.04493.x

8. Petschnigg J, Snider J, Stagljar I (2011) Inter-active proteomics research technologies: recent applications and advances. Curr Opin Biotechnol 22:50–58

9. Alexander SPH, Christopoulos A, Davenport AP et al (2019) The concise guide to pharmacology 2019/20: G protein-coupled receptors. Br J Pharmacol 176:S21–S141. https://doi.org/10.1111/bph.14748

10. Schöneberg T, Schulz A, Biebermann H et al (2004) Mutant G-protein-coupled receptors as a cause of human diseases. Pharmacol Ther 104:173–206

11. Weis WI, Kobilka BK (2018) The molecular basis of G protein–coupled receptor activation. Annu Rev Biochem 87:897–919. https://doi.org/10.1146/annurev-biochem-060614-033910

12. Gurevich VV, Gurevich EV (2019) GPCR signaling regulation: the role of GRKs and arrestins. Front Pharmacol 10. https://doi.org/10.3389/fphar.2019.00125

13. Rosen H, Stevens RC, Hanson M et al (2013) Sphingosine-1-phosphate and its receptors: structure, signaling, and influence. Annu Rev Biochem 82:637–662. https://doi.org/10.1146/annurev-biochem-062411-130916

14. Cahalan SM, Gonzalez-Cabrera PJ, Sarkisyan G et al (2011) Actions of a picomolar short-acting S1P 1 agonist in S1P 1-eGFP knock-in mice. Nat Chem Biol 7:254–256. https://doi.org/10.1038/nchembio.547

15. Maggio R, Innamorati G, Parenti M (2007) G protein-coupled receptor oligomerization provides the framework for signal discrimination. J Neurochem 103:1741–1752

Chapter 16

NanoLuc-Based Methods to Measure β-Arrestin2 Recruitment to G Protein-Coupled Receptors

Xiaoyuan Ma, Rob Leurs, and Henry F. Vischer

Abstract

Cytosolic β-arrestins are key regulators of G protein-coupled receptors (GPCRs) by sterically uncoupling G protein activation, facilitating receptor internalization, and/or acting as G protein-independent signaling scaffolds. The current awareness that GPCR ligands may display bias toward G protein signaling or β-arrestin recruitment makes β-arrestin recruitment assays important additions to the drug discovery toolbox. This chapter describes two NanoLuc-based methods to monitor β-arrestin2 recruitment to the human histamine H_1 receptor by measuring bioluminescence resonance energy transfer and enzyme-fragment complementation in real-time on living cells with reasonable high throughput. In addition to the detection of agonism, both assay formats can be used to qualitatively evaluate the binding kinetics of antihistamines on the human histamine H_1 receptor.

Key words GPCR, β-Arrestin, Bioluminescence resonance energy transfer (BRET), Enzyme-fragment complementation (EFC), Luciferase, Protein-protein interaction (PPI)

1 Introduction

G protein-coupled receptors (GPCRs) are cell surface receptors that induce intracellular signaling upon binding of an extracellular ligand and are involved in regulation of nearly all cellular processes in our body [1, 2]. This central role in (patho)physiology has made the GPCR family an attractive target for therapeutic interventions with approximately one-third of the currently marketed drugs targeting this receptor family [1, 2].

GPCRs predominantly signal via heterotrimeric G proteins that consist of Gα and Gβγ subunits. Most GPCRs preferentially couple to one of the four Gα subfamilies (i.e., $Gα_s$, $Gα_{i/o}$, $Gα_{q/11}$, and $Gα_{12/13}$) resulting in activation of different downstream signaling pathways. Subsequent phosphorylation of the C-terminal tail of activated GPCRs by G protein-coupled receptor kinases (GRKs) triggers the recruitment of β-arrestin1 and/or β-arrestin2, which

Sofia Aires M. Martins and Duarte Miguel F. Prazeres (eds.), *G Protein-Coupled Receptor Screening Assays: Methods and Protocols*, Methods in Molecular Biology, vol. 2268, https://doi.org/10.1007/978-1-0716-1221-7_16,
© Springer Science+Business Media, LLC, part of Springer Nature 2021

desensitizes the GPCR by preventing further G protein coupling due to steric hindrance [3]. In addition, GPCR-bound β-arrestin facilitates receptor internalization by scaffolding components of the cellular internalization machinery (e.g., clathrin and the AP2 adaptor complex) [4], but can also facilitate mitogen-activated protein kinase signaling (e.g., ERK1/2, p38, and c-Jun) by acting as adaptor protein to compartmentalize the involved kinases [4]. Traditionally in drug discovery, GPCR activity has been predominantly focused on G protein-mediated second messenger formation and associated activation of downstream transcription factors using well-established biochemical and reporter gene assays, respectively. However, in the last two decades genetically engineered biosensor constructs have been developed to monitor the interaction between GPCRs and β-arrestins in living cells using multilabel microplate readers. The availability of these functional assays to measure these protein–protein interactions in a (reasonably) high-throughput format has contributed to the discovery of biased GPCR ligands that preferentially activate G protein- or β-arrestin-mediated responses at the relative expense of the other [5]. Biased ligands are considered to have better therapeutic value for a number of GPCRs as compared to unbiased ligands as they can separate desirable from adverse on-target responses (Fig. 1) [6].

Bioluminescence resonance energy transfer (BRET) and enzyme-fragment complementation (EFC) technologies have been most frequently applied to measure ligand-induced interactions between GPCRs and β-arrestin2 [7]. BRET relies on the resonance energy transfer from a bioluminescent donor luciferase to a compatible acceptor fluorescent protein that is in very close proximity (<10 nm) upon luciferase-substrate conversion and resulting in light emission from the fluorescent acceptor. In particular, the Renilla luciferase (Rluc) variant Rluc8 in combination with green-fluorescent protein (GFP) variants eYFP/Venus have been widely applied as BRET donor-acceptor pair in engineered fusion proteins to measure protein–protein interactions [7, 8].

The EFC method relies on the dissection of an enzyme into two inactive fragments that are genetically fused to proteins of interest. These fragments can refold into an active enzyme if brought in close proximity by the interaction of the proteins to which they are fused [7]. In particular, split β-galactosidase, firefly, and click beetle luciferase have been used to monitor GPCR interactions with β-arrestin2.

More recently, a novel luciferase Nanoluc (Nluc) is increasingly used in both BRET- and EFC-based assays (i.e., NanoBRET and Nluc Binary Technology (NanoBIT), respectively) to study protein–protein interaction [9, 10]. Nluc was engineered from the small subunit of the heteromeric luciferase from deep sea shrimp *Oplophorus gracilirostris* to increase its stability in mammalian cells [11]. In combination with its codeveloped novel

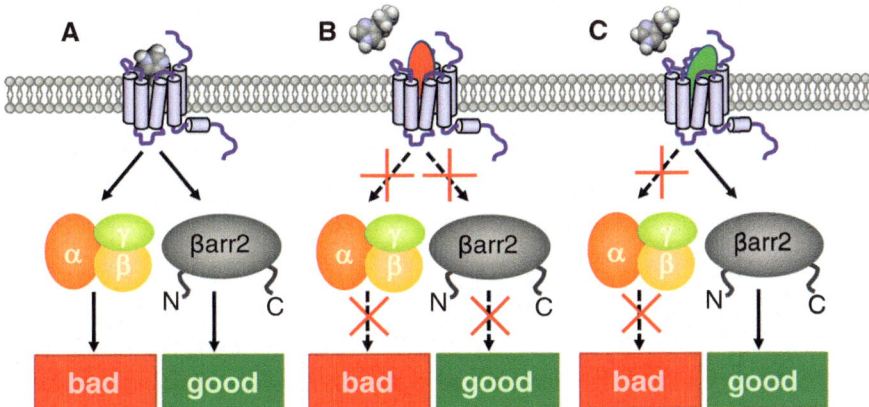

Fig. 1 Biased signaling at GPCRs. (**a**) Unbiased endogenous agonist induces equal G protein- and β-arrestin2-mediated signaling that in this hypothetical example results in physiologically detrimental and beneficial responses, respectively. (**b**) Blocking binding of the unbiased endogenous agonist by an antagonist silences both receptor-mediated adverse and desired responses. (**c**) The β-arrestin2-biased agonist selectively activates β-arrestin2-mediated responses but not G protein signaling. Consequently, it can antagonize detrimental G protein signaling by the endogenous unbiased agonist while inducing beneficial β-arrestin2-mediated responses by itself

coelenterazine-based substrate furimazine, Nluc emits 150-fold brighter bioluminescence with a ~20% narrower spectrum as compared to Rluc8, allowing detection at lower protein expression levels [11, 12]. In addition, the smaller size of Nluc (19 kDa) as compared to Rluc8 (36 kDa) and firefly luciferase (61 kDa) is preferable in fusion proteins to minimize its effect on structure and function of the protein of interest. The split Nluc fragments LgBiT (18 kDa) and SmBiT (VTGYRLFEEIL; 1.3 kDa) have been optimized by random mutagenesis for increased structural stability and reduced intrinsic affinity ($K_D = 190$ μM) between the complementary fragments, respectively, allowing rapid and reversible detection of protein–protein interactions in living mammalian cells at 37 °C [8].

In BRET-based β-arrestin2 recruitment assays the luciferase donor is generally fused to the GPCR C-tail, while the fluorescent protein acceptor has been fused to either the N- or C-terminus of β-arrestin2 [13]. The NanoBiT-based β-arrestin2 recruitment assay generally prefers fusion of the complementary LgBiT and SmBiT fragments via flexible (GAQ-GNS)- GSSGGGGSGGGGSSG - (GAQ-GNS) linker sequences to the C-tail of a GPCR and N-terminus of β-arrestin2, respectively [14–18]. However, for some GPCR subtypes the opposite NanoBiT configuration was reported [10, 19, 20], and found to be more sensitive to detect β-arrestin2 [14]. Fusion of NanoBiT fragments to the N-terminus of β-arrestin2 resulted in better EFC-assay sensitivity for reported GPCRs as compared to C-terminal fusion [10, 14–20].

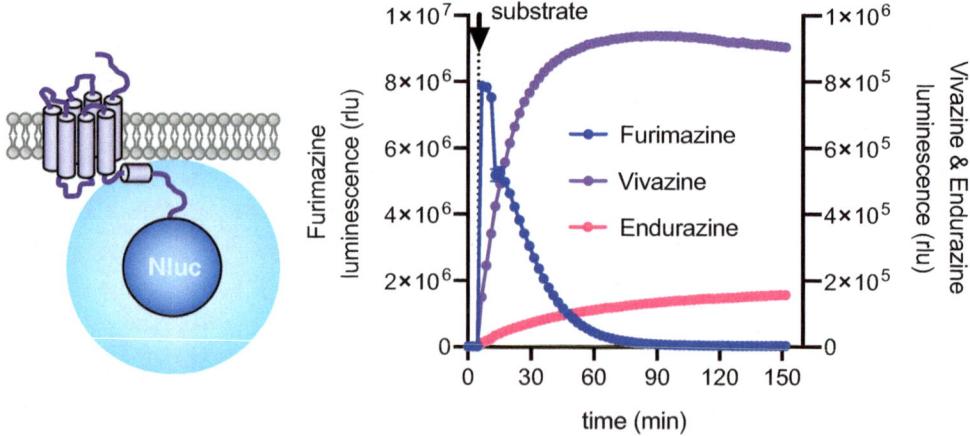

Fig. 2 Bioluminescence of H_1R-Nluc transiently expressed in HEK293T cells is measured as a function of time upon incubation with Furimazine (1:300), Vivazine (1:100), and Endurazine (1:100) in HBSS. Mean $\pm$ SD is shown from representative experiments

Furimazine is rapidly degraded and can only be used for relatively short time traces (Fig. 2). While NanoBRET can ratiometrically correct to a reasonable extent for this loss in donor luminescence, the interpretation of NanoBiT assay readouts might be more difficult due to substrate degradation. The recently marketed substrates Vivazine and Endurazine are more suitable for extended real-time traces, as these caged versions of furimazine are steadily hydrolyzed by intracellular esterases, resulting in a continuous release of furimazine during the course of the assay (Fig. 2) [8].

In this chapter, real-time β-arrestin2 recruitment to the histamine H_1R upon histamine stimulation is measured in transiently transfected HEK239T cells using both NanoBRET and NanoBiT technology. Under similar transient transfection conditions in HEK293T cells, the histamine-induced BRET signal was larger between hH_1R-Nluc and Venus-β-arrestin2 as compared to β-arrestin2-mVenus (Fig. 3a, b). Similar to most other GPCRs, H_1R-LgBiT in combination with SmBiT-β-arrestin2 was found to be the optimal NanoBiT configuration to detect histamine-induced β-arrestin2 recruitment to H_1R as compared to the other three possible configurations, if transiently transfected under similar conditions in HEK293T cells (Fig. 3c, d).

The factor time is important to consider in the quantification of agonist-biased signaling between different signaling pathways; therefore real-time quantification of agonist-induced responses is preferred above an end-point quantification after a fixed time interval [21]. In addition, the real-time β-arrestin2 recruitment assay can be used to qualitatively evaluate the H_1R binding kinetics of antihistamines in competition with histamine, as in the absence of

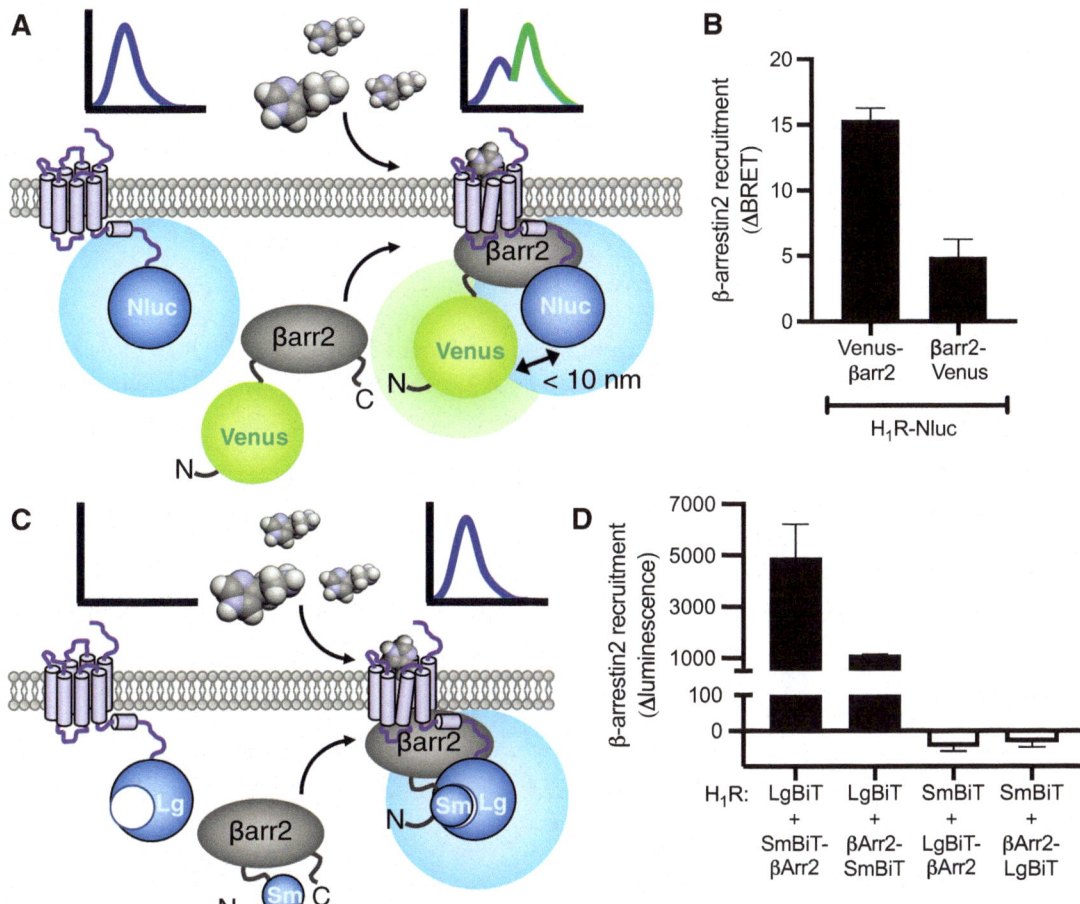

Fig. 3 Detection of histamine-induced β-arrestin2 recruitment to the H_1R in transiently transfected HEK293T cells using Nluc-based technologies. (**a**) NanoBRET-based β-arrestin2 recruitment assay in which Nluc was fused to the C-terminus of the H_1R and Venus to the N-terminus of β-arrestin2. Stimulation with histamine recruits Venus-β-arrestin2 to the H_1R-Nluc bringing BRET donor and acceptor in close proximity (<10 nm) resulting in BRET signal in the presence of furimazine substrate. (**b**) ΔBRET upon stimulation with 10 μM histamine is higher in HEK293T if H_1R-Nluc (1 μg plasmid/dish) is co-transfected with Venus-β-arrestin2 (4 μg plasmid/dish) as compared to β-arrestin2-mVenus (4 μg plasmid/dish). (**c**) NanoBiT-β-arrestin2 recruitment assay in which LgBiT was fused to the C-terminus of the H_1R and SmBiT to the N-terminus of β-arrestin2. Histamine-induced SmBiT-β-arrestin2 recruitment to the H_1R-LgBiT brings the two EFC-fragments in close proximity, allowing functional reconstitution of the Nluc. (**d**) Exchange of LgBiT and SmBiT between H_1R and β-arrestin2, or fusing the Nluc fragments to the C-terminus of β-arrestin2 impairs NanoBiT-based detection of histamine-induced β-arrestin2 recruitment to H_1R in HEK293T cells under similar transfection conditions (0.75 μg H_1R-LgBiT plasmid/dish, 1.5 μg SmBiT-βarrestin2 plasmid/dish, and 2.75 μg empty plasmid/dish). Mean ± SD is shown from representative experiments

receptor reserve recruited β-arrestin2 is a measure of bound histamine to the receptor [22]. In the presence of an antihistamine with a short target residence time (e.g., mepyramine), histamine-induced β-arrestin2 recruitment to the H_1R gradually increases in time up to a steady-state level in which binding of histamine and antihistamine to the H_1R is in equilibrium. In contrast, in the presence of a long target residence antihistamines (e.g., levocetirizine) an initial overshoot is observed after which the histamine-induced β-arrestin2 recruitment returns to a steady-state level. These kinetic signatures are comparable to competition association binding curves of these antihistamines for the H_1R using a labeled tracer ligand and can provide insight in the target residence time of antagonists [22].

2   Materials

All buffers and solutions are prepared using Millipore water and all other chemicals were of analytical grade and obtained from standard commercial sources.

2.1 Cell Culture and Transfection

1. Human embryonic kidney 293 T (HEK293T) cells (CRL-3216) from ATCC (*see* **Note 1**).

2. Cell culture medium: Dulbecco's Modified Eagles Medium (DMEM), 0.04 mM phenol red, 1.0 mM sodium pyruvate, 4 mM L-glutamine, 4.5 g/L into glucose, supplemented with 10% fetal bovine serum, 50 IU/mL penicillin, 50 mg/mL streptomycin.

3. Trypsin (0.05%) EDTA solution with phenol red.

4. Dulbecco's Phosphate Buffered Saline (DPBS).

5. 10-cm Sterile cell culture dishes.

6. Automated cell counter.

7. Filter (0.2 μm) sterilized 25 kDa linear polyethyleneimine (PEI) in dH_2O (1 μg/μL), pH 7.4 (*see* **Note 1**).

8. Filter (0.2 μm) sterilized 150 mM NaCl in dH_2O.

9. Mammalian expression plasmids such as, for example, pcDNA3, pcDEF3, and NanoBiT starter kit plasmids to express the NanoBRET/NanoBiT fusion protein in HEK293T cells (*see* **Note 2**): H_1R-Nluc/pcDEF3, H_1R-LgBiT/pcDEF3, H_1R-SmBiT/pcDEF3, Venus-β-arrestin2/pcDNA3 and β-arrestin2-mVenus/pcDEF3, LgBiT-β-arrestin2/pBiT1.1-N, SmBiT-β-arrestin2/pBiT2.1-N, βarrestin2-LgBiT/pBiT1.1-C, and β-arrestin2-SmBiT/pBiT2.1-C (*see* **Note 3**).

2.2 Assay and Detection

1. Sterile poly-L-lysine solution 0.01% (w/v) in H_2O.

2. Sterile white 96-well flat-bottom cell culture plates (*see* **Note 4**).

3. Multichannel vacuum aspirator.

4. 12-Channel pipets: 2–20 μL range and 20–200 μL range.

5. (Deep) 96-well polypropylene U- or V-bottom plates.

6. Hanks' Balanced Salt Solution (HBSS: 1.26 mM $CaCl_2$, 0.49 mM $MgCl_2$-$6H_2O$, 0.41 mM $MgSO_4$-$7H_2O$, 5.33 mM KCl, 0.44 mM KH_2PO_4, 4.17 mM $NaHCO_3$, 137.93 mM NaCl, 0.34 mM Na_2HPO_4, 5.56 mM D-glucose) without phenol red (*see* **Note 5**).

7. Ligand stocks: 1 mM in 100% dimethylsulfoxide (DMSO). Histamine is soluble in H_2O at 100 mM stock concentration.

8. NanoLuc substrates: Furimazine (i.e., Nano-Glo® Luciferase Assay Substrate), Vivazine™, and Endurazine™.

9. Plate reader to sequentially or simultaneously detect luminescence at 460–480 nm and 540 nm for NanoBRET or luminescence at 460–480 nm for NanoBiT.

3 Methods

3.1 Day 1: Seeding HEK293T Cells for Transfection

1. HEK293T cells are maintained in log-phase growth in culture medium in a humidified incubator at 37 °C with 5% CO_2. All cell culture steps should be performed under sterile conditions in a laminar flow hood.

2. Aspirate the medium from an 80–90% confluent 10-cm dish and wash the cells with 3 mL pre-warmed DPBS.

3. Replace the DPBS with 1 mL pre-warmed trypsin-EDTA and incubate for 2 min.

4. Add 3 mL pre-warmed culture medium to collect the detached cells and transfer the suspension to a 12-mL centrifuge tube and centrifuge at $260 \times g$ for 3 min at room temperature.

5. Aspirate the supernatant and resuspend the cells in 4 mL culture medium.

6. Count the cells by using an (automated) cell counter.

7. Seed 2×10^6 cells in 8 mL cell culture medium in a new 10-cm culture dish and culture 24 h in a humidified incubator at 37 °C and 5% CO_2 (Fig. 4).

3.2 Day 2: Transient Transfection HEK293T Cells

1. The transfection procedure should be performed under sterile conditions in a laminar flow hood.

2. Combine 5 μg DNA plasmids in 250 μL NaCl (150 mM) in a 1.5-mL vial: 1 μg H_1R-Nluc is combined with 4 μg Venus-β

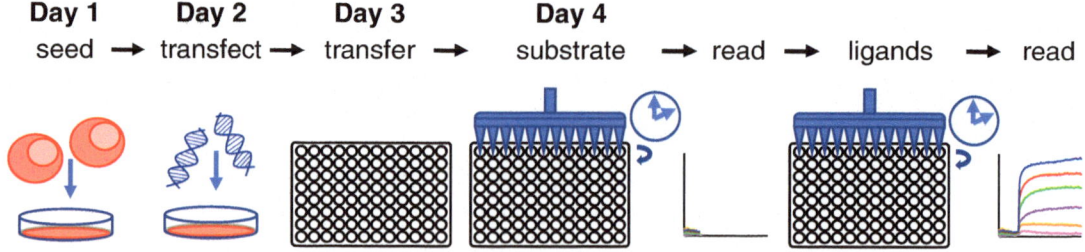

Day 1   **Day 2**   **Day 3**   **Day 4**
seed → transfect → transfer →   substrate → read → ligands → read

Fig. 4 Standard protocol for real-time NanoBRET and NanoBiT-based β-arrestin2 recruitment assays to GPCRs in transiently transfected HEK293T cells in response to agonist stimulation (see main text for step-by-step explanation)

arrestin2 for the NanoBRET assay, whereas 0.75 µg hH_1R-LgBiT is combined with 1.5 µg SmBiT-βarrestin2 and 2.75 µg empty expression plasmid for the NanoBiT assay (*see* **Note 6**).

3. Prepare 20 µg PEI solution per 250 µL NaCl (150 mM) and mix by vortexing.

4. Add 250 µL PEI/NaCl solution to the vial with 250 µL DNA/NaCl solution (PEI:DNA ratio is 4:1), directly mix by vortexing and incubate for 30 min at room temperature.

5. Aspirate in the meantime the cell culture medium from the HEK293T cells and carefully add 6 mL pre-warmed cell culture medium to the cells.

6. Gently resuspend the PEI/DNA mixture and pipette dropwise in a spiral form on the cell culture medium in the dish.

7. Gently swirl the dish and culture 24 h in a humidified incubator at 37 °C and 5% CO_2 (Fig. 4).

3.3 Day 3: Transfer Transfected HEK293T Cells into Assay Microplates

1. All cell culture steps should be performed under sterile conditions in a laminar flow hood.

2. Coat a white 96-well flat-bottom plate with 0.01% poly-L-lysine solution by adding 100 µL poly-L-lysine solution to each well and incubate 10 min at room temperature (*see* **Note 7**).

3. Aspirate the poly-L-lysine using a multichannel vacuum aspirator and wash twice with 150 µL sterile H_2O per well.

4. Carefully aspirate all H_2O and air-dry the plate for at least 30 min at room temperature.

5. Collect and count the transfected cells in cell culture medium as described before (*see* Subheading 3.1: **steps 2–6**).

6. Add 100 µL of cell suspension (5×10^5 cells/mL) per well in coated white 96-well flat-bottom plate and culture 24 h in a humidified incubator at 37 °C and 5% CO_2 (Fig. 4) (*see* **Note 8**).

3.4 Day 4: Prepare Ligand Dilutions

3.4.1 Agonist Assay

1. Dilute agonist stock to 10x the desired final concentration in HBSS, and prepare logarithmic serial dilution for concentration response curves. Ensure to mix at each dilution step and that DMSO concentrations are similar in all dilutions (<1%), including the vehicle.

2. Transfer the ligand dilutions to U- or V-bottom 96-well plate in the same layout as the assay plate, so that ligands can be added (*see* Subheading 3.5, **step 6**) row by row to the assay plate by using a multichannel pipet.

3.4.2 Agonist in Competition with Antagonist Assay

1. Dilute agonist and antagonist stocks to 20× the desired final concentration (here 10 μM for histamine and 100 nM for both antihistamines) in HBSS. Ensure that DMSO concentrations are similar in all dilutions (<1%), including the vehicle.

2. Mix equal volumes of agonist and antagonist dilutions and transfer into U- or V-bottom 96-well plate in the same layout as the assay plate, so that agonist and antagonist can be simultaneously added (*see* Subheading 3.5, **step 6**) row by row to the assay plate by using a multichannel pipet (Fig. 4) (*see* **Note 9**).

3.5 Day 4: NanoBRET and NanoBiT Assay Procedure

1. Carefully aspirate culture medium from the cells to avoid cell loss and add 50 μL pre-warmed HBSS buffer in each well using multichannel pipette.

2. Optional for the NanoBRET assay: measure fluorescence (excitation and emission at: 485 nm and 535 nm, respectively) to check Venus-β-arrestin2 expression (*see* **Note 10**).

3. Dilute the substrate stock in pre-warmed HBSS directly prior to the measurements and keep protected from light: furimazine (150×) and Vivazine (50×). Furimazine can be used for short-term (<60 min) NanoBRET (Fig. 5a, b) and NanoBiT (Fig. 6a, b) assays, while Vivazine should be used for real-time assays that measure longer than 60 min.

4. Dispense 50 μL/well of the substrate dilution to the first row of white 96-well plate with cells by using a 12-channel pipette, and stagger the addition of substrate to each of the subsequent rows to account for time it takes the microplate reader to measure luminescence and BRET per row of the 96-well plate (*see* **Note 11**) (Fig. 4).

5. Measure a few cycles bioluminescence and BRET at 460–480 and 540 nm (*see* **Note 10**), respectively, in a microplate reader at 37 °C (*see* **Note 12**) (Fig. 4).

6. After 5 min or 45 min incubation with furimazine or Vivazine, respectively, ligands are added. Pause the measurements and add 11 μL/well ligand to the first row of the 96-well plate by using a 12-channel pipette, and continue with adding ligands per row at the same pace as dispensing of substrate at **step 4** (*see* **Note 11**) (Fig. 4).

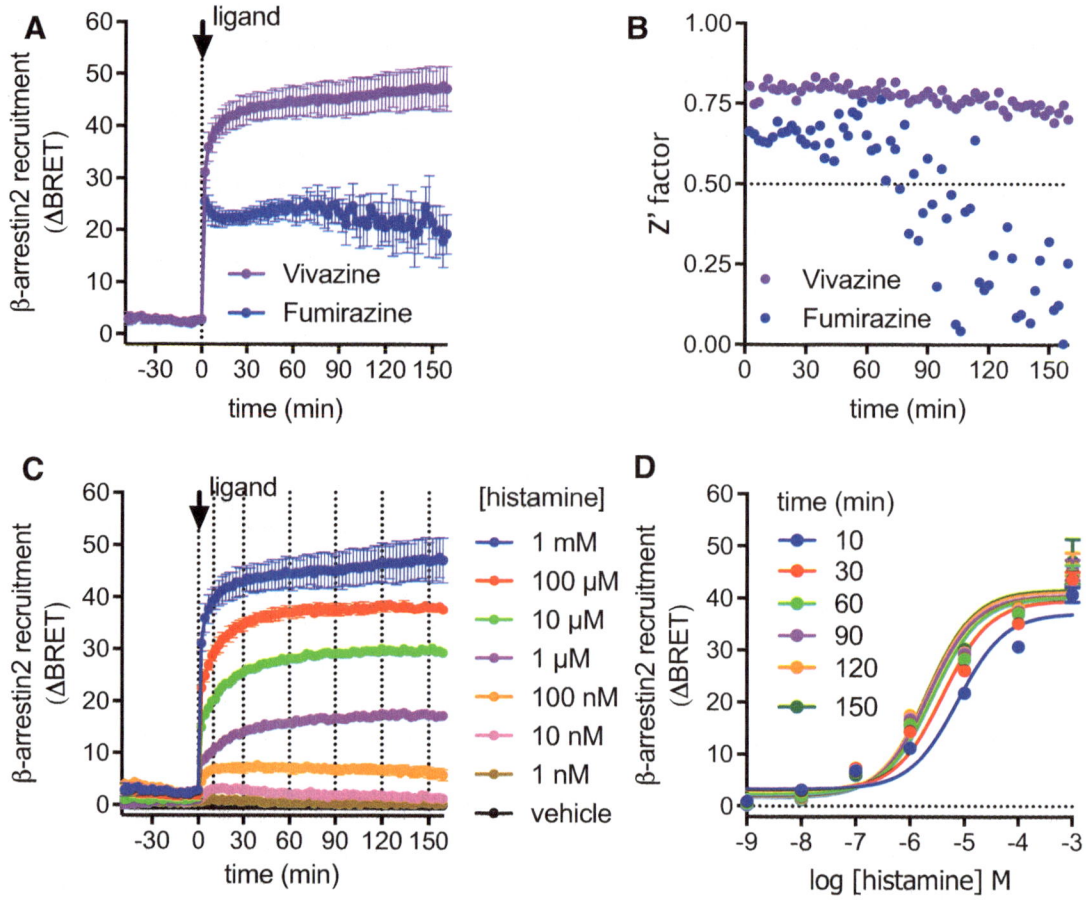

Fig. 5 NanoBRET-based β-arrestin2 recruitment assay in HEK293T cells transiently expressing H₁R-Nluc and Venus-β-arrestin2. (**a**) BRET-ratio over vehicle-stimulated (ΔBRET) in time upon stimulation with 1 mM histamine, which was preceded by 5 min or 45 min pre-incubation of the cells with furimazine or Vivazine, respectively. (**b**) Z′ factor in time for furimazine and Vivazine to detect histamine-induced β-arrestin2 over vehicle by NanoBRET. (**c**) β-arrestin2 recruitment to H₁R in response to increasing concentrations histamine was measured following 45 min pre-incubation with Vivazine. (**d**) Concentration-response curves at different time points following histamine stimulation in panel **c**. Mean ± SD is shown from representative experiments

7. Continue the bioluminescence and BRET measurement cycles at 460–480 and 540 nm, respectively, in a microplate reader for the desired amount of time (Fig. 4).

3.6 Data Analysis

1. Export measurement data from plate reader to a spreadsheet program (e.g., Excel or Numbers).

2. For NanoBRET-based β-arrestin2 recruitment assay, calculate the BRET ratios by dividing the BRET signal at 540 nm by the Nluc signal at 460–480 nm for each individual well if this is not automatically reported by the plate reader software.

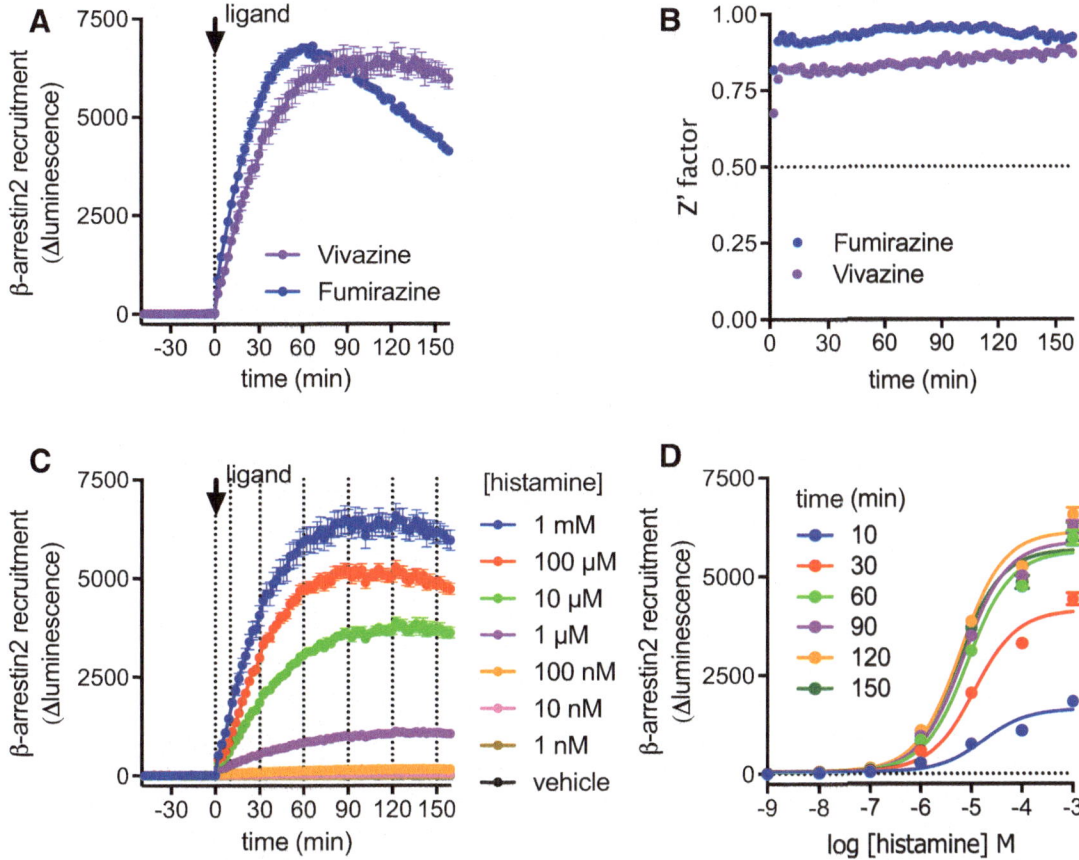

Fig. 6 NanoBiT-based β-arrestin2 recruitment assay in HEK293T cells transiently expressing H₁R-LgBiT and SmBiT-β-arrestin2. (**a**) Luminescence over vehicle-stimulated (Δluminescence) in time upon stimulation with 1 mM histamine, which was preceded by 5 min or 45 min pre-incubation of the cells with furimazine or Vivazine, respectively. (**b**) Z′ factor in time for furimazine and Vivazine to detect histamine-induced β-arrestin2 over vehicle by NanoBiT. (**c**) β-arrestin2 recruitment to H₁R in response to increasing concentrations histamine was measured following 45 min pre-incubation with Vivazine. (**d**) Concentration-response curves at different time points following histamine stimulation in panel **c**. Mean ± SD is shown from representative experiments

3. To quantify agonist-induced changes in BRET ratio over vehicle-stimulated, the ΔBRET is calculated for each well as ((BRET ratio$_{agonist}$ − BRET ratio$_{vehicle}$)/BRET ratio$_{vehicle}$) × 100, by using the average BRET ratio of vehicle-stimulated wells for corresponding timepoints. Optimal BRET configuration was determined using furimazine as substrate and at the peak response of 10 μM histamine, 50 min after stimulation (Fig. 3a, b) (*see* **Note 13**).

4. ΔBRET values are graphically plotted as function of time (Fig. 5c) or agonist concentration (Fig. 5d), and concentration response curves are analyzed by nonlinear regression for dose response stimulatory fit (three independent parameters) using GraphPad Prism 8.0.

5. For NanoBiT-based β-arrestin2 recruitment assays quantify agonist-induced changes in NanoBiT complementation over vehicle-stimulated as Δluminescence for each well as ((luminescence$_{agonist}$ − luminescence$_{vehicle}$)/luminescence$_{vehicle}$) × 100, by using average luminescence of vehicle-stimulated wells for corresponding timepoints. Optimal NanoBiT configuration was determined using furimazine as substrate and at the peak response of 10 μM histamine, 50 min after stimulation (Fig. 3c, d).

6. Δluminescence values are graphically plotted and/or analyzed as function of time (Fig. 6c) or agonist concentration (Fig. 6d) as described for the NanoBRET assay (*see* Subheading 3.6, **step 4**).

7. Calculate Z′-factor to evaluate the assay window between vehicle and agonist stimulation, as $Z' = 1 − ((3\sigma[agonist] + 3\sigma[vehicle])/(|\mu[agonist] − \mu[vehicle]|))$, where σ represents standard deviation and μ represents the mean (*see* **Note 14**).

8. Visually inspect agonist-induced β-arrestin2 recruitment kinetics in the presence of an antagonist for an initial overshoot signature (Fig. 7), which is indicative for relatively slower binding kinetics of the antagonist (e.g., levocetirizine) as compared to the used agonist [22]. The absence of this overshoot suggests that the antagonist (e.g., mepyramine) binds the receptor with similar or faster kinetics than the agonist (Fig. 7).

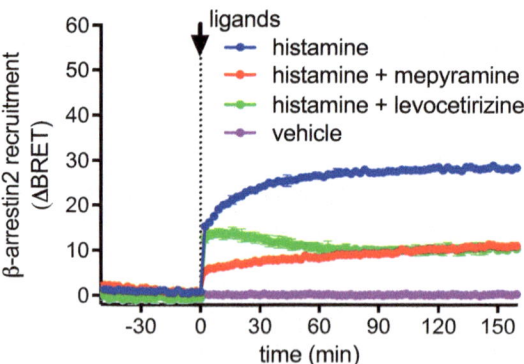

Fig. 7 Histamine-induced β-arrestin2 recruitment to H$_1$R in time to reveal binding kinetic profile of antihistamines. HEK293T cells transiently expressing H$_1$R-Nluc and Venus-β-arrestin2 incubated for 45 min with Vivazine, followed by stimulation with 10 μM histamine in the absence or presence of 0.1 μM mepyramine or 0.1 μM levocetirizine. BRET-ratio over vehicle-stimulated (ΔBRET) is subsequently measured in time. Co-incubation with the slow-residence time antihistamine levocetirizine results in the characteristic overshoot pattern of the histamine-induced β-arrestin2 recruitment time-trace, whereas short-residence time antihistamine mepyramine does not affect the shape of the histamine trace [22]. Mean ± SD is shown from representative experiment

The steady state indicates the level of receptor blockade by the used antagonist concentration (e.g., 0.1 μM) when binding of both agonist and antagonist is in equilibrium (*see* **Note 15**).

4   Notes

1. We generally use HEK293T cells with PEI for transient transfections as this combination yields reproducible high transfection efficiency at relatively affordable cost. Other cell lines and/or transfection reagents can be used after optimization of transfection conditions.

2. The pcDNA3 variants and pcDEF3 contain human cytomegalovirus (CMV) and human elongation factor-1α (EF1α) promoter sequences to drive constitutive gene expression in mammalian cells, whereas pBiT1.1 and pBiT2.1 plasmids contain a herpes simplex virus thymidine kinase (HSV-TK) promoter for lower level constitutive gene expression. In general, low expression levels are recommended to measure protein–protein interactions [12].

3. Nluc and Venus are fused via short linker sequences TSAAA and SGLKSRRALDS to H_1R and β-arrestin2, respectively, whereas LgBiT and SmBiT are fused via a longer linker sequence to H_1R C-tail (TSS-GSSGGGGSGGGGSSG), and the N- (GSSGGGGSGGGGSSGGAQGNS) and C-terminus (GNSGSSGGGGSGGGGSSG) of β-arrestin2. Fusion of Nluc or NanoBiT fragments to the C-tail of GPCRs is generally well tolerated with respect to its expression and signaling properties. We routinely verify proper receptor folding and functioning by radioligand binding and G protein signaling assays. The linker sequence might be optimized to minimize detrimental effects of protein fusion on receptor functioning or increase nanoBRET or NanoBiT assay window [23]. Fusions to the C-tail of β-arrestin might affect its positioning in N-domain cavity upon and facilitates its basal interaction with clathrin and/or AP2 [4, 13].

4. White plates are recommended for BRET and bioluminescent assays as they reflect light and consequently maximize detection of bioluminescence.

5. Phenol red interferes with detection of bioluminescence and BRET and should therefore be avoided.

6. The ratio between complementary NanoBRET or NanoBiT plasmids used for transfection can be further optimized to increase the response window. Total plasmid DNA should be kept constant for PEI transfections and should be supplemented with empty plasmid as carrier DNA if the combined amount of fusion protein-encoding plasmids is less than 5 μg.

7. Coating 96-well plates with poly-L-lysine is recommended to enhance attachment of HEK293T cells and consequently reduce cell loss while replacing media.

8. Cells can be seeded in parallel in a poly-L-lysine-coated transparent 96-well flat-bottom plate to monitor cell growth and viability using light microscopy.

9. If functional determination of antagonist binding kinetics is not the scope of the experiment, the preparation of an agonist-antagonist pre-mix is not necessary. In that case, the antagonist can be added first to the cells, subsequently followed by the agonist. Stagger the addition of agonist to account measurement times of the plate-reader.

10. One of the advantages of NanoBRET assay is that the expression of both donor and acceptor fusion constructs can be easily measured as luminescence and fluorescence, respectively, so that intra- and inter-assay variation can be easily monitored.

11. Add substrate and ligands in the same direction as the plate reader is measuring the 96-well plate, and stagger the addition of both substrate and ligand by the time it takes the plate reader to measure the same number of wells.

12. Kinetic β-arrestin2 recruitment can also be performed at room temperature (25 °C) to slow down both ligand binding and β-arrestin2 recruitment processes. Pre-equilibrate the cells in HBSS for 15 min at 25 °C before adding substrate and ligands.

13. Transfection of 4 μg Venus-β-arrestin2 plasmid in combination with 1 μg H_1R-Nluc yielded ~2.5-fold more fluorescence as compared to the same transfection conditions for β-arrestin2-mVenus, which may contribute to the observed difference in histamine-induced BRET. Further optimization by changing the ratio between receptor-donor and β-arrestin2-acceptor plasmids might be executed if desired.

14. Z' factor > 0.5 indicates that assay allows statistical separation of ligand-induced effects from control treatment.

15. Assay might also be performed in an end-point format instead of real-time. First stimulate with 100 μL agonist in the absence or presence of antagonist for the desired time period to obtain steady-state β-arrestin recruitment followed by addition of 11 μL furimazine (1:30 pre-dilution in HBSS) with a multi-channel pipet. Measure bioluminescence and BRET after 5 min. Stagger addition of furimazine between each row to account for the plate reader measurement time of a row. Alternatively, substrate might be injected prior to measurements per well if the plate reader is equipped with an injection system. This assay setup is useful in compound library screens in both agonist (i.e., direct stimulation) and antagonist (i.e., in competition with EC_{60-80} concentration of agonist) mode.

Acknowledgments

The NanoBiT starter kit plasmids were kindly provided by Promega Corporation (Madison, Wisconsin, USA). The β-arrestin2 Nano-BiT constructs were kindly provided by Dr. J.Y. Seong (Korea University, Seoul, Republic of Korea). The Venus-β-arrestin2-pcDNA3 construct was kindly provided by Dr. V.V. Gurevich (Vanderbilt University School of Medicine, Nashville, Tennessee, USA). X.M. is supported by a CSC Chinese scholarship grant (201703250074).

References

1. Insel PA, Sriram K, Gorr MW et al (2019) GPCRomics: an approach to discover GPCR drug targets. Trends Pharmacol Sci 40:378–387. https://doi.org/10.1016/j.tips.2019.04.001

2. Hauser AS, Attwood MM, Rask-Andersen M et al (2017) Trends in GPCR drug discovery: new agents, targets and indications. Nat Rev Drug Discov 16:829–842. https://doi.org/10.1038/nrd.2017.178

3. Gurevich VV, Gurevich EV (2019) GPCR signaling regulation: the role of GRKs and arrestins. Front Pharmacol 10:125. https://doi.org/10.3389/fphar.2019.00125

4. Peterson YK, Luttrell LM (2017) The diverse roles of arrestin scaffolds in G protein-coupled receptor signaling. Pharmacol Rev 69:256–297. https://doi.org/10.1124/pr.116.013367

5. Smith JS, Lefkowitz RJ, Rajagopal S (2018) Biased signalling: from simple switches to allosteric microprocessors. Nat Rev Drug Discov 17:243–260. https://doi.org/10.1038/nrd.2017.229

6. Kenakin T (2019) Biased receptor signaling in drug discovery. Pharmacol Rev 71:267–315. https://doi.org/10.1124/pr.118.016790

7. Hattori M, Ozawa T (2015) Bioluminescent tools for the analysis of G-protein-coupled receptor and arrestin interactions. RSC Adv 5:12655–12663. https://doi.org/10.1039/C4RA14979C

8. Dale NC, Johnstone EKM, White CW, Pfleger KDG (2019) NanoBRET: the bright future of proximity-based assays. Front Bioeng Biotechnol 7:587–513. https://doi.org/10.3389/fbioe.2019.00056

9. Machleidt T, Woodroofe CC, Schwinn MK et al (2015) NanoBRET—a novel BRET platform for the analysis of protein–protein interactions. ACS Chem Biol 10:1797–1804. https://doi.org/10.1021/acschembio.5b00143

10. Dixon AS, Schwinn MK, Hall MP et al (2016) NanoLuc complementation reporter optimized for accurate measurement of protein interactions in cells. ACS Chem Biol 11:400–408. https://doi.org/10.1021/acschembio.5b00753

11. Hall MP, Unch J, Binkowski BF et al (2012) Engineered luciferase reporter from a deep sea shrimp utilizing a novel imidazopyrazinone substrate. ACS Chem Biol 7:1848–1857. https://doi.org/10.1021/cb3002478

12. White CW, Vanyai HK, See HB et al (2017) Using nanoBRET and CRISPR/Cas9 to monitor proximity to a genome-edited protein in real-time. Sci Rep 7:3187. https://doi.org/10.1038/s41598-017-03486-2

13. Donthamsetti P, Quejada JR, Javitch JA et al (2015) Using Bioluminescence Resonance Energy Transfer (BRET) to characterize agonist-induced arrestin recruitment to modified and unmodified G protein-coupled receptors. Curr Protoc Pharmacol 70:2.14.1–2.1414. https://doi.org/10.1002/0471141755.ph0214s70

14. Cannaert A, Storme J, Franz F et al (2016) Detection and activity profiling of synthetic cannabinoids and their metabolites with a newly developed bioassay. Anal Chem 88:11476–11485. https://doi.org/10.1021/acs.analchem.6b02600

15. Cannaert A, Vasudevan L, Friscia M et al (2018) Activity-based concept to screen biological matrices for opiates and (synthetic) opioids. Clin Chem 2018:289496. https://doi.org/10.1373/clinchem.2018.289496

16. Storme J, Cannaert A, Van Craenenbroeck K, Stove CP (2018) Molecular dissection of the human A3 adenosine receptor coupling with β-arrestin2. Biochem Pharmacol

148:298–307. https://doi.org/10.1016/j.bcp.2018.01.008

17. Reyes-Alcaraz A, Lee Y-N, Yun S et al (2018) Conformational signatures in β-arrestin2 reveal natural biased agonism at a G-protein-coupled receptor. Commun Biol 1:128–112. https://doi.org/10.1038/s42003-018-0134-3

18. Littmann T, Buschauer A, Bernhardt G (2019) Split luciferase-based assay for simultaneous analyses of the ligand concentration- and time-dependent recruitment of β-arrestin2. Anal Biochem 573:8–16. https://doi.org/10.1016/j.ab.2019.02.023

19. Dupuis N, Laschet C, Franssen D et al (2017) Activation of the orphan G protein–coupled receptor GPR27 by surrogate ligands promotes β-arrestin 2 recruitment. Mol Pharmacol 91:595–608. https://doi.org/10.1124/mol.116.107714

20. Szpakowska M, Nevins AM, Meyrath M et al (2018) Different contributions of chemokine N-terminal features attest to a different ligand binding mode and a bias towards activation of ACKR3/CXCR7 compared with CXCR4 and CXCR3. Br J Pharmacol 175:1419–1438. https://doi.org/10.1111/bph.14132

21. Klein Herenbrink C, Sykes DA, Donthamsetti P et al (2016) The role of kinetic context in apparent biased agonism at GPCRs. Nat Commun 7:10842. https://doi.org/10.1038/ncomms10842

22. Bosma R, Moritani R, Leurs R, Vischer HF (2016) BRET-based β-arrestin2 recruitment to the histamine H1 receptor for investigating antihistamine binding kinetics. Pharmacol Res 111:679–687. https://doi.org/10.1016/j.phrs.2016.07.034

23. Chen X, Zaro JL, Shen W-C (2013) Fusion protein linkers: property, design and functionality. Adv Drug Deliv Rev 65:1357–1369. https://doi.org/10.1016/j.addr.2012.09.039

Chapter 17

Luciferase Complementation Approaches to Measure GPCR Signaling Kinetics and Bias

Nicola C. Dijon, Desislava N. Nesheva, and Nicholas D. Holliday

Abstract

An understanding of the kinetic contributions to G protein-coupled receptor pharmacology and signaling is increasingly important in compound profiling. Nonequilibrium conditions are commonly present *in vivo*, for example, as the drug competes with dynamic changes in hormone or neurotransmitter concentration for the receptor. Under such conditions individual binding kinetic properties of the ligands can influence duration of action, local ligand concentration, and functional properties such as the degree of insurmountable inhibition. Mapping the kinetic patterns of GPCR signaling events elicited by agonists, rather than a peak response at a single timepoint, is often key to predicting their functional impact. This is also a path to a better understanding of the origins of ligand bias, and whether such ligands demonstrate their effects through selection of distinct GPCR conformations, or via their kinetic properties. Recent developments in complementation approaches, based on a small bright shrimp luciferase Nanoluc, provide a new route to kinetic analysis of GPCR signaling in living cells that is amenable to the throughput required for compound profiling. In the NanoBiT luciferase complementation system, GPCRs and effector proteins are tagged with Nanoluc fragments optimized for their low interacting affinity and stability. The interactions brought about by GPCR recruitment of the effector are reproduced by a rapid and reversible increase in NanoBiT luminescence, in the presence of its substrate furimazine. Here we discuss the methods for optimizing and validating the GPCR NanoBiT assays, and protocols for their application to study endpoint and kinetic aspects of agonist and antagonist pharmacology. We also describe how timecourse families of agonist concentration response curves, derived from a single NanoBiT assay experiment, can be used to evaluate the kinetic components in operational model derived parameters of ligand bias.

Key words G protein-coupled receptors, Luciferase complementation, Binding kinetics, Signaling kinetics, Ligand bias, G protein, Arrestin, Operational model

Nicola C. Dijon and Desislava N. Nesheva contributed equally to this work.

Sofia Aires M. Martins and Duarte Miguel F. Prazeres (eds.), *G Protein-Coupled Receptor Screening Assays: Methods and Protocols*, Methods in Molecular Biology, vol. 2268, https://doi.org/10.1007/978-1-0716-1221-7_17,
© Springer Science+Business Media, LLC, part of Springer Nature 2021

1   Introduction

1.1   Kinetic Context in G Protein-Coupled Receptor (GPCR) Compound Profiling

In defining ligand pharmacology at GPCRs, we are typically focused on deriving the key parameters of binding affinity and a measure of efficacy—the ability of the ligand to activate the receptor once bound. GPCRs are allosteric proteins that form a ternary complex with both ligand and effector—for example, G protein or β-arrestin—and so affinity and efficacy measurements are also dependent on the nature of the engaged effector protein and signaling pathway, and cellular context [1–3]. Biased GPCR ligands were initially revealed because they display preferential interaction with a subset of signaling pathways from a single receptor relative to a reference agonist, and for some examples, there is now strong structural evidence in favor of ligand-specific conformations of receptors (for example, angiotensin, opioid receptors, and adrenoceptors [4–7]) that generate bias through being functionally distinct.

However, many of our approaches to predict ligand pharmacology and bias are underpinned by an assumption of equilibrium conditions—such as our use of the dissociation constant K_D (the concentration of ligand required to occupy 50% of the target receptor population *at equilibrium*) in ranking affinity, or the Black and Leff operational model in analysis of concentration response curves [8, 9]. Increasingly, better quantitative understanding of the kinetics of ligand binding and receptor signaling is essential to consider effects at a GPCR target in physiological context, where the interplay between the ligand and receptor is often highly dynamic in vivo [10–15]. The concentration of unbound drug in the local target vicinity is continually changing, dependent on pharmacokinetic properties regulating plasma concentration and distribution to local tissue reservoirs. The drug may also be in competition with native messengers for the GPCRs, whose concentrations—for example, at a neuronal synapse—can alter greatly over a short space of time. Rates of ligand association with and dissociation from GPCR complexes then have a significant impact on functional behavior. For example, slow dissociating antagonists generate insurmountable effects under nonequilibrium conditions, and so potentially improved inhibition of receptor responses, even in the presence of high neurotransmitter or hormone concentrations [16, 17]. Dissociation and association rates of binding can also influence duration of action and local ligand concentration, particularly when target rebinding is enhanced within tissue reservoirs (such as synapses) with restricted exchange with the systemic circulation [11, 14, 16, 18, 19].

For GPCR agonists, both binding and signaling pathway kinetics are key to understanding the overall functional consequences. The timescales, as well as subcellular organization, of GPCR signaling events vary dramatically from second messenger transients and oscillations over seconds, to alterations in gene expression and phenotypic cellular changes occurring over hours to days [20, 21]. This can lead to discrepancies in aligning agonist efficacies at different signaling endpoints. For example, the antiproliferative effects of different Gs-coupled receptor agonists in lung fibroblasts are poorly correlated with their abilities to elevate peak levels of cAMP, suggesting that the spatiotemporal kinetics of such second messenger signals is a critical driver in the actual phenotypic effects required for treating idiopathic pulmonary fibrosis [22]. In assessment of ligand bias at the dopamine D2 receptor, Klein Herenbrink et al. [23] demonstrated time-dependent variation in agonist concentration-response curves for different signaling endpoints, influenced both by the time profile of the pathway (for example, signal desensitization) and variation in the receptor binding kinetics of the ligands. As a result, the observed pathway bias for some D2 agonists changed substantially according to assay timepoint [23]. Thus, a robust screening analysis to identify ligands with conformation bias should also consider kinetic response factors and map the spatiotemporal dynamics of pathway responses.

The clear relevance of kinetic context poses a question for selecting GPCR screening assays for functional analysis of agonists and antagonists. Many common biochemical cellular assays are optimized to maximize assay window for robust profiling, using dissimilar timepoints—for example, a peak intracellular calcium response (10–20 s), the integrated accumulation of cAMP (30 min–1 h), reporter gene transcription (4–5 h), or proliferation/differentiation (24–48 h) [20, 21]. Most are also endpoint in nature, in that one well provides a single timepoint of measurement, and so incorporating kinetic analysis into routine iterative screening can lead to significant reductions in throughput. The pioneering development of fluorescence based small molecule indicators [24] has now been followed by an array of genetically encoded imaging biosensors for GPCR signaling [25–30]. These biosensors provide opportunities to develop kinetic analysis of the effects of compounds in living cells, located at highly defined receptor–effector interactions or pathway nodes. The use of bioluminescence and fluorescence resonance energy transfer (BRET/FRET) sensors has been reviewed elsewhere [25, 26, 28–30] and so the application of complementation strategies will be specifically considered here.

1.2 Luciferase Complementation to Detect Real-Time GPCR Signaling Events

The basis of a complementation assay method is to employ a reporter protein that can be split into nonfunctioning fragments, which are then used to tag the proteins of interest. Interaction between these proteins then brings the fragments together to reconstitute the activity of the reporter, and this activity is detected as a readout of the target protein–protein interaction. In principle, the proximity constraints imposed by complementation provide a similar degree of specificity to BRET/FRET. The original complementation assays used split-enzyme reporters including dihydrofolate reductase [31] and beta-galactosidase [32] (for example, DiscoverX), but the detection and development steps involved, for example, using a colorimetric substrate, mean that single-well kinetic reads using such systems are impractical. In contrast, bimolecular fluorescence complementation (BiFC), involving split fluorescent protein (FP) reporters, generates protein–protein interaction signals capable of being monitored in live cells, since the fragments refold and generate a mature fluorescent protein without further reagent addition [33–35]. However, the delayed chromophore maturation, and stable formation of the beta-barrel fluorescent protein after fragment refolding, limits the ability of BiFC to measure fast and transient GPCR signaling interactions.

The use of split-luciferase as a reporter combines the advantages of a live cell, imaging platereader-based approach, with the rapid and reversible fragment domain interaction of enzyme systems. GPCR complementation assays have previously been described using other luciferases [36], but the subsequent development of an engineered shrimp luciferase Nanoluc has offered much improved reporter characteristics [37]. Nanoluc is a small 19 kD luciferase that generates a long duration, high amplitude, luminescence in the presence of its substrate furimazine, so providing signal sensitivity and stability over assay timecourses. Different strategies are possible for generating split-Nanoluc fragments for complementation, including the Promega NanoBiT system described here [38, 39]. Dixon et al. (2016) used a split point in Nanoluc to generate a complementary LgBiT fragment (157 aa) and a 11 amino acid SmBiT peptide that offers size advantages to preserve the tagged protein function [39]. Furthermore point mutations introduced into LgBiT increased its pre-folded stability, while the SmBiT peptide sequence was varied to produce a variety of partners with ~270,000-fold range in LgBiT affinity [39]. The optimization of the low-affinity SmBiT sequence enabled the generation of a complementation system that is both rapid and reversible, with the signal dependent only on the interaction between the protein partners under study [39].

We and others have applied the NanoBiT system to measure the interactions between GPCRs and their effector proteins [39–45], including β-arrestins and mini-Gα proteins that preserve the receptor-G protein binding interface [40, 46] (Fig. 1). This technology provides a real time, simple luminescence readout for GPCR signaling in living cells, from which the kinetics of GPCR-effector interactions can be derived from single-well analysis using appropriate platereaders. This can reveal the time-dependence of both antagonist insurmountability and relative agonist efficacies for the monitored pathways, within single assay plates. Here we describe

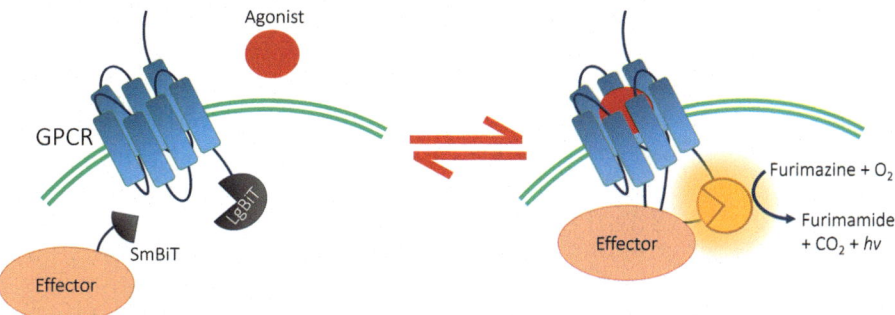

Fig. 1 NanoBiT luminescence complementation assay measuring effector recruitment at GPCRs The receptor is C-terminally tagged with the LgBiT fragment and the effector is tagged with the SmBiT at the N terminus. The receptor recruits the effector constitutively or upon agonist stimulation which results in enzyme complementation, and the production of luminescence on addition of the furimazine substrate

optimization steps to validate the NanoBiT complementation approach and kinetic analysis, followed by examples of its use to examine agonists, antagonists, and modulators of GPCRs. Finally we outline the application of the Black and Leff operational model [8, 9, 47] to these data in the analysis of agonist efficacy and bias across the signaling timecourse.

1.3 Assay Optimization

The success of GPCR NanoBiT complementation assays first depends on careful construct plasmid design focusing on a number of factors. Routinely we use GPCR fusion proteins tagged at the C terminus with the LgBiT fragment, but the positioning of the SmBiT/LgBiT tags between receptor and effector interactors should be validated as necessary. For many GPCRs with extended C termini, a minimal linker region between receptor and LgBiT is sufficient. However, some receptors are characterized by short C terminal regions consisting only of juxtamembrane helix 8 followed by the palmitoylation site. As an example, for the dopamine D2 receptor, a linker of 15–20 amino acids is required to position the LgBiT fragment away from the membrane and allow flexibility for efficient complementation. The linker length needs to be tested empirically, and while Ser-Gly-based linkers are routine, some consideration on their composition is useful, for example, if considering functional modifications such as receptor phosphorylation. The position of the SmBiT-linker fusion on the effector protein can also be critical; for example, mini Gα subunits must be tagged at the N terminus [40], and G protein receptor kinases at the C terminus to preserve receptor interaction [48]. Given the small size of the SmBiT fragment, its internal insertion flanked by suitable 10–15 amino acid linkers into receptor or effector protein (such as full length Gα [49]) is feasible. It is also useful to monitor expression of the partner proteins, and for the GPCR-LgBiT fusions, we incorporate an extracellular N terminal SNAP tag to allow selective detection of cell surface expression by immunofluorescence microscopy [34, 50, 51] (*see* **Note 1**). A final consideration is the choice of promoter for the expression constructs. In contrast to BRET, which is a ratiometric measurement, both the basal and stimulated NanoBiT luminescence signals increase with higher levels of partner protein expression and this allows great sensitivity [39], including for proteins under endogenous levels of promotion [45]. In recombinant systems, it is possible to consider both high strength promoters such as cytomegalovirus (CMV, e.g., pcDNA3.1) and more nuanced choices such as thymidine kinase (e.g., pBiT, Promega) in optimizing both expression levels and stoichiometry of the partners of interest. Given the need to test a variety of construct combinations in a matrix format, the use of a transient transfection approach as described first below is favored during optimization, and also enables the incorporation of controls to validate the selectivity of the receptor signaling interaction under study. For example, in establishing GPCR-mini Gα recruitment assays [40] it is useful to test the receptor of interest across the mini Gα panel to ensure preferential recruitment of the expected G protein effector. Following the optimization of construct design, it is expected that stably transfected cells will be generated for routine assays. The most straightforward approach is to use mixed populations of stably co-transfected cells, and so the plasmids encoding GPCR-LgBiT and SmBiT effector proteins should encode different eukaryotic resistance markers (e.g., neomycin, zeocin, hygromycin) to avoid shuttle cloning to new vectors.

2   Materials

1. Mammalian expression vectors pcDNA3.1(+) (neomycin, hygromycin or zeocin resistance) or pBiT 1.1 N/C containing (1) C-X-C motif chemokine receptor 2 (CXCR2) cDNA (human; GenBank accession NM_001557.3); (2) beta 2 (β2) adrenoceptor cDNA (human; GenBank accession NM_000024), modified with an N-terminal SNAP tag and with the LgBit sequence at the C terminus; (3) β arrestin2 (human; GeneBank accession NM_004313) and mini $G\alpha_s$ and mini $G\alpha_{oA}$ [46] with an N terminal SmBiT sequence followed by Ser/Gly linker.

2. HEK 293 cells (ATCC CRL-1573).

3. Dulbecco's modified Eagle's medium (DMEM): 4500 mg/L glucose, L-glutamine, sodium pyruvate, and bicarbonate, 10% fetal bovine serum (FBS) added on the day of experiments.

4. Opti-MEM reduced serum cell culture medium without Phenol Red: L-glutamine, 0.1% bovine serum albumin (BSA), added on the day of experiment.

5. Lipofectamine 3000 cationic lipid transfection reagent.

6. Poly-D-lysine hydrobromide Mr-70–150 kDa, dissolved in PBS (5 mg/mL), stored in 40 μL aliquots at −20 °C. For coating plates, a 40 μL aliquot is diluted in 20 mL ultrapure water (18 MΩ cm) and filter sterilized (0.2 μm).

7. 96-Well flat-bottomed, white plates.

8. 6-Well flat-bottomed plates.

9. HEPES-buffered saline solution (HBSS): 147 mM NaCl, 24 mM KCl, 1.3 mM $CaCl_2$, 1 mM $MgSO_4$, 1 mM Na pyruvate, 1 mM $NaHCO_3$, 10 mM HEPES, pH 7.4, and sterilized by autoclave. D-Glucose added to 10 mM before first use. 0.1% BSA added on the day of experiment.

10. β2 Adrenoceptor ligands: isoprenaline, salmeterol, salbutamol, formoterol, propranolol, ICI118551, dissolved in ultrapure water to a stock concentration of 10 mM and kept in single-use aliquots at −20 °C; CXCR2 peptide agonist CXCL8 (aa 28–99), powder diluted to a stock concentration of 10 μM in ultrapure water, stored in single-use aliquots −20 °C; navarixin (synthesized by School of Pharmacy, University of Nottingham), SB265610 diluted in DMSO to a stock concentration of 10 mM and stored in single-use aliquots at −20 °C.

11. Furimazine substrate, NanoGlo Live Cell Assay system.

12. Benchtop hotbox-style 37 °C incubator.

13. Incubator at 37 °C, gassed with 5% CO_2 for cell culture; incubator at 37 °C without CO_2 for experiments using HBSS.

14. Vacuum Pump with 1 L waste trap.

15. BMG PHERAstar FS or FSX, standard luminescence filter.

16. BMG Mars data analysis software.

17. Graphpad Prism v 7 or v 8 data analysis and manipulation software.

---

3 Methods

3.1 Assay Optimization: Transient Transfection of NanoBiT-Tagged DNAs

As an enzyme, the luminescence signal from complemented luciferase is influenced by substrate concentration. It is therefore vital that assay conditions are tested to avoid depletion of furimazine across the timecourse, so that interpretation of kinetic signatures relate to the underlying changes in protein–protein interactions. The remaining optimization protocols describe how to exclude furimazine concentration effects empirically, using an example NanoBiT assay detecting the transient recruitment of β-arrestin2 to the β2-adrenoceptor (Fig. 2). The first of these protocols simply

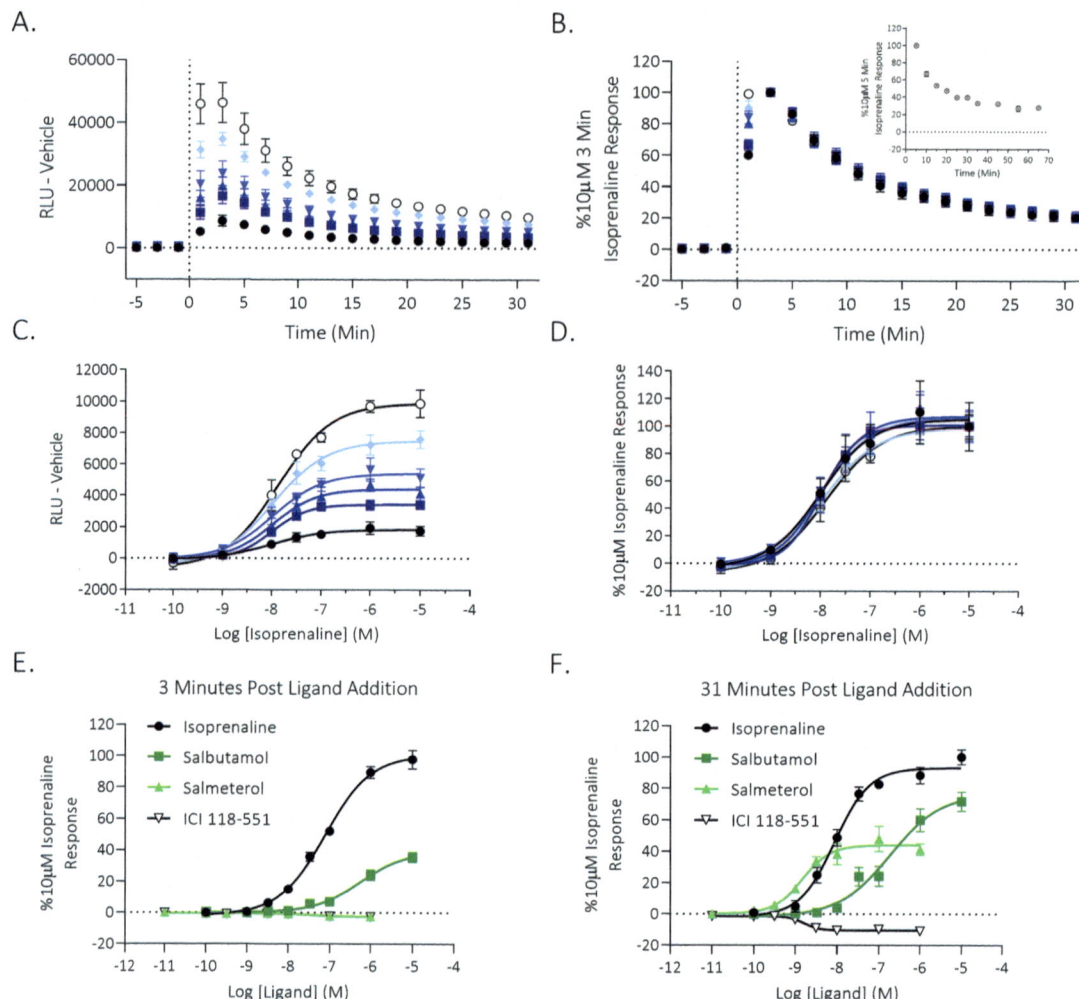

Fig. 2 Testing furimazine concentration and ligand pharmacology in the β2 adrenoceptor-β-arrestin2 NanoBiT assay. (**a**) Timecourse of β-arrestin2 recruitment by the β2 adrenoceptor stimulated with 10 μM isoprenaline in the presence of a range of furimazine concentrations (dilutions from manufacturer's stock), overlaid by normalization to the peak isoprenaline response in (**b**). The inset for (**b**) demonstrates an equivalent time-profile obtained when cells are stimulated with agonist for different times, followed by furimazine added to all wells and a single endpoint read. (**c**) Equivalent concentration-response curves of isoprenaline stimulated arrestin recruitment by the β2 adrenoceptor were obtained with a range of furimazine concentrations, shown normalized in (**d**). (**e** and **f**) Examples of the concentration response curves obtained to three representative agonists and an inverse agonist (ICI118551), at the peak (3 min, **e**) and equilibrium (31 min, **f**) timepoints. All data points represent mean ± standard error of the mean ($n \geq 3$), with luminescence responses relative to 10 μM isoprenaline at the indicated timepoint

detects the furimazine concentration dependence of kinetic reads, but it is also useful to collect and compare kinetic data constructed via an endpoint assay, in which the furimazine incubation conditions are then kept short and constant for all time-points. This comparison also controls for other potential artifacts, such as changes in stability of the active luciferase enzyme over time.

The monitoring of the luciferase complementation signal is possible using a wide variety of platereaders capable of monitoring luminescence emission at the maximum of 460 nm [37]. Our experiments use the Labtech BMG Pherastar FS/FSX, which offers a number of advantages regarding 37 °C temperature control, sensitivity, speed, and precision useful for kinetics. In the section on platereader optimization, some guidance is provided regarding Pherastar settings to optimize assay window, temporal resolution, and decrease background signals; and these considerations are more broadly applicable to other measurement systems.

1. Add 2 mL of poly-D-lysine (*see* **Note 2**) to each well of clear 6-well plate and incubate for 30 min.

2. Wash with 2 mL of DMEM media, with wash replaced with a further 2 mL of DMEM media supplemented with 10% fetal calf serum (DMEM/10% FCS).

3. Passage a flask of HEK293 cells at 70–80% confluency, using 1–2 mL trypsin-EDTA solution to lift the cells from flask. Inactivate the cell/trypsin suspension in 10 mL serum containing DMEM, in a 30 mL universal tube, and centrifuge at 200 rcf for 5 min.

4. Resuspend pellet in 10 mL of DMEM/10% FCS and estimate number of cells using a hemocytometer.

5. Seed 500,000 cells/well into prepared 6-well plate and incubate in humidified incubator at 37 °C and 5% CO_2.

6. After 24 h, prepare DNA/lipofectamine 3000 reagent mixes for each well as detailed below (*see* **Note 3**):
   (a) Add 3.75 μL of lipofectamine 3000 reagent to 125 μL reduced serum OptiMEM medium.
   (b) Add 2.5 μg of cDNA mix (*see* **Note 4**) to 125 μL of OptiMEM, then add 5 μL of P3000 reagent.
   (c) Combine DNA and lipofectamine mixes and incubate at room temperature for 5 min (*see* **Note 5**).

7. Wash 6-well plates with 1.5 mL OptiMEM and replace with 1 mL OptiMEM.

8. After 5 min DNA/lipofectamine incubation (*see* **Note 5**), add each 250 μL DNA/lipofectamine mix to each well of the prepared 6-well plate and incubate in humidified incubator at 37 °C and 5% CO_2 for 24 h.

9. To split cells from a 6-well plate transfection, lift cells using trypsin and agitation through pipetting DMEM/10% FCS. Pellet cells through centrifugation at 200 rcf, resuspend pellet in 3 mL DMEM/10% FCS and count cells using hemocytometer.

10. Prepare poly-D-lysine-coated white-walled 96-well plates by adding 50 μL poly-D-lysine and incubate for 30 min at room temperature.

11. Remove the poly-D-lysine and wash wells with DMEM. Aspirate DMEM wash from all wells. A second plate could be seeded at the same time to test for SNAP-receptor expression (*see* **Note 6**).

12. Seed cells at 35–40,000 cells/well into poly-D-lysine-coated 96-well plate and incubate in humidified incubator at 37 °C and 5% CO_2.

13. After 24 h, perform the NanoBiT assay as described in Subheading 3.5. We recommend the initial buffer choice of HBSS/0.1% BSA (*see* **Notes 7** and **8**) and to employ a screening paradigm that analyzes a full agonist concentration response curve across the transfection conditions (platemap, Fig. 3a). Although single point screens are a possibility for higher throughout (evaluating basal/maximal agonist concentration) differences in complementation efficiencies between construct partners of different design may also be revealed through changes in agonist potency.

3.2 Assay Optimization: Substrate Concentration

1. Prepare cells from a T75 flask stably co-transfected with receptor and effector NanoBiT DNAs (*see* **Note 9**), as described above (**steps 2** and **3**).

2. Seed into poly-D-coated 96-well flat-bottom white plates at 35,000–40,000 cells per well and incubate in humidified incubator at 37 °C and 5% CO_2 for 24 h.

3. Conduct luminescence assay as described below (Subheading 3.5, **steps 1–5**) with increasing concentrations of furimazine substrate used across the assay plate columns (as shown in Fig. 3a).

4. Resulting raw luminescence responses should increase with furimazine concentration (Fig. 2a, c). However, normalized timecourse profiles and agonist potencies should be unaltered (as shown in Fig. 2b, d).

3.3 Assay Optimization: Staggered Kinetics with Endpoint Measurement

1. Prepare assay plate as described above.

2. On day of assay, aspirate DMEM/10% FCS and wash cells with 50 μL/well of warmed HBSS/0.1% BSA buffer and replace with 40 μL/well of HBSS/0.1% BSA buffer.

3. Over a timecourse of 65 min, stagger the addition a maximal concentration of a receptor agonist or vehicle to assay wells (Fig. 3b) in duplicate, incubating the plate in a 37 °C/0% CO_2 benchtop incubator. Each well representing different agonist treatment time should therefore finish incubation at the same timepoint.

4. At timepoint 5, add the final agonist/vehicle addition followed by 10 μL of furimazine mix (dilution dependent, refer to Subheading 3.2), incubate for 5 min before reading luminescence on BMG PHERAstar platereader.

A. Concentration Dependence

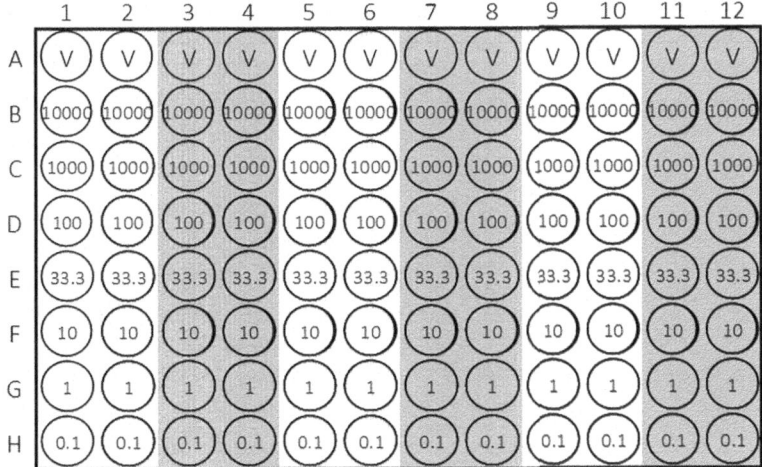

B. Reverse Timecourse

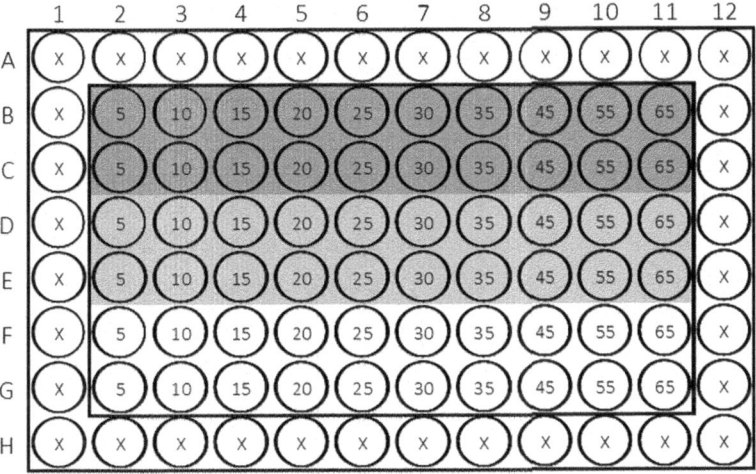

Fig. 3 Example plate maps for NanoBiT luminescence complementation assay for studying (**a**) the action of six different agonists/conditions in duplicate and (**b**) reverse timecourse experiments. In (**a**) each set of two duplicate columns represent concentration response curves with different ligands, or to the same agonist in the presence of varying assay conditions (e.g., furimazine concentration), or in the absence and presence of different concentrations of antagonist or modulator. In (**b**), numbers indicate agonist pretreatment timings of additions of receptor agonists (rows BC and DE) and vehicle (rows FG) in duplicate, before luminescence read

5. Compare agonist timecourses constructed in kinetic and end-point methods (Fig. 2b), with the expectation that these should closely overlay.

3.4 Assay Optimization: Luminescence Platereader Measurement Settings

1. The Pherastar FS/FSX platereader stage should be preheated to 37 °C, and this temperature can be monitored throughout the assay. Steps should be taken to minimize plate cooling during compound/furimazine addition steps. The adoption of this physiological temperature is most suitable both for GPCR-effector recruitment assays and measured luciferase activity.

2. As the microplate is moved to a read position of a well, there is a positioning delay to allow for settlement and stabilization of the well liquid, typically defaulted 0.5 s. In kinetic reads this setting can be minimized to decrease the overall reading time for each well and therefore the cycle time of the entire plate (e.g., 96 wells). This can increase temporal resolution, for example, of a whole plate read kinetically. The PHERAstar offers a "check timing" feature to estimate this minimal cycle time required to read the plate.

3. The measurement interval time defines the length of time that the PHERAstar reads each well (e.g., 1 s), integrating the luminescence produced. An increased interval time provides higher luminescence counts, but also increases plate cycle time and reduces kinetic resolution.

4. If increased temporal resolution is critical, a subset of wells can be read within each full kinetic read, and the order of well-read configuration (e.g., column by column versus row by row) can be adjusted on the PHERAstar.

5. To ensure the best signal-to-noise ratio for a new assay, a focal height and gain adjustment is required for candidate wells (e.g., agonist stimulated), and this may be implemented using an automated protocol or manually. These settings, which must be kept constant across the plate conditions, are a key element of optimizing assay window and background. It is also important that luminescence detection is kept within the linear range, for example, a high set gain does not saturate the signal. Robust assay luminescence measurements with optimized settings indicate the potential to reduce the measurement interval time (**step 3** above) and so allow more frequent kinetic reads.

6. For certain assay configurations (endpoint reads), the availability of online injection may allow rapid substrate addition at optimal concentration. Otherwise, manual additions of compound and substrate may be programmed at the appropriate cycle interval. In Figs. 2 and 4, kinetic reads were collected from a full 96-well cycle read time at 2 min intervals; after substrate addition, 3 baseline read cycles were required to monitor equilibration, after which receptor ligands were added (at time 0) and the resulting effect on the luciferase complementation was observed.

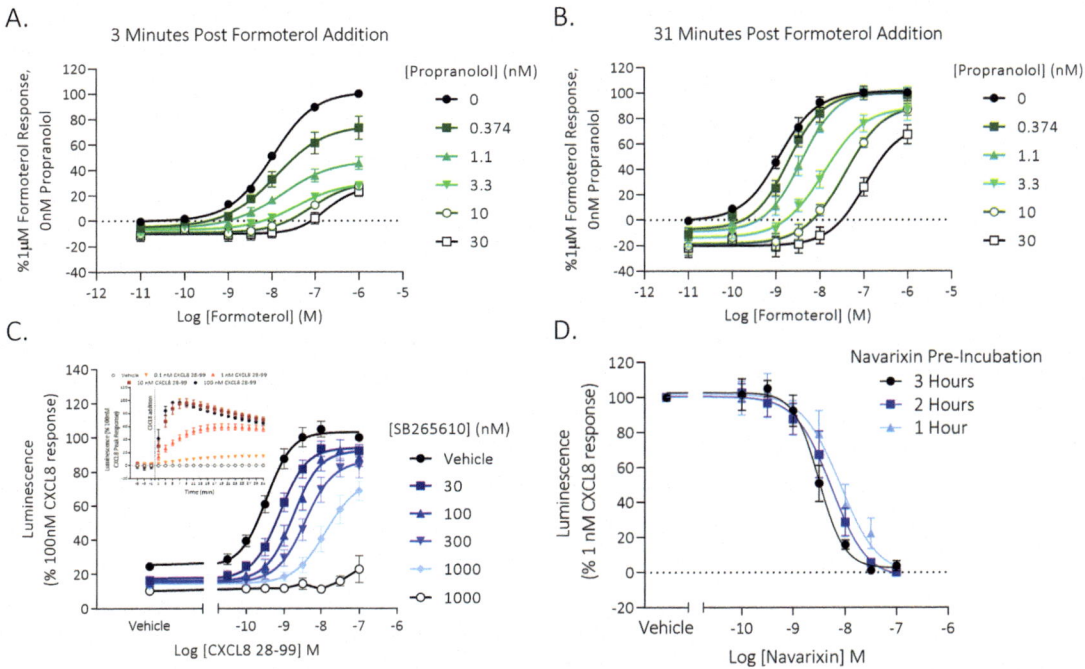

Fig. 4 Kinetic aspects of antagonist pharmacology evaluated by GPCR luminescence complementation assays. Following preincubation with propranolol (10 min), concentration-response curves for β-arrestin2 recruitment by the formoterol stimulated β2 adrenoceptor were constructed at peak (3 min, **a**) and 31 min (**b**) post agonist addition—revealing a time-dependent change in non-surmountable to surmountable antagonism. (**c**) Analysis of the effect of the negative allosteric modulator SB265610 on CXCL8 stimulated recruitment of β-arrestin2 to the CXCR2 receptor, measured at 31 min post-agonist addition. Note the more sustained pattern of arrestin recruitment in response to CXCL8 stimulation (inset), compared to the β2 adrenoceptor-arrestin assay (Fig. 2a, b). (**d**) The effects of the negative allosteric modulator navarixin on the association of preformed CXCR2—mini mini $G\alpha_{oA}$ NanoBiT complexes, measured as IC_{50} curves after 1 h 1 nM CXCL8 pretreatment followed by 1–3 h navarixin treatment. All data points represent mean ± standard error of the mean ($n \geq 3$), with luminescence responses relative to (**a**, **b**) 10 μM isoprenaline, in the absence of propranolol, (**c**) 100 nM CXCL8, in the absence of SB265610, (**d**) 1 nM CXCL8 in the absence of navarixin

3.5 NanoBiT Assays: Direct Analysis of Ligand Signaling and Pharmacology

These generic protocols outline NanoBiT kinetic measurement of direct ligand effects on GPCR-effector interactions, for example, agonists, and also the effects of antagonists or modulators on agonist function. In each case, timecourse reads from single assay plates allow analysis of concentration response data across multiple timepoints.

If available, it is advantageous to first test the assay against known agonists of varying efficacy (Fig. 2e, f), together with potential inverse agonists (ligands which stabilize the inactive GPCR conformation). Comparison with literature data for each test ligand provides a number of useful indicators. One impact that is less recognized for direct assays of GPCR: effector recruitment is the continuing pharmacological relevance of receptor reserve, despite limited signal amplification, if receptor expression exceeds the pool of available effector molecules for complementation. This stoichiometry allows maximal complementation responses to be obtained without fully occupying the receptor population—leading to increased agonist potencies relative to binding affinity (for full agonists), and relative maximal responses compared to a reference (for partial agonists). Adjustments to the transfection co-expression levels may then be required. Well-characterized inverse agonists, if available, can be used to inhibit basal levels of complementation between receptor and effector, demonstrating reversibility in the system (see also antagonists below) and also the levels of complementation driven by constitutive activity of the receptor. Typically, antagonist and modulator effects are best assessed through a pretreatment paradigm. An antagonist preincubation time is used to promote equilibrium conditions before addition of the stimulating agonist. These stimulatory effects can then be followed over time, during which binding equilibrium between agonist, antagonist, and receptor may be re-established. One of the factors that defines this rate of transition from hemi-equilibrium conditions, for competitive antagonists, is the rate of dissociation of the antagonist from the receptor—since antagonists with slow dissociation rates will release receptor availability for agonist activation more slowly [11]. This can be monitored experimentally, within the same assay, by the change from insurmountable to surmountable behavior at later timepoints (Fig. 4a, b). Since the luciferase complementation assay is rapidly reversible, an inverse treatment protocol (agonist followed by antagonist) is also beneficial for considering the relative kinetics of agonists and antagonists, particularly where agonist-stimulated recruitment of the effector and the complementation response is sustained. The example in Fig. 4c, d shows such a comparison for the analysis of negative allosteric modulation of β-arrestin2 or mini $G\alpha_{OA}$ recruitment by the chemokine receptor CXCR2.

1. Prepare assay plate as described above (Subheading 3.2; *see* **Note 10**).

2. On day of assay, aspirate DMEM/10% FCS, wash cells with 50 μL/well of warmed HBSS/0.1% BSA assay buffer, and replace with 40 μL/well of HBSS/0.1% BSA buffer (*see* **Notes 7** and **8**).

3. Add 10 μL/well furimazine substrate diluted in HBSS/0.1% BSA to the optimized concentration and incubate for 5 min. During this period, transfer the plate to the PHERAstar platereader and allow equilibration to 37 °C. Read luminescence for 3 × 2 min cycles using the optimized protocol (Subheading 3.4; *see* **Note 11**).

4. Prepare a compound plate containing the ligands to be tested in sufficient volume at 6× final assay concentration, in HBSS/0.1% BSA, arranged in the desired concentration format (e.g.,

7 concentrations plus vehicle; Fig. 3a; *see* **Note 12**). Following the basal reads, rapidly transfer compounds (10 μL/well) to the center of the assay plate wells using a multichannel pipette, taking care to avoid carryover or disturbing the adherent cells. Immediately restart read cycles on the platereader.

5. Collect the assay kinetics using multiple PHERAstar reads for the desired number of timepoints (*see* **Notes 13** and **14**).

3.6 NanoBiT Assays: Antagonist and Modulator Pretreatment

1. Prepare assay plate as described above (*see* **Note 10**).

2. On day of assay, aspirate DMEM/10% FCS, wash cells with 50 μL/well of warmed HBSS/0.1% BSA assay buffer.

3. Replace with 40 μL/well of HBSS/0.1% BSA buffer containing vehicle or antagonist at the desired concentrations and incubate for 1 h at 37 °C (*see* **Note 15**). For example, Fig. 3a illustrates a suggested 96-well platemap option to run duplicate concentration response curves for vehicle and five antagonist concentrations.

4. Add furimazine substrate and stimulate with agonists as described in the protocol above (Subheading 3.5).

3.7 NanoBiT Assays: Stimulation with Agonist Followed by Antagonist

1. Prepare assay plate as described in Subheading 3.2 (*see* **Note 10**).

2. On day of assay, aspirate DMEM/10% FCS, wash cells with 50 μL/well of warmed assay buffer, and replace with 60 μL/well of assay buffer (*see* **Notes 7, 8,** and **16**).

3. Add 10 μL/well of agonist to the desired single final assay concentration (for example, EC_{80}) or vehicle according to the plate plan and incubate for desired time (e.g., 60 min; *see* **Note 17**) in a 37 °C incubator.

4. Add a range of concentrations of the antagonist/NAM or vehicle to the wells pretreated with an agonist or vehicle and incubate at 37 °C for the desired time (*see* **Note 18**).

5. Add 10 μL/well furimazine substrate diluted in assay buffer to the optimized concentration and incubate for 5 min. Read a single cycle of luminescence on the PHERAstar.

3.8 Analyses of Agonist and Antagonist Action: Normalizing Agonist Timecourse Data

General data normalization and analysis considerations for agonist and antagonist pharmacology are discussed here, prior to a further option for analysis of bias using the operational model across timepoints. The two distinct representations of NanoBiT assay data come in the form of timecourse traces and concentration response curves. In either representation, it is important to select a consistent reference ligand throughout investigations to both normalize data between experiments and to minimize system influences in bias calculations when comparing two pathways (*see* **Note 19**).

1. To remove background luminescence, subtract the luminescence in vehicle-treated wells at each timepoint for all agonist-treated wells.

2. To allow for pooling of replicate experiments, normalize data to a peak response of the top concentration of a reference ligand as 100% (*see* **Note 19**) with the remaining timecourse data points represented as a percentage of the peak data point (Fig. 2b).

3. Ligand, receptor, and effector-specific timecourse profiles have been observed using GPCR NanoBiT complementation assays. For example, the small molecule isoprenaline stimulated transient recruitment of β-arrestin2 at the β2 adrenoceptor, peaking at 3 min (Fig. 3a, b). However, the chemokine CXCL8 (aa 28–99) stimulated a more sustained arrestin recruitment response to CXCR2 in comparison, with the luminescence peaking between 3 and 5 min (Fig. 4c).

4. The kinetic profile of different responses can be fitted empirically using Graphpad Prism (for example, as sustained recruitment or rise and fall to plateau), with complex underlying contributions of ligand binding kinetics, effector association, and regulation of the receptor–effector complex (e.g., desensitization) (*see* **Note 20**).

3.9 Ligand Concentration Response Curves

1. Concentration response relationships are described using a nonlinear regression model, with agonist curves defined by the 4 parameter Eq. (1):

$$Y = \text{Min} + \left(\frac{\text{Max} - \text{Min}}{1 + 10^{(\text{LogEC}_{50} - X) * \text{Hill Slope}}}\right) \quad (1)$$

where Min and Max represent the minimum and maximum plateau of each ligand response, and the Hill Slope describes the steepness of the fit. The relative maximal response (R_{max}) is defined as "Max − Min," while EC_{50} is the concentration of ligand to produce the response half way between Min and Max, representing potency of the agonist.

2. Concentration response data at each timepoint are routinely normalized to the vehicle (0%) and reference agonist (100%) control (*see* **Note 21**), which allows data to be pooled for presentation. However, log EC_{50} and R_{max} values are pooled from estimates using each individual experiment. Figure 2e, f illustrates how timepoint analysis can reveal changes in agonist potency and relative R_{max} from a nonequilibrium peak timepoint for the β2-adrenoceptor-β-arrestin2 response, to a later timepoint following equilibration.

3. Neither the agonist EC_{50} or R_{max} provide sole indicators of agonist efficacy, and to consider bias requires application of further analysis, such as the operational model (*see* Modelling signaling bias using timecourse data).

3.10 Analysis of Antagonist Action

1. Analyses of antagonist action at GPCRs is well established, for example, in the use of Schild regression in the classification of reversible antagonists [52]. Here, we describe considerations in Schild analysis or obtaining IC_{50} values using the antagonist complementation experiments outlined above.

2. Schild analysis requires that the antagonist be competitive and reversible, and that the functional system be at equilibrium—enabling surmountable antagonism (no change in agonist R_{max} in the presence of the antagonist). Collecting kinetic data from antagonist assay plates enables the degree of surmountability to be monitored over time (Fig. 4a, b) and the appropriate time-point chosen. The concentration ratio (CR) is usually calculated as the EC_{50} for the antagonist in the presence of each antagonist concentration [B], divided by the control agonist EC_{50}. A plot of log [CR-1] against log [B] should be linear with slope of 1, with an intercept providing an estimate of the antagonist dissociation constant (log K_D). Global analysis using Graphpad Prism v8 often circumvents the need to calculate and draw the Schild plot directly, but this is still a useful graphical representation to reveal nonlinearity and slopes other than 1, which suggest noncompetitive, slowly reversible or nonequilibrium effects.

3. The NanoBiT assays are capable of revealing inverse agonist activity (reduction in constitutive receptor activation and basal complementation) of GPCR ligands that otherwise behave as antagonists (Fig. 2f). In this case, it is important to take an equi-effective concentration of agonist from each point of the curve (using Prism's interpolation function, rather than the individual EC_{50} values), when calculating CR values for Schild analysis.

4. Effects of antagonists can also be described as IC_{50} values, the antagonist concentration for 50% inhibition, when the antagonist concentration is varied against a single concentration of agonist (Fig. 4d). Depending on the mode of inhibition, the antagonist IC_{50} depends both on the antagonist affinity for the receptor, and the agonist concentration used in the assay. Typically, the agonist concentration for IC_{50} determinations is selected to fit within the mid-range of the control concentration response curve (EC_{25}–EC_{75}).

3.11 Agonist Pharmacology and Bias Across Timepoints Using the Operational Model: Calculating Transduction Coefficients for Use in Bias Calculations

The operational model of agonism, originally described by Black and Leff (1983), derives measures of ligand affinity and efficacy using functional data, with parameters being agonist concentration [A], the agonist receptor dissociation constant (K_A), maximal response of the system (E_{max}), and a measure of coupling efficacy (τ) (Eq. 2) [1, 2, 8, 9, 47].

$$E = \frac{E_{max}\tau[A]}{K_A + [A](1 + \tau)} \tag{2}$$

τ is still a combined measurement of efficacy, influenced by agonist intrinsic efficacy and system coupling (combined as parameter K_E) and receptor expression level [2, 9]. However τ and K_A provide an improved empirical framework to consider ligand efficacy compared to R_{max} and EC_{50} alone, and this can then be applied to calculate estimates of ligand bias. One disadvantage is that single concentration response curves cannot be fitted unambiguously with the operational model to define all the parameters. Typically, a global fit of multiple concentration response curve data will be employed, using the presence of a full reference agonist to define E_{max}. This allows τ of all compared agonists and K_A values of partial agonists (only) to be defined. A modified approach is to fit a single transduction ratio $R = (\tau/K_A)$ as the key parameter, which is then used as in bias calculations (*see* **Note 22**) [47]. Agonist values for log R are first normalized to the reference ligand within a single pathway (Δlog R), to control for system-dependent influences on the parameter including receptor number and signal implication. Δlog R for each ligand are then subtracted between compared signaling pathways ($\Delta\Delta$ log R); if significantly different from 0, $\Delta\Delta$ log R therefore indicates "bias" between ligands. In the complementation experiments described above, $\Delta\Delta$ log R values can be readily calculated from the families of agonist concentration response curves at each timepoint, providing powerful time-dependent tracking of the observed bias calculation (*see* **Notes 23** and **24**).

1. Obtain complementation assay concentration response curve data for the ligand panel, with sufficient concentration ranges which unambiguously represent R_{max} and EC_{50} of each ligand and ensure the reference ligand is present as the positive control on all assay plates. Normalize the data as outlined in Subheading 3.9 for each timepoint. Typical kinetic complementation assay plates, with 2 min cycle read time over a 30 min time-course, therefore yield a family of 15 concentration response curve sets that can be analyzed for bias at each timepoint.

2. Following van der Westhuizen et al. [47], input the user-defined equation of nonlinear regression for the adapted operational model of agonism into GraphPad PRISM software (*see* **Note 25**).

3. Input concentration response data of each ligand into an *XY data table*, with the reference ligand in column A, other full agonist data in columns B to O, and partial agonists in columns P onward.

4. Using the user-defined equation, analyze concentration response data to obtain $LogR$ values for each agonist (*see* **Note 26**). Repeat for each experimental repeat and take the mean $LogR$ value for each agonist.

5. To minimize the system contributions to $LogR$ (*see* **Note 22** and above), first obtain $\Delta LogR$ values for each experimental agonist by subtracting $LogR$ of the reference agonist:

$$\Delta LogR_{\text{Experimental Agonist}} = LogR_{\text{Experimental Agonist}} - LogR_{\text{Reference Agonist}}$$

6. To compare signaling between two pathways, subtract $\Delta LogR$ values of an experimental ligand obtained in different complementation signaling assays (i.e., β-arrestin2 recruitment vs. mini Gα recruitment):

$$\Delta\Delta LogR = \Delta LogR_{\text{Pathway 1}} - \Delta LogR_{\text{Pathway}}$$

Positive $\Delta\Delta LogR$ values indicating bias toward pathway 1 and negative to pathway 2, relative to the reference agonist.

4 Notes

1. Routine detection of plasma membrane SNAP-GPCR-LgBiT constructs in transient or stably transfected cells is performed by incubation of live cells in DMEM containing 0.1 μM SNAP-surface AlexaFluor488, which is membrane impermeant. Cells may be imaged using appropriate excitation /emission filter settings (FITC) by widefield or confocal microscopy. Adaptions of this method enable agonist-stimulated receptor internalization to be quantified to provide a further readout of pathway pharmacology [50].

2. The use of poly-D-lysine (specifications in Subheading 2, **item 6**), as a positively charged polymer coating for flask and plate plastic, is essential for maintaining HEK293 cell adherence during procedures with wash steps during the transfection and experimental assay.

3. A method based on lipofectamine 3000 is highlighted, but a variety of other cationic transfection reagents can also be used for high-efficiency transfection in HEK293 systems. Times and DNA: reagent mixtures need to be independently optimized. Transient transfections were optimized at 3.75 μL of lipofectamine 3000 reagent, see manufacturers protocol (Invitrogen) for further optimization steps.

4. The amount of DNA shown represents the total DNA mix, which always should be kept constant within a matrix format to explore co-transfection of different construct combinations. The ratio of receptor:effector:carrier (empty vector, such as pcDNA3.1) DNA should also be optimized—for example, to evaluate SNAP-β2-adrenoceptor-LgBiT:SmBiT-β-arrestin2 ratios (keeping receptor transfection constant), 0.8 µg GPCR DNA can be mixed with 0, 0.2, 0.4, 0.8, or 1.6 µg arrestin plasmid, with carrier DNA added to the same total of 2.5 µg in each case.

5. DNA:lipofectamine 3000 mixes should be used following 5 min incubation, as longer incubation times result in a decline in transfection efficiency.

6. There is potential for the incorporation of C terminal LgBiT, as well as N terminal SNAP tag if used, to impact on receptor cell surface expression. If an imaging platereader system is available, seeding a duplicate black-walled clear-bottomed plate for SNAPsurface AF488 labeling and imaging (*see* **Note 1**) provides an opportunity to monitor receptor expression during assay optimization. Functional assays (for example, second messenger measurements) could also be carried out to validate receptor coupling. It should be noted that due to the sensitivity of the NanoBiT assay, we have seen robust complementation luminescence responses for receptors with only minimal levels of cell surface expression detected by imaging.

7. HBSS/0.1% BSA is a commonly used buffer to monitor cell-based assays investigating GPCR pharmacology, and we have found it suitable for assay preincubations and reads of up to 2 h. The use of reduced serum OptiMEM media, lacking phenol red, is an alternative for longer complementation assay incubations, but, for example, may require a platereader system with environmental (CO_2) as well as temperature control. We have found the inclusion of BSA improves well by well measurement consistency in live cell assays described here. Alternatively, the use of the proprietary reagent provided with the Promega NanoGlo live cell assay kit is an option to reduce autoluminescence and background.

8. The use of 0.1% BSA also limits ligand binding to plastic surfaces such as vial and plate walls, for example, particularly relevant for peptide agonists such as the CXCL8 example described here. There should be an awareness of the potential for some compounds to bind BSA directly, and this may require testing by its removal from assay media if suspected.

9. While it is assumed that a move to stably co-transfected cells will be made for convenience, routine assays can be based on solely transient systems. The use of independent plasmids for

receptor and effector in a transient system (compared to bicistronic vectors or dual stable lines) provides some continued control over receptor:effector expression stoichiometry.

10. Cell seeding density, as well as substrate concentration and measurement settings, is a further factor in optimizing the assay window size without saturation. The quoted seeding density of 35,000–40,000 cells per well is designed to generate well confluence of 80–90% at the 24 h assay timepoint.

11. It is important that the membrane permeable substrate furimazine equilibrates in the cells prior to initiating agonist stimulation. Repeat baseline luminescence reads, for example, at 2 min intervals provide an opportunity to judge when this has been achieved.

12. The plate layouts for compound addition can be optimized for the desired assay throughput (e.g., single or duplicate, 7 or 10 point concentration response curves), and the addition of multiple basal and reference agonist controls at different well positions to calculate both assay window and Z′. An initial test using basal and maximal agonist-stimulated wells across all plate rows and columns is a useful control to analyze intraplate variability.

13. Typical assay timecourses run for 1 h under the conditions described here. If much longer timecourses are required, the choice of buffer should be considered (*see* **Note 7**), while endurazine and vivazine (Promega) are available as esterified furimazine pro-substrates. Cellular esterase activity promotes extended steady release of furimazine for complementation measurements over several hours.

14. Recent evidence indicates that GPCR-effector signaling complexes are spatially and temporally organized within cells [25]. Developments in the sensitivity of luminescence microscopy (e.g., Olympus LV200) provide an opportunity for complementation signals to be monitored at a subcellular level in single living cells [39], in addition to the well-based platereader approaches described here.

15. Picking the appropriate antagonist or modulator preincubation time is essential to ensure their equilibrium binding to the receptor prior to agonist stimulation. This timepoint can be established using an antagonist pretreatment timecourse in the complementation assay, or inferred from other sources, for example, direct receptor binding data under similar assay buffer conditions and temperature. The antagonist or modulator ligand should first be tested for direct agonist or inverse agonist effects in the assay.

16. For longer-term experiments, attention should be paid to the assay well volume as well as the buffer choice (*see* **Note 7**). Increased assay volumes will help reduce time-dependent artifacts resulting from evaporation. If compounds have very high affinity for the receptor under investigation, increased assay volumes may also reduce the likelihood of effects from ligand depletion.

17. This assay protocol determines an IC_{50} curve for the ability of an antagonist or modulator to disturb preformed effector–receptor complexes. In the absence of known inverse agonists for the GPCR of interest, it is also useful as a further demonstration of the reversibility of NanoBiT detection for the protein partners of interest.

18. For example, agonist pretreatment time can be set at 60 min for equilibrating CXCL8-CXCR2 interaction and effector recruitment (Fig. 4d), prior to treatment with the negative allosteric modulator for 30 min–3 h.

19. Commonly, the native ligand of the investigated receptor is chosen as a reference ligand, though this is by no means essential. For example, some native messengers may have reduced chemical stability in the assay, which may itself modulate kinetic signatures, compared to high efficacy pharmacological agonists. Regardless of the choice, the key consideration is that any estimate of ligand "bias" within a pathway is always made relative to the reference ligand of choice in the assay—and so for bias analysis across pathways it is paramount that this ligand should be kept consistent.

20. Mathematical models have been put forward to use functional timecourse data in obtaining estimates of ligand binding affinity and kinetics. For example, Hoare et al. (2018) provide options for different types of signaling kinetics to estimate both agonist affinity, and efficacy in the form of a transduction rate constant. Simpler options for this model assume rapidly equilibrating agonists prior to signal generation, although versions incorporating slower binding kinetics can also be considered [53].

21. Commonly, vehicle-treated wells are used as the negative control and thus the basal measure of luminescence response. However, in cases where pretreatments alter basal measurements (e.g., inverse agonism), 0% may equally be defined as the absence of any luminescent signal.

22. In generating complementation assay sets to analyze bias profiles, care should be taken to ensure equivalence in receptor expression levels and host cell context (e.g., HEK293) for the different pathway endpoints. This minimizes system-dependent differences in the calculation of $\Delta\log R$ for each

pathway. The focus on direct receptor–effector recruitment in complementation assays has benefits in reducing the number of intervening system amplification steps—for example, low efficacy agonists are more likely to appear partial, with reduced R_{max} compared to reference ligand.

23. The operational model was framed under the assumption of equilibrium conditions (hence its use of the agonist equilibrium dissociation constant K_A). Thus, time-dependent calculations of transduction ratios under this method provide important means to track observed bias at different timepoints across signaling pathways, and reveal if there are kinetic influences on this calculation through nonequilibrium binding kinetics [23]. Observation of consistent bias which persists across timepoints, as equilibrium is approached, provides enhanced evidence for a basis in functionally distinct ligand–receptor conformations.

24. Other methods for bias determination can also be considered [2], with commonalities in data normalization to the reference ligand to minimize system coupling effects in the comparison. Bias factors, for example, use independent agonist K_i values in the operational model calculations, obtained from binding experiments [54]. The highlighted method to fit R directly has the advantage of requiring no other data from additional experiments where assay and time conditions need to be matched.

25. To input adapted operational model of agonism into GraphPad PRISM, the "user-defined equation" is reproduced here from van der Westhuizen et al. [47].

Select Analyze tab, Non-linear regression (curve fit), New and Create new equation.

EQUATION TAB.

Equation Type – Explicit Equation: Y = a function of X and parameters.

Name – Example: "Operational Model for Bias".

Definition –.

A = 10^X.

operate1 = ((1 + A)/((10^LogR)*A))^n.

operate2 = ((1 + A/(10^LogKA))/((10^LogR)*A))^n.

Y1 = basal+(Emax-basal)/(1 + operate1).

Y2 = basal+(Emax-basal)/(1 + operate2).

<A:O > Y=Y1.

<~A:O > Y=Y2.

RULES FOR INITIAL VALUES TAB.

LogR   -1.0   *(Value of X at YMID)

n   1.0

LogKA   1.0   *(Value of X at YMID)

basal   1.0   *YMIN

Emax   1.0   *YMAX
DEFAULT CONSTRAINTS TAB.
LogR   No constraint
n   Shared value for all datasets
LogKA   No constraint
basal   Shared value for all datasets
Emax   Shared value for all datasets

26. The user-defined equation will also provide K_A values for partial agonists. Using the $\text{Log}R$ and $\text{Log}K_A$ values, $\text{Log}\tau$ values for partial agonists can be obtained: $\text{Log}R = \text{Log}(\tau/K_A) = \text{Log}(\tau) - \text{Log}(K_A)$.

Acknowledgments

Nicola C. Dijon and Desislava N. Nesheva contributed equally to this work.

References

1. Kenakin T (2019) Biased receptor signaling in drug discovery. Pharmacol Rev 71:267–315. https://doi.org/10.1124/pr.118.016790

2. Stott LA, Hall DA, Holliday ND (2016) Unravelling intrinsic efficacy and ligand bias at G protein coupled receptors: a practical guide to assessing functional data. Biochem Pharmacol 101:1–12. https://doi.org/10.1016/j.bcp.2015.10.011

3. Wootten D, Christopoulos A, Marti-Solano M et al (2018) Mechanisms of signalling and biased agonism in G protein-coupled receptors. Nat Rev Mol Cell Biol 19:638–653. https://doi.org/10.1038/s41580-018-0049-3

4. Wingler LM, Elgeti M, Hilger D et al (2019) Angiotensin analogs with divergent bias stabilize distinct receptor conformations. Cell 176:468–478.e11. https://doi.org/10.1016/j.cell.2018.12.005

5. Masureel M, Zou Y, Picard L-P et al (2018) Structural insights into binding specificity, efficacy and bias of a beta2AR partial agonist. Nat Chem Biol 14:1059–1066. https://doi.org/10.1038/s41589-018-0145-x

6. Lee Y, Warne T, Nehmé R et al (2020) Molecular basis of β-arrestin coupling to formoterol-bound β(1)-adrenoceptor. Nature. https://doi.org/10.1038/s41586-020-2419-1

7. Manglik A, Lin H, Aryal DK et al (2016) Structure-based discovery of opioid analgesics with reduced side effects. Nature 537:185–190. https://doi.org/10.1038/nature19112

8. Black JW, Leff P, Shankley NP, Wood J (1985) An operational model of pharmacological agonism: the effect of E/[A] curve shape on agonist dissociation constant estimation. Br J Pharmacol 84:561–571. https://doi.org/10.1111/j.1476-5381.1985.tb12941.x

9. Black JW, Leff P (1983) Operational models of pharmacological agonism. Proc R Soc London: Biol Sci. https://doi.org/10.1098/rspb.1983.0093

10. Vauquelin G (2016) Effects of target binding kinetics on in vivo drug efficacy: koff, kon and rebinding. Br J Pharmacol 173:2319–2334. https://doi.org/10.1111/bph.13504

11. Charlton SJ, Vauquelin G (2010) Elusive equilibrium: The challenge of interpreting receptor pharmacology using calcium assays. Br J Pharmacol 161:1250–1265. https://doi.org/10.1111/j.1476-5381.2010.00863.x

12. van der Velden WJC, Heitman LH, Rosenkilde MM (2020) Perspective: implications of ligand-receptor binding kinetics for therapeutic targeting of G protein-coupled receptors. ACS Pharmacol Transl Sci 3:179–189. https://doi.org/10.1021/acsptsci.0c00012

13. Brunner HR, Goenaga J, Wittek R et al (2002) Role of prohormone convertases in pro-neuropeptide Y processing: coexpression and in vitro kinetic investigations. Biochemistry 36:16309–16320. https://doi.org/10.1021/bi9714767

14. de Witte WEA, Danhof M, van der Graaf PH, de Lange ECM (2016) In vivo target residence time and kinetic selectivity: the association rate

constant as determinant. Trends Pharmacol Sci 37:831–842. https://doi.org/10.1016/j.tips.2016.06.008

15. Schuetz DA, de Witte WEA, Wong YC et al (2017) Kinetics for drug discovery: an industry-driven effort to target drug residence time. Drug Discov Today 22:896–911. https://doi.org/10.1016/j.drudis.2017.02.002

16. Sykes DA, Moore H, Stott L et al (2017) Extrapyramidal side effects of antipsychotics are linked to their association kinetics at dopamine D2 receptors. Nat Commun 8:763. https://doi.org/10.1038/s41467-017-00716-z

17. Sykes DA, Bradley ME, Riddy DM et al (2016) Fevipiprant (QAW039), a slowly dissociating CRTh2 antagonist with the potential for improved clinical efficacy. Mol Pharmacol 89:593–605. https://doi.org/10.1124/mol.115.101832

18. Sykes DA, Dowling MR, Leighton-Davies J et al (2012) The Influence of receptor kinetics on the onset and duration of action and the therapeutic index of NVA237 and tiotropium. J Pharmacol Exp Ther 343:520–528. https://doi.org/10.1124/jpet.112.194456

19. Scimemi A, Beato M (2009) Determining the neurotransmitter concentration profile at active synapses. Mol Neurobiol 40:289–306. https://doi.org/10.1007/s12035-009-8087-7

20. Grundmann M, Kostenis E (2017) Temporal bias: time-encoded dynamic GPCR signaling. Trends Pharmacol Sci 38:1110–1124. https://doi.org/10.1016/j.tips.2017.09.004

21. Bdioui S, Kenakin T, Pierre N et al (2018) Equilibrium assays are required to accurately characterize the activity profiles of drugs modulating Gq-protein-coupled receptors. Mol Pharmacol 94:992–1006. https://doi.org/10.1124/mol.118.112573

22. Roberts MJ, Broome RE, Kent TC et al (2018) The inhibition of human lung fibroblast proliferation and differentiation by Gs-coupled receptors is not predicted by the magnitude of cAMP response. Respir Res 19:56. https://doi.org/10.1186/s12931-018-0759-2

23. Klein Herenbrink C, Sykes DA, Donthamsetti P et al (2016) The role of kinetic context in apparent biased agonism at GPCRs. Nat Commun 7:10842. https://doi.org/10.1038/ncomms10842

24. Paredes RM, Etzler JC, Watts LT et al (2008) Chemical calcium indicators. Methods 46:143–151. https://doi.org/10.1016/j.ymeth.2008.09.025

25. Halls ML, Canals M (2018) Genetically encoded FRET biosensors to illuminate compartmentalised GPCR signalling. Trends Pharmacol Sci 39:148–157. https://doi.org/10.1016/j.tips.2017.09.005

26. Haider RS, Godbole A, Hoffmann C (2018) To sense or not to sense-new insights from GPCR-based and arrestin-based biosensors. Curr Opin Cell Biol 57:16–24. https://doi.org/10.1016/j.ceb.2018.10.005

27. Maryu G, Miura H, Uda Y et al (2018) Live-cell imaging with genetically encoded protein kinase activity reporters. Cell Struct Funct 43:61–74. https://doi.org/10.1247/csf.18003

28. Sanford L, Palmer A (2017) Recent advances in development of genetically encoded fluorescent sensors. Methods Enzymol 589:1–49. https://doi.org/10.1016/bs.mie.2017.01.019

29. Dale NC, Johnstone EKM, White CW, Pfleger KDG (2019) NanoBRET: the bright future of proximity-based assays. Front Bioeng Biotechnol 7:56. https://doi.org/10.3389/fbioe.2019.00056

30. El Khamlichi C, Reverchon-Assadi F, Hervouet-Coste N et al (2019) Bioluminescence resonance energy transfer as a method to study protein–protein interactions: application to g protein coupled receptor biology. Molecules 24. https://doi.org/10.3390/molecules24030537

31. Michnick SW, Ear PH, Manderson EN et al (2007) Universal strategies in research and drug discovery based on protein-fragment complementation assays. Nat Rev Drug Discov 6:569–582. https://doi.org/10.1038/nrd2311

32. Carter AA, Hill SJ (2005) Characterization of isoprenaline- and salmeterol-stimulated interactions between beta2-adrenoceptors and beta-arrestin 2 using beta-galactosidase complementation in C2C12 cells. J Pharmacol Exp Ther 315:839–848. https://doi.org/10.1124/jpet.105.088914

33. Kerppola TK (2008) Bimolecular fluorescence complementation (BiFC) Analysis as a probe of protein interactions in living cells. Annu Rev Biophys 37:465–487. https://doi.org/10.1146/annurev.biophys.37.032807.125842

34. Kilpatrick LE, Humphrys LJ, Holliday ND (2015) A G protein-coupled receptor dimer imaging assay reveals selectively modified pharmacology of neuropeptide Y Y1/Y5 receptor heterodimers. Mol Pharmacol 87:718–732. https://doi.org/10.1124/mol.114.095356

35. Kilpatrick LE, Briddon SJ, Hill SJ, Holliday ND (2010) Quantitative analysis of neuropeptide y receptor association with β-arrestin2 measured by bimolecular fluorescence complementation. Br J Pharmacol 160:892–906. https://doi.org/10.1111/j.1476-5381.2010.00676.x

36. Takakura H, Hattori M, Takeuchi M, Ozawa T (2012) Visualization and quantitative analysis of G protein-coupled receptor-β-arrestin interaction in single cells and specific organs of living mice using split luciferase complementation. ACS Chem Biol 7:901–910. https://doi.org/10.1021/cb200360z

37. Hall MP, Unch J, Binkowski BF et al (2012) Engineered luciferase reporter from a deep sea shrimp utilizing a novel imidazopyrazinone substrate. ACS Chem Biol 7:1848–1857. https://doi.org/10.1021/cb3002478

38. Yano H, Cai NS, Javitch JA, Ferré S (2018) Luciferase complementation based-detection of G-protein-coupled receptor activity. BioTechniques 65:9–14. https://doi.org/10.2144/btn-2018-0039

39. Dixon AS, Schwinn MK, Hall MP et al (2016) NanoLuc complementation reporter optimized for accurate measurement of protein interactions in cells. ACS Chem Biol 11:400–408. https://doi.org/10.1021/acschembio.5b00753

40. Wan Q, Okashah N, Inoue A et al (2018) Mini G protein probes for active G protein-coupled receptors (GPCRs) in live cells. J Biol Chem 293:7466–7473. https://doi.org/10.1074/jbc.RA118.001975

41. Reyes-Alcaraz A, Lee Y-N, Yun S et al (2018) Conformational signatures in beta-arrestin2 reveal natural biased agonism at a G-protein-coupled receptor. Commun Biol 1:128. https://doi.org/10.1038/s42003-018-0134-3

42. Storme J, Cannaert A, Van Craenenbroeck K, Stove CP (2018) Molecular dissection of the human A3 adenosine receptor coupling with beta-arrestin2. Biochem Pharmacol 148:298–307. https://doi.org/10.1016/j.bcp.2018.01.008

43. Storme J, Tosh DK, Gao Z-G et al (2018) Probing structure-activity relationship in β-arrestin2 recruitment of diversely substituted adenosine derivatives. Biochem Pharmacol 158:103–113. https://doi.org/10.1016/j.bcp.2018.10.003

44. Inoue A, Raimondi F, Kadji FMN et al (2019) Illuminating G-protein-coupling selectivity of GPCRs. Cell 177:1933–1947.e25. https://doi.org/10.1016/j.cell.2019.04.044

45. White CW, Caspar B, Vanyai HK et al (2020) CRISPR-mediated protein tagging with nanoluciferase to investigate native chemokine receptor function and conformational changes. Cell Chem Biol 27:499–510.e7. https://doi.org/10.1016/j.chembiol.2020.01.010

46. Carpenter B, Tate CG (2016) Engineering a minimal G protein to facilitate crystallisation of G protein-coupled receptors in their active conformation. Protein Eng Des Sel 29:583–594. https://doi.org/10.1093/protein/gzw049

47. van der Westhuizen ET, Breton B, Christopoulos A, Bouvier M (2013) Quantification of ligand bias for clinically relevant 2-adrenergic receptor ligands: implications for drug taxonomy. Mol Pharmacol 85:492–509. https://doi.org/10.1124/mol.113.088880

48. Jorgensen R, Holliday ND, Hansen JL et al (2008) Characterization of G-protein coupled receptor kinase interaction with the neurokinin-1 receptor using bioluminescence resonance energy transfer. Mol Pharmacol 73:349–358. https://doi.org/10.1124/mol.107.038877

49. Galés C, Rebois RV, Hogue M et al (2005) Real-time monitoring of receptor and G-protein interactions in living cells. Nat Methods 2:177–184. https://doi.org/10.1038/nmeth743

50. Watson S-JS-J, Brown AJH, Holliday ND (2012) Differential signaling by splice variants of the human free fatty acid receptor GPR120. Mol Pharmacol 81:631–642. https://doi.org/10.1124/mol.111.077388

51. Valentin-Hansen L, Groenen M, Nygaard R et al (2012) The arginine of the DRY motif in transmembrane segment III functions as a balancing micro-switch in the activation of the beta2-adrenergic receptor. J Biol Chem 287:31973–31982. https://doi.org/10.1074/jbc.M112.348565

52. ARUNLAKSHANA O, SCHILD HO (1959) Some quantitative uses of drug antagonists. Br J Pharmacol Chemother 14:48–58. https://doi.org/10.1111/j.1476-5381.1959.tb00928.x

53. Hoare SRJ, Pierre N, Moya AG, Larson B (2018) Kinetic operational models of agonism for G-protein-coupled receptors. J Theor Biol 446:168–204. https://doi.org/10.1016/j.jtbi.2018.02.014

54. McPherson J, Rivero G, Baptist M et al (2010) mu-opioid receptors: correlation of agonist efficacy for signalling with ability to activate internalization. Mol Pharmacol 78:756–766. https://doi.org/10.1124/mol.110.066613

Chapter 18

Gradient Tracking by Yeast GPCRs in a Microfluidics Chamber

Sara Kimiko Suzuki, Joshua B. Kelley, Timothy C. Elston, and Henrik G. Dohlman

Abstract

Cells typically exist in a highly dynamic environment, which cannot easily be recreated in culture dishes or microwell plates. Microfluidic devices can provide precise control of the time, dose, and orientation of a stimulus, while simultaneously capturing quantitative single-cell data. The approach is particularly powerful when combined with the genetically tractable yeast model organism. The GPCR pathway in yeast is structurally conserved and functionally interchangeable with those in humans. We describe the implementation of a microfluidic device to investigate morphological and transcriptional responses of yeast to a gradient or pulse administration of a GPCR ligand, the peptide mating pheromone α-factor.

Key words G protein-coupled receptor, Microfluidics, Gradient, Cell biology, Yeast

1 Introduction

Studies of GPCR signaling are usually done with spatially uniform and unchanging concentrations of a ligand. In a physiological setting however, cells are likely to be exposed to a directed and fluctuating stimulus. In this regard, an emerging strategy is to use microfluidic chambers to monitor responses to a gradient or pulsatile stimulus. While technically challenging, this is the only method that can assess the ability of an individual cell to properly track a gradient over time.

Haploid yeast cells use pheromone-binding GPCRs to detect the presence of a potential mating partner (Fig. 1a). As a consequence of activation, the cells will elongate in the direction of the pheromone gradient in order to increase the likelihood of successful mating [1, 2]. Traditional methods, such as the partner

Electronic supplementary material: The online version of this chapter (https://doi.org/10.1007/978-1-0716-1221-7_18) contains supplementary material, which is available to authorized users.

Sofia Aires M. Martins and Duarte Miguel F. Prazeres (eds.), *G Protein-Coupled Receptor Screening Assays: Methods and Protocols*, Methods in Molecular Biology, vol. 2268, https://doi.org/10.1007/978-1-0716-1221-7_18,

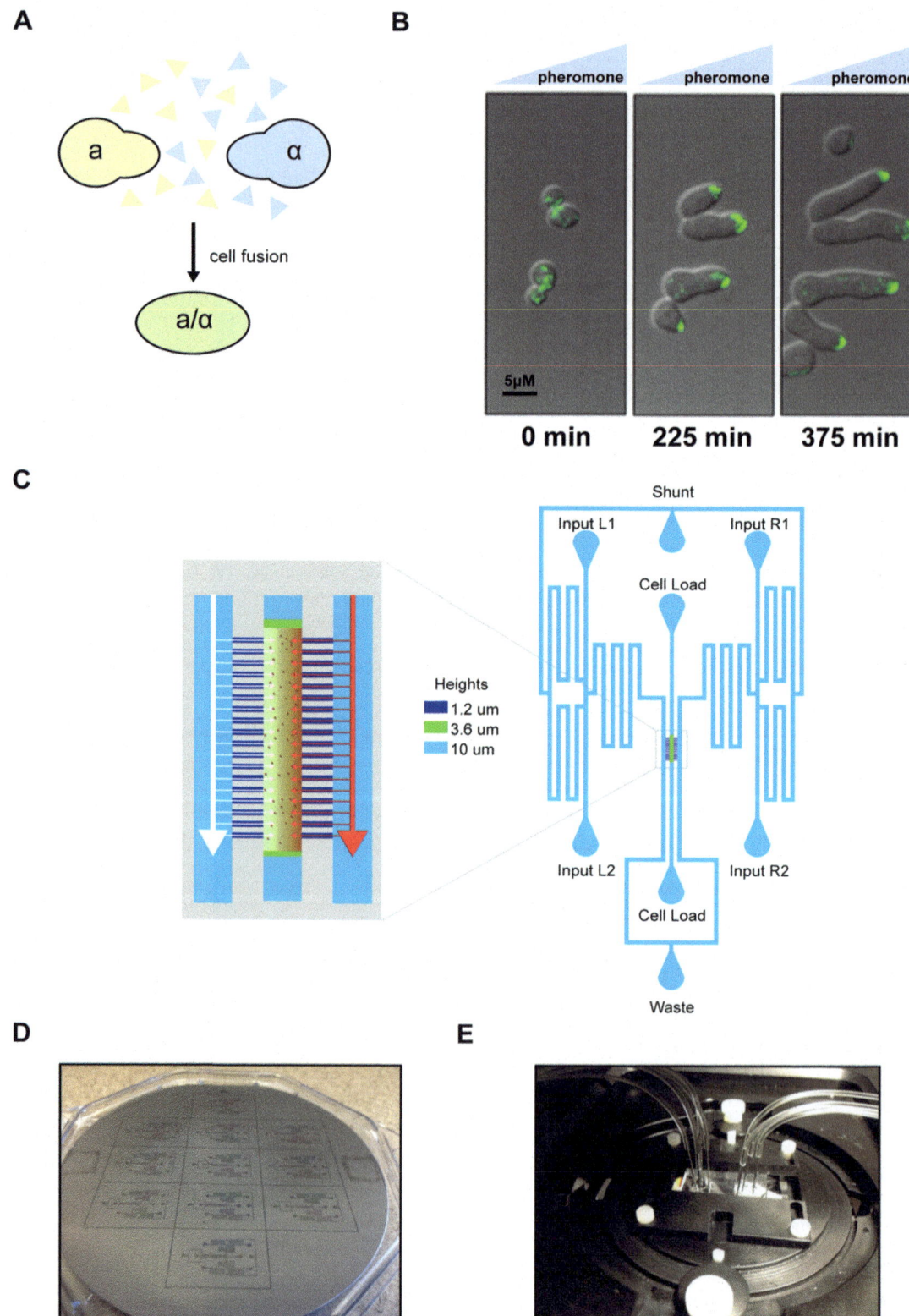

Fig. 1 Microfluidic device for tracking single-cell response to pulses or gradients of stimuli. (**a**) a yeast cell and an α yeast cell chemotropically grow in the direction of the opposite mating pheromone. Once the cells mate,

discrimination and pheromone confusion assays, provide indirect measures of gradient tracking, do not allow for rapid changes in growth conditions and do not provide single-cell resolution (reviewed in [3]). Other technologies such as flow cytometry allow the acquisition of single-cell data but cannot be used to track individual cells over time. Imaging cytometers can track individual cells over time, but cannot be used to increase or decrease the stimulus, in either space or time [4]. The purpose of this chapter is to share our experiences using microfluidics devices to investigate GPCR-mediated signal regulation in yeast. This is meant to be a practical guide. For a more detailed explanation of chip design, theory, and optimization we refer readers to a comprehensive review published by the developers of the technology used by us and described below [5].

In yeast, G protein-coupled receptors respond to mating pheromones produced by cells of the opposite mating type. When fully activated, these receptors stimulate MAPK phosphorylation, transcriptional regulation, cell cycle arrest, and formation of a pear-shaped projection or "shmoo." In addition to dividing and shmooing cells, a third morphogenic state is evident at intermediate pheromone concentrations, where cells have stopped dividing but continue to grow in the direction of a weak pheromone gradient [6, 7]. We refer to this morphology as "elongated" growth. When cells are expanding specifically in the direction of a pheromone gradient, we refer to this as "chemotropic" growth (Fig. 1b). The chemotropic growth state is functionally analogous to important GPCR-mediated processes in animals, such as neuronal outgrowth and migration of macrophages in pursuit of a bacterial invader.

The microfluidic-based methods described below have allowed investigators to identify new proteins and processes involved in gradient tracking (*versus* elongated growth, which occurs in response to a uniform stimulus and does not require the use of a microfluidic device). We use a custom-built microfluidics chamber made of polydimethylsiloxane (PDMS), originally developed in the laboratory of Jeff Hasty, capable of producing a linear concentration gradient of pheromone (Fig. 1c) [2, 8, 9]. This device is a variation on the Hasty Lab's Dial-A-Wave design and is capable of generating a gradient in two different directions and switching between them [10–13]. PDMS is optically transparent, permeable to biologically important gases, chemically and thermally stable,

Fig. 1 (continued) they form an **a/**α diploid cell. **(b)** BY4741 cells expressing Bem1-GFP grow in the direction of the pheromone (α-factor) gradient. Bem1 localizes to the polar cap in the direction of cell growth. **(c)** Schematic of the gradient chamber with inputs and outputs labeled. Inset shows the pheromone gradient. **(d)** Image of the gradient chamber wafer mold. **(e)** Image of the gradient chamber positioned on the microscope. The eight needles are inserted into the ports of the chamber

and it does not absorb water. The gradient is achieved by passive diffusion between two parallel channels, one containing standard growth medium and the other growth medium plus pheromone. Because there is no direct flow within the growth chamber, the cells remain stationary throughout the experiment. These devices, or ones of similar design, have been used to identify proteins that are essential for gradient tracking; these include the regulator of G protein signaling (RGS) protein Sst2 [6, 8, 9], the effector MAPK Fus3 [7, 14–16], and the MAPK scaffold protein Ste5 [2]. Sst2 accelerates G protein inactivation, thereby focusing the region of GPCR-initiated signaling to the region of highest pheromone concentration. Sst2 is also required for proper assembly of the septin collar, which further limits the wandering of activated G proteins and receptors from the site of polarized cell expansion [9, 17].

Finally, the pheromone pathway has been adapted for the systematic identification of ligands for human GPCRs expressed in yeast [18]. Thus, analysis of G protein signaling in yeast positions us to understand more complex systems (such as humans) and eventually predict which interventions (such as pharmaceuticals) and physiological circumstances (such as pulses and gradients) afford the most effective treatments with the fewest side effects.

2   Materials

1. SYLGARD 184 Silicone Elastomer, supplied as two-part liquid (base + curing agent) component kits.

2. Mold template (Fig. 1d).

3. 0.5-mm Puncher.

4. Plasma cleaner.

5. Tygon tubing, 0.020 in. inner diameter.

6. 23-gauge Luer stub.

7. 21-gauge Luer stub.

8. 5 mL Syringes.

9. Pheromone (α-factor).

10. Alexa Fluor 647 dye.

11. Yeast growth media (SCD medium).

12. Small alligator clips.

13. Petri dishes.

14. Vaccum chamber.

15. Razor blades.

16. Scotch Tape.

17. ACS grade methanol ($\geq$99.8%).

18. Microscope with fluorescence power source.

19. 70% Ethanol made with filtered water.

20. Yeast cells (e.g., BY4741) [19], available from ATCC (www. atcc.org).

3 Methods

The complete fabrication of a microfluidic chip consists of three steps: (i) creation of a patterned wafer mold, by photolithography, (ii) using the mold to create a silicon rubber chamber, through the process of soft lithography, and (iii) bonding of the silicon chamber to a glass coverslip to produce a functional microfluidic device. The photolithography step creates reusable master molds through the use of photoresists, which are viscous chemicals spun on silicon wafers to very precise heights. When exposed to ultraviolet (UV) light, the photoresist becomes resistant to solvent producing a negative image with device features. We limit our discussion to the last two steps, which can be carried out in a typical lab space and in a standard chemical fume hood. The first (photolithography) step requires a specialized cleanroom available at major research universities, and entails considerable trial-and-error (the height of the trap is a critical feature; at the optimal height of about 3.5 μm the cells will be wedged between the ceiling and the glass cover slip. If it is too high, cells will not be trapped. If too low, the cells will never enter the trap). The mask used to print the wafer mold is produced on a polyester Mylar sheet coated with photographic silver, a process that is best done commercially using a freely distributed design (prepared by AutoCAD and transmitted as a *.dwg file, *see* Electronic Supplementary file).

3.1 Pouring Chambers

1. To prepare PDMS, combine 36 g of SYLGARD 184 silicone elastomer base and 4 g of SYLGARD 184 silicone elastomer curing agent in a plastic weigh boat (*see* Video S1 and **Note 1**).

2. Using a plastic pipette or clean glass rod, mix the liquids thoroughly for >2 min (*see* **Note 2**).

3. Place the mold in a second weigh boat and pour the PDMS mixture on top of the mold, transferring as much mixture as possible by squeezing the weigh boat.

4. Place the mold in a vacuum chamber for 30–60 min to remove bubbles (*see* **Note 3**).

5. Using a pipette tip, remove any remaining bubbles and any dust particles near the features of the mold.

6. Heat for ~1 h at 68–70 °C, or according to manufacturer's instructions.

7. Gently cut the corners of the weigh boat and remove the mold and PDMS (*see* Video S2 and **Note 4**).

8. Use a razor blade to separate any PDMS from the bottom of the mold.

9. Use gravity to slowly detach the PDMS from the mold.

10. Store the mold for future reuse.

3.2 Fusing the Chamber

1. Cover the feature side of the chambers with clear tape (*see* Video S3).

2. Using a razor blade cut out one chamber.

3. Hold the chamber to the light to check that the features are intact.

4. Use a permanent ink pen or marker and mark the tape to indicate the location of the ports.

5. Flip the chamber so the tape is on the bottom.

6. Push the 21-gauge Luer stub through the PDMS and tape, on top of the port marking. Remove any PDMS that is inside the puncher.

7. Hold the chamber down with forceps and pull out the 21-gauge Luer stub.

8. Repeat for all of the remaining ports.

9. Check that there is overlap between the punched holes and ports.

10. Remove tape. Fill a syringe with filtered deionized water and push the water through each port to make sure it is clear. Dry the chamber with a Kimwipe.

11. Place adhesive tape on both sides of the chamber and press the feature side to remove any dust. Repeat 3–4 times. Remove the tape.

12. Wash the chamber with ACS grade methanol.

13. Wash the chamber with 70% ethanol made with filtered water.

14. Wash the chamber with filtered deionized water.

15. Blow the chamber dry using filtered compressed air and place in a clean petri dish with the feature side facing up.

16. Clean a glass slide with methanol, ethanol, and water (as described above), dry, and place in the petri dish.

17. Place the glass slide and chamber (feature side facing upward) in the plasma cleaner (*see* **Notes 5, 6** and **7**).

18. With the chamber sealed and the valve closed, turn on plasma cleaner and pump.

19. Turn radiofrequency (RF) to "High." Plasma cleaner should display a dark purple color. Open the valve until the plasma cleaner glows a bright magenta/pink color. Allow to clean for 30–45 s. Treatments longer than a minute can result in decreased efficiency of fusion.

20. Turn the RF to "off," turn off the pump, and gently open the vacuum seal.

21. Using forceps, place the slide glass on top of the petri dish, making sure not to touch the surface of the slide.

22. Place the chamber on top of the slide with features face down, making sure not to touch the face of the chamber (*see* Video S4 and **Notes 8** and **9**).

3.3 Setting Up Yeast Cultures

1. Use freshly streaked cells (<2 weeks). The night before an experiment, inoculate for overnight growth in filter-sterilized SCD medium (*see* **Notes 10** and **11**).

2. The following day dilute 1:50 or 1:100 into 5 mL of fresh SCD medium and grow to an optical density (OD 600 nm) below 0.5.

3.4 Prepare Syringes

1. Turn on microscope objective heater and set to 30 °C to allow time for it to reach the target temperature.

2. Prepare 8 lines (4× ~1 m long and 4× ~0.6 m long) by inserting a 23-gauge needle (made from a 23-gauge Luer stub) into one end of the tubing and attaching another 23-gauge Luer stub to the other end. Cells will be added to one of the syringes with a shorter tubing (~0.6 m long) (*see* Video S5 and **Note 12**).

3. Prepare a conical tube with SCD containing pheromone plus a 1:10,000 dilution of stock Alexa Fluor 647 dye, to visualize the presence of pheromone in the mating chamber (*see* **Note 13**).

4. Using a pipetter or small syringe, fill the Luer stub with media and remove any bubbles (*see* Video S6).

5. Connect a syringe to the Luer stub and add 4.5–5 mL of SCD medium with pheromone + dye to a syringe with long tubing. Mark this line to distinguish it from the others. Add 4.5–5 mL SCD medium to the 7 remaining syringes.

6. Using the syringe plunger, push enough medium through the tubing to remove any air bubbles and clamp with an alligator clip. Any remaining air bubbles can be dislodged by knocking the Luer stub at an upward angle.

7. While the tubing is still clamped, sterilize the needle-end of the tubing by passing it though a flame, insert the needle into the top of the syringe, and secure it with a plunger.

8. Repeat for all syringes.

3.5 Microscope Set Up

1. Turn all equipment on (power source, laser, microscope, camera, and perfect focus system) and open image acquisition software (Metamorph or NIS-Elements, for example). Set the microscope to 40× objective and to DIC imaging.

2. Place the chamber in a chamber adapter (3D printed to hold the chamber and fit into a 60 mm Petri Dish mechanical stage insert) and center the slide. Place a bumper on each corner of the slide and position the sliders on top of the bumpers. The sliders should be placed just on the edge of the PDMS. Make sure the slide is centered and lock the sliders into position (*see* Video S7).

3. Place the chamber holder on the stage and lock in position.

4. Begin to fill the chamber with media by connecting tubing that contains media without dye (*see* Video S8). Pull the tubing tip out of the top of the syringe and plunge ~1 mL media through the tubing to clear any air. Locate the port by using the tubing tip to "scan" the surface of the chamber, then gently push the tip into the chamber (*see* **Note 14**). The plunger should still be in the syringe, so media will be coming out. Absorb any media with a Kimwipe to avoid dripping any media and damaging the microscope.

5. Once a needle is in its port, push some media through. Media should emerge from the other ports. Clamp the line and remove the needle.

6. Using the microscope, check that all channels are clear of air bubbles. Especially vulnerable areas are ports and the microchannels that lead to the imaging chamber. If air bubbles are present, remove them by gently pushing media through the port with a syringe.

7. Once the last port has been checked for bubbles, make sure there is a drop of media above each port. This ensures that new bubbles are not introduced into the channels. Clamp the line and remove the needle.

8. Remove the plunger from the clamped line and check that media flows freely from the line using gravity. Connect the needle to the chamber while the syringe is above the chamber, to ensure no bubbles are introduced.

9. Repeat Subheading 3.5, **step 8** for all ports, plunging at least 1 mL media through the line before checking that media flows freely using gravity (*see* **Note 15**). Once all of the needles have been connected, the chamber should now look like Fig. 1e.

10. Check that air bubbles were not introduced into the lines. If bubbles are present, return to Subheading 3.5, **step 7** for that line.

11. To alter the mixing ratio between two source reservoirs, we use a stepper-motor to regulate (by gravity) the hydrostatic pressure of the inlets [5] (*see* **Note 16**). The syringes are moved to a height that provides the desired flow to the cell trap. When the syringes are at the same height each provides 50% of the input. They are then moved in equal and opposite directions until there is only media from one source or the other; this becomes the 100% point for a single input.

12. If using the motor, calibrate the motor and load the motor program. Ensure that the chamber switches from stimulus to no stimulus and vice versa. Switching time should be 30 s or less.

13. Set the chamber to no stimulus (non-stimulus syringes are higher than those with stimulus) and begin the cell-loading process.

14. Change the objective to the imaging objective (usually $60\times$ or $100\times$). If this is an oil objective, place enough oil on the objective. Gently place the chamber on top of the objective and lock it in place.

15. With a pipette remove most of the media from the cell-loading syringe. Leave just enough to fill the Luer stub.

16. Pour cell culture into the cell-loading syringe.

17. Raise the cell-loading syringe above the shunt to begin loading. The higher above the shunt, the faster the loading will occur. Begin with ~20 cm above the shunt. This will create flow along the chamber. It can take a few minutes to see cells flowing into the chamber depending on cell density and flow rate (*see* **Note 17**).

18. Wait for about 2 rows of cells to collect above the cell trap. Then lower the syringe to a little above the syringe connected to the top shunt. When the pressure is correct, small movement of the tubing will cause the cells to briefly bounce up, and rapidly return to the edge of the cell trap. Holding the tubing between the thumb and forefinger of one hand, gently flick the tubing into the palm of that hand using the other. Loading is an art.

19. Repeat 3–4 times until the trap contains 100–200 evenly dispersed cells. Fewer flicks are better (*see* **Notes 18** and **19**).

20. Once loaded, equalize the cell load and cell waste syringes and move both well below the shunt.

21. Tape tubing to the stage, each to its own side so that tubing is not crossing the device. This will reduce the chances, in a multi-position experiment, that tension on the tubing causes a shift in the XY-position of the device.

3.6 Image Acquisition

1. Set Köhler illumination. Make sure that the condenser does not touch the tubing. Depending on the working distance of your condenser, you may have to sacrifice perfect Köhler illumination to avoid hitting the tubing (*see* **Notes 20–26**).

2. During the first acquisition, make sure all stage positions are in focus.

3. On the second or third acquisition, begin stimulus by raising the appropriate inputs to a higher elevation.

3.7 Clean Up

1. Disconnect all tubing from the chamber and close the plungers on the tips.

2. Discard the chamber, syringes, media, and waste in the appropriate manner.

3. Connect all tubing to a vacuum manifold. Place the tips in 1 L of filtered deionized water and turn on the vacuum. Follow with 15 mL 70% EtOH, to sterilize (*see* **Note 27**).

4. Make sure all liquid has been vacuumed and the tubing is completely dry. If not, push air through the tubing with a 30 mL syringe until it is dry.

5. Alternatively, tubing can be cleaned with a 30 mL syringe by washing with water, followed by 70% EtOH, and finally air.

4   Notes

1. Use nitrile instead of latex gloves during all stages of pouring and working with the mold. Latex gloves will prevent PDMS from curing. Objects touched with latex gloves will retain sufficient latex to stop PDMS from curing.

2. Inadequate mixing will cause uneven curing. The liquid should appear foamy if mixed properly.

3. To accelerate removal of the bubbles, periodically open the vacuum chamber valve for 10 s and close it again. The influx of air will cause bubbles at the surface to pop.

4. If final PDMS is tacky or sticky, something is wrong. Discard and either try again or use a new PDMS kit. If final PDMS is too hard, try a 1:10 ratio of curing agent to base for a softer PDMS or bake for less time. If there are frequently tears in the ports, the PDMS may be too hard.

5. The purpose of the plasma cleaner is to replace Si–CH$_3$ bonds with Si–OH bonds.

6. Both the chamber and the glass slide must be very clean and completely dry to ensure complete fusion.

7. Lint and other contaminants will prevent fusion of the chamber and slide. Ensure that gloves remain free of lint. Ensure that the chamber is clear and without smudges. Powder from powdered gloves is problematic, and should be avoided or washed off.

8. The slide and chamber should fuse instantly. If this does not happen, press down the corners of the chamber. Be careful not to press down on the features at the center of the chamber.

9. Check that the slide and chamber fused properly. If not, place the petri dish containing the unfused chamber in the 58–60 °C oven for 15–30 min until fusion is complete. Overbaking can cause the device to become brittle, and a short time is generally sufficient to set the fusion.

10. Use filter-sterilized medium (using a 0.22 μm filtration system) rather than autoclaved medium, to reduce particulate matter that could clog the microfluidic device. For experiments using fluorescence, use Synthetic Complete media to minimize autofluorescence.

11. We use strains that contain integrated Bem1-GFP to visualize the polar cap. Bem1 binds to activated Cdc42 [20] and promotes actin polymerization at, and polarized exocytosis to, the site of cell expansion [21] (*see* Fig. 1b).

12. Label the lines for reuse with the same ports. Lines that contain treatment or cells should only be reused for the same input solutions.

13. Use pheromone concentrations matched to the sensitivity of the individual strains; for example: 0–150 nM for wild type, 0–50 nM for cells lacking the pheromone protease Bar1.

14. Be careful not to push the tip into the PDMS where there is not a hole already punched. Too much force can break the slide.

15. Look for any signs of leakage (media leaking from the sides of the chamber or from the ports). If a small amount of media is leaking from the ports, it is possible to stop the leakage by taping a strategically positioned Kimwipe to the stage or to superglue the tear. If using superglue, first clamp the line to stop flow and use the minimum amount of glue. The glue can enter into the chamber and ruin it and the tubing.

16. The actuator system for the syringes could consist of a manually operated guide and pulley system. However, we use the automated Multiple Dial-A-Wave system, developed by Jeff Hasty and colleagues. This system features two linear actuator controllers, which connect the actuators to a single communication gateway module used for communication to a computer. A recent review lists all the required parts for building the system and links to the software [5].

17. Do not push in the cells using the syringe; this will force the cells into the channels on either side of the chamber.

18. If the chamber becomes overloaded, start over with a new chamber.

19. For a more even distribution of cells throughout the chamber, switch the cell load syringe with the cell waste syringe and load from the bottom cell trap.

20. We use a Nikon Ti-E inverted fluorescence microscope with Perfect Focus, coupled with a Hamamatsu Orca-flash 4.0 digital camera and an XLED Multi-Triggering LED Illumination System (Excelitas). Images are taken using a Nikon Plan Apo VC X60 oil immersion objective (NA 1.40 WD 0.17 MM).

21. We take images every 10 min in the brightfield, far-red, and green channels.

22. We use the lowest LED intensity setting, to prevent photobleaching and phototoxicity.

23. Our images are registered using the descriptor-based series registration (2d/3d + t) plugin based on the DIC images in ImageJ [22].

24. Focus on cells residing in the region of the chamber with the largest linear difference in pheromone concentration, as evaluated by the intensity of an inert dye in the pheromone solution.

25. To track the movement of the polar cap, we use Bem1-GFP in conjunction with the manual tracking plugin in ImageJ. Plots of the single polar cap traces and polar histograms of the angle of the traces can be generated in Python. Cells are segmented based on the GFP images in ImageJ, and kymographs of the GFP intensity around the edge of a cell are generated in MATLAB using code available at https://github.com/aeallen/pher-response-quantification. Segmentation is checked manually.

26. To quantify gradient tracking we trace the "angle of orientation," defined as the position of the polar cap relative to the direction of the gradient source, as a function of time, and is plotted as a kymograph [9]. Other time-dependent behaviors are "frequency of turning" (i.e., frequency of turns $>60°$), "memory" (the time period for which the current angle of orientation is correlated with future angles of orientation), and "persistence" (the difference between the position of the polar cap at the beginning and end of a fixed time interval, divided by the total length of the path travelled by the polar cap during that interval) [9].

27. Tubing is the most expensive part of these experiments and can be reused multiple times if it is thoroughly cleaned. Thus, tubing should be cleaned immediately to prevent contamination.

Acknowledgments

This work was supported by National Institutes of Health (NIH) T32 GM067553-12 (SKS), R15 GM128026 (JBK), R35 GM127145 (TCE), and R35 GM118105 (HGD).

References

1. Erdman S, Lin L, Malczynski M, Snyder M (1998) Pheromone-regulated genes required for yeast mating differentiation. J Cell Biol 140:461–483

2. Hao N, Nayak S, Behar M, Shanks RH, Nagiec MJ, Errede B, Hasty J, Elston TC, Dohlman HG (2008) Regulation of cell signaling dynamics by the protein kinase-scaffold Ste5. Mol Cell 30:649–656

3. Stone DE, Arkowitz RA (2016) In situ assays of chemotropism during yeast mating. Methods Mol Biol 1407:1–12

4. Shellhammer JP, Pomeroy AE, Li Y, Dujmusic L, Elston TC, Hao N, Dohlman HG (2019) Quantitative analysis of the yeast pheromone pathway. Yeast 36:495–518

5. Ferry MS, Razinkov IA, Hasty J (2011) Microfluidics for synthetic biology: from design to execution. Methods Enzymol 497:295–372

6. Segall JE (1993) Polarization of yeast cells in spatial gradients of alpha mating factor. Proc Natl Acad Sci U S A 90:8332–8336

7. Erdman S, Snyder M (2001) A filamentous growth response mediated by the yeast mating pathway. Genetics 159:919–928

8. Dixit G, Kelley JB, Houser JR, Elston TC, Dohlman HG (2014) Cellular noise suppression by the regulator of G protein signaling Sst2. Mol Cell 55:85–96

9. Kelley JB, Dixit G, Sheetz JB, Venkatapurapu SP, Elston TC, Dohlman HG (2015) RGS proteins and septins cooperate to promote chemotropism by regulating polar cap mobility. Curr Biol 25:275–285

10. Nayak S (2013) Investigating the single cell dynamics of Saccharomyces cerevisiae using microfluidics. PhD University of California, San Diego

11. Cookson S, Ostroff N, Pang WL, Volfson D, Hasty J (2005) Monitoring dynamics of single-cell gene expression over multiple cell cycles. Mol Syst Biol 1(2005):24

12. Bennett MR, Hasty J (2009) Microfluidic devices for measuring gene network dynamics in single cells. Nat Rev Genet 10:628–638

13. Jin M, Errede B, Behar M, Mather W, Nayak S, Hasty J, Dohlman HG, Elston TC (2011) Yeast dynamically modify their environment to achieve better mating efficiency. Sci Signal 4:ra54

14. Errede B, Vered L, Ford E, Pena MI, Elston TC (2015) Pheromone-induced morphogenesis and gradient tracking are dependent on the MAPK Fus3 binding to Galpha. Mol Biol Cell 26:3343–3358

15. Hegemann B, Unger M, Lee SS, Stoffel-Studer I, van den Heuvel J, Pelet S, Koeppl H, Peter M (2015) A cellular system for spatial signal decoding in chemical gradients. Dev Cell 35:458–470

16. Conlon P, Gelin-Licht R, Ganesan A, Zhang J, Levchenko A (2016) Single-cell dynamics and variability of MAPK activity in a yeast differentiation pathway. Proc Natl Acad Sci U S A 113: E5896–E5905

17. McClure AW, Minakova M, Dyer JM, Zyla TR, Elston TC, Lew DJ (2015) Role of polarized G protein signaling in tracking pheromone gradients. Dev Cell 35:471–482

18. Dowell SJ, Brown AJ (2009) Yeast assays for G protein-coupled receptors. Methods Mol Biol 552:213–229

19. Brachmann CB, Davies A, Cost GJ, Caputo E, Li J, Hieter P, Boeke JD (1998) Designer deletion strains derived from Saccharomyces cerevisiae S288C: a useful set of strains and plasmids for PCR-mediated gene disruption and other applications. Yeast 14:115–132

20. Madden K, Snyder M (1992) Specification of sites for polarized growth in Saccharomyces cerevisiae and the influence of external factors on site selection. Mol Biol Cell 3:1025–1035

21. Bi E, Park HO (2012) Cell polarization and cytokinesis in budding yeast. Genetics 191:347–387

22. Preibisch S, Saalfeld S, Schindelin J, Tomancak P (2010) Software for bead-based registration of selective plane illumination microscopy data. Nat Methods 7:418–419

Chapter 19

Monitoring Intracellular Calcium in Response to GPCR Activation: Comparison Between Microtiter Plates and Microfluidic Assays

Sofia Aires M. Martins, Duarte Miguel F. Prazeres, Virginia Chu, and João P. Conde

Abstract

Microfluidic strategies combined with transduction and electronic integration have the promise of enabling miniaturized, combinatorial assays at higher speeds and lower costs, while at the same time mimicking the local chemical concentrations and force fields of the cellular in vivo environment. In this chapter we introduce a microfluidic structure with hydrodynamic cell traps and a culture volume in the nanoliter range (50 nL), to quantitatively evaluate the transient calcium response of the endogenous Muscarinic type 1 receptor (M1) in HEK 293 T cells. The microfluidic fabrication protocol is described as well as a methodology to monitor the cell response in real time, after stimulation with M1 agonists (e.g., carbachol) and antagonists (e.g., pirenzepine).

Key words G-protein-coupled receptors, Endogenous muscarinic receptor, Calcium signaling, Microfluidics, Fluorescence

1 Introduction

G-protein-coupled receptors (GPCR) comprise one of the largest groups of membrane proteins involved in the translation of extracellular stimulus into intracellular responses [1, 2]. These receptors are involved in different biological processes including sensorial perception, cell proliferation, immune response, or neuronal activity [3], and their importance is reflected by the fact that approximately 30% of all marketable drugs are GPCR-acting compounds. Nevertheless, these drugs address only up to 35 members of the entire GPCR repertoire [4]. Moreover, endogenous ligands have

Electronic supplementary material: The online version of this chapter (https://doi.org/10.1007/978-1-0716-1221-7_19) contains supplementary material, which is available to authorized users.

Sofia Aires M. Martins and Duarte Miguel F. Prazeres (eds.), *G Protein-Coupled Receptor Screening Assays: Methods and Protocols*, Methods in Molecular Biology, vol. 2268, https://doi.org/10.1007/978-1-0716-1221-7_19,
© Springer Science+Business Media, LLC, part of Springer Nature 2021

yet to be identified for ~150 GPCRs (orphan GPCRs). Thus, screening for new targets remains an active and challenging task [5, 6].

A variety of compounds like hormones, neurotransmitters, proteins, photons, and small molecules activate GPCRs and initiate specific downstream signaling pathways such as cyclic adenosine monophosphate (cAMP), calcium (Ca^{2+}), or β-arrestin mediated signaling. Monitoring those effectors through functional cell-based assays constitutes an established strategy to identify GPCR targets. Typically, cells overexpressing the receptors of interest are challenged with candidate test compounds and the response signals are analyzed using fluorescent or luminescent read-outs. Conventional tissue culture protocols have been translated into microtiter plates, enabling these assays to run in high-throughput (HTS) or high-content (HCS) screening formats and test, in parallel, hundreds of compounds [7, 8].

Nevertheless, the existence of differences between the microtiter plate assays and the natural cell microenvironment is well recognized. A first downside concerns to the use of engineered cell lines, overexpressing the target receptor. Alternatively, the assaying of cells endogenously expressing the GPCR of interest would offer the advantage of studying the receptor at its normal expression levels, while replicating the interaction of the receptor with its natural regulatory partners [9]. A second drawback of plate assays lies on the discrete delivery of compounds to the cultured cells, which contrasts with the dynamic fluxes that cells experience in their natural environment.

Microfluidic-based GPCR screening assays are an attractive alternative to classical microplate assays. One important advantage of microfluidic technology is the ability to precisely control fluid flow with spatial and temporal resolution. By adopting proper designs and flow rates, shear stress and solute concentrations can be fine-tuned to mimic in vivo conditions [10, 11]. Additional advantages include the compatibility with automation and reduced assay footprints to save on handling requirements and reagent costs.

In this chapter, we explore the ability of using microfluidics to quantitatively evaluate the transient calcium response of endogenous Muscarinic type 1 (M1) receptor in HEK 293 T cells, induced by the action of the agonist carbachol. M1 receptors are usually coupled to the Gαq/11 type protein, which activates phospholipase C with the consequent transient rise of intracellular calcium (iCa^{2+}). The specificity of the assay was evaluated by stimulating the cells with pirenzepine, a well-known neutral antagonist of the M1 receptor. The addition of pirenzepine to the cell culture is expected to specifically block M1 activation, inhibiting agonist-induced signaling in a dose response manner [7]. The microfluidic device (Fig. 1) comprises a cell chamber (~50 nL cell culture volume) that is shifted

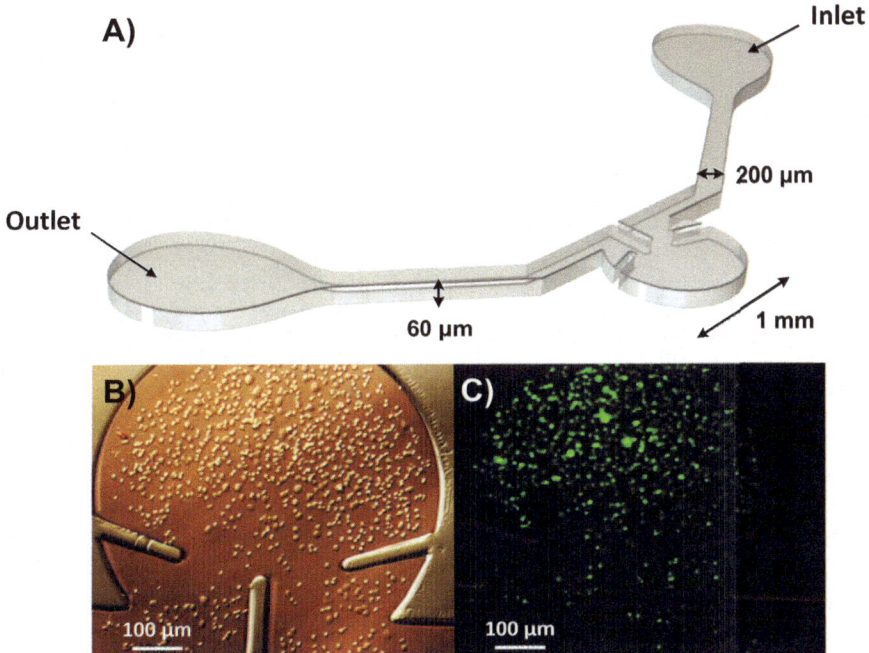

Fig. 1 Microfluidic hydrodynamic cell traps. (**a**) Schematic diagram of microfluidic cell trap with inlet and outlet channels. Key dimensions are highlighted. (**b**) Micrograph showing adherent HEK 293 T cells on cell trapping chamber. (**c**) Fluorescence micrograph showing the same region as (**b**). The cells were incubated with Fluo4 and the luminescence results from the presence of iCa^{2+}

from the main flow in the device to provide a low-velocity niche with hydrodynamic conditions suitable for cell settling and adhesion. Cell response was monitored by following, in real-time, the iCa^{2+} dynamics using fluorescence microscopy.

2 Materials

2.1 Cell Assay

1. Human Embryonic kidney (HEK) 293 T: American type culture collection, ATTC (ATTC CRL-11268).

2. Sterile T75 cell culture flasks.

3. Sterile Falcon Tubes of 15 mL.

4. Sterile 96-well microtiter plates.

5. Disposable hemocytometers.

6. CO_2 incubator.

7. Disposable micropipettes for cell culture handling.

8. Inverted microscope with fluorescence filters compatible with excitation (*Ex.*) at 490 nm and emission (*Em.*) at 520 nm.

9. Camera enabling real-time imaging acquisition.

10. Disposable vacuum filters with the capacity to filter 1 L solutions.

11. Water bath with temperature control.

2.2 Microfluidics Fabrication

1. Glass substrates.

2. Silicon substrates.

3. Microscopy glass slides.

4. Lithography equipment (*see* **Note 1**).

5. UV lamp.

6. Ultrasonic bath with temperature control.

7. Vacuum exicator.

8. Hot-plate with temperature control up to 150 °C.

9. UV ozone cleaner.

10. Oven with temperature control up to 70 °C.

11. Laurel Spin-coater (*see* **Note 2**).

12. Forceps to manipulate the substrates and channels.

13. Polyethylene (PE) capillary tubing for microfluidics infusion, 20 Gauge diameter.

14. Blunt end syringe tips, Luer Stub 20 Gauge.

15. 1 mL syringes.

16. Infusion pump.

2.3 Chemicals and Reagents

1. Acetone 99.5%.

2. Isopropyl alcohol 99%.

3. Positive photoresist: PFR7790G (*see* **Note 3**).

4. Positive photoresist developer: TMA238WA.

5. Aluminum etchant solution.

6. Negative photoresist: SU8-50 (*see* **Note 4**).

7. Propylene glycol methyl ether acetate (PGMEA) stock solution.

8. Sylgard® 184 polydimethyl siloxane (PDMS) polymer and corresponding curing agent.

9. Hank's balanced solution (HBSS): 26 mM $CaCl_2$, 0.493 mM $MgCl_2$, 0.407 mM $MgSO_4$, 5.33 mM KCl, 137 mM NaCl, 0.338 mM Na_2HPO_4, 5.5 mM D-glucose.

10. Stock solution (1 M) of 4-(2-hydroxyethyl)-1-piperazineethanesulfonic acid (HEPES).

11. Dulbecco's Modified Eagle medium (DMEM), high glucose + pyruvate + glutamine.

12. Fetal bovine serum (FBS).

13. Phosphate buffer saline (PBS): 1 mM KH_2PO_4, 155 mM NaCl, 3 mM Na_2HPO_4 filter sterilized.

14. Antibiotic-antimycotic solution (*see* **Note 5**).

15. Trypsin-0.05% EDTA solution.

16. 0.4% Trypan Blue solution.

17. Fibronectin (powder).

18. Carbachol (powder).

19. Probenecid (water soluble) solution: 250 mM in 1× HBSS. Store at −20 °C.

20. Assay buffer: 20 mM HEPES, 2.5 mM probenecid, 1× HBSS.

21. Pirenzepine (powder).

22. Fluor-4-Direct calcium dye (*see* **Note 6**).

23. Ethanol 96%.

2.4 Software

1. AutoCAD or other computer-aided design software.

2. Image acquisition software.

3. ImageJ or equivalent software for the analysis of fluorescence intensities.

4. GraphPad Prism.

---

3 Methods

3.1 Preparation of Working Cell Banks

1. Prepare a stock solution of complete cell culture medium: for 1 L of cell culture medium, add 890 mL of DMEM medium, 100 mL of FBS, and 10 mL of antibiotic-antimycotic solution (DMEM, 10% FBS, 1% antibiotic-antimycotic). Filter using a disposable vacuum filter. Perform all steps inside a laminar flow hood.

2. Warm the cell-containing cryovial (*see* **Note 7**), in the water bath at 37 °C for approximately 5 min to defrost the cell pellet.

3. Pipette 6 mL of complete cell culture medium (DMEM, 10% FBS, 1% antibiotic-antimycotic) into a 15 mL sterile Falcon tube. Add the cell pellet and mix well and with care, by inverting the tube several times.

4. Centrifuge at $200 \times g$ for 5 min, at 20 °C. Discard the supernatant and add 1 mL of complete cell culture medium. Homogenize the pellet by careful pipetting steps.

5. Add 9 mL of complete cell culture medium to a T75 culture flask and finally add the 1 mL of the previous prepared cell suspension solution. Mix gently by shaking the T75 flask.

6. Incubate the cells under a 5% CO_2 atmosphere at 37 °C. Change the culture medium after 48 h or when cells reach 80% confluency (*see* **Note 8**).

3.2 Microfluidics Fabrication

The microfluidic devices were fabricated using Soft Lithography techniques and UV Lithography. The fabrication methodology is described in this section.

1. Design the microfluidic 2D pattern in AutoCAD or other equivalent software.

2. For the fabrication of the hard mask, metalize the total area of a clean glass substrate (5 cm × 5 cm) with a 300 nm aluminum film (*see* **Note 9**). Spin coat a 1.5 μm film of positive photoresist using the spin coater (*see* **Note 2**).

3. Transfer the 2D design to the hard mask by exposing at 405 nm. We used the Heidelberg DWL II direct-write laser and the respective exposure protocol.

4. After exposure, solubilize the positive photoresist by incubating briefly with the photoresist developer until the aluminum film of the exposed areas (channels) is visible. Rinse with dH_2O and dry.

5. Remove the aluminum film in the exposed areas by wet-etch using an aluminum etchant solution (*see* **Note 10**). When the channels become visible in the glass substrate, stop the etching with dH_2O.

6. Remove the remaining, nonexposed photoresist by washing the substrate in acetone for 5 min, followed by a rinse with dH_2O. Dry under a stream of compressed air. In a correct hard mask, aluminum is removed from the glass only at the exposed areas, which in turn correspond to the 2D design of the channels (Fig. 2).

7. Clean a silicon substrate (5 cm × 5 cm) with Alconox detergent, using an ultrasonic bath at 65 °C and for 20 min.

8. Rinse with isopropyl alcohol and finish with one rinse with dH_2O. Dry the substrate under a stream of compressed air.

9. Perform a final cleaning step in the UV Ozone cleaner, programming one cycle of 10 min under UV light and 5 min for ventilation (*see* **Note 11**).

10. Spin-coat the SU8-50 negative photoresist with a thickness of 60 μm over the clean silicon substrate (*see* **Note 2**). Be careful not to the touch the SU8-50 film. Use forceps to handle the substrate.

11. Making use of the hot-plate, pre-bake the SU8-50-coated substrate at 65 °C for 3 min and at 95 °C for 8 min to evaporate the solvent. Let the substrate cool down to room temperature.

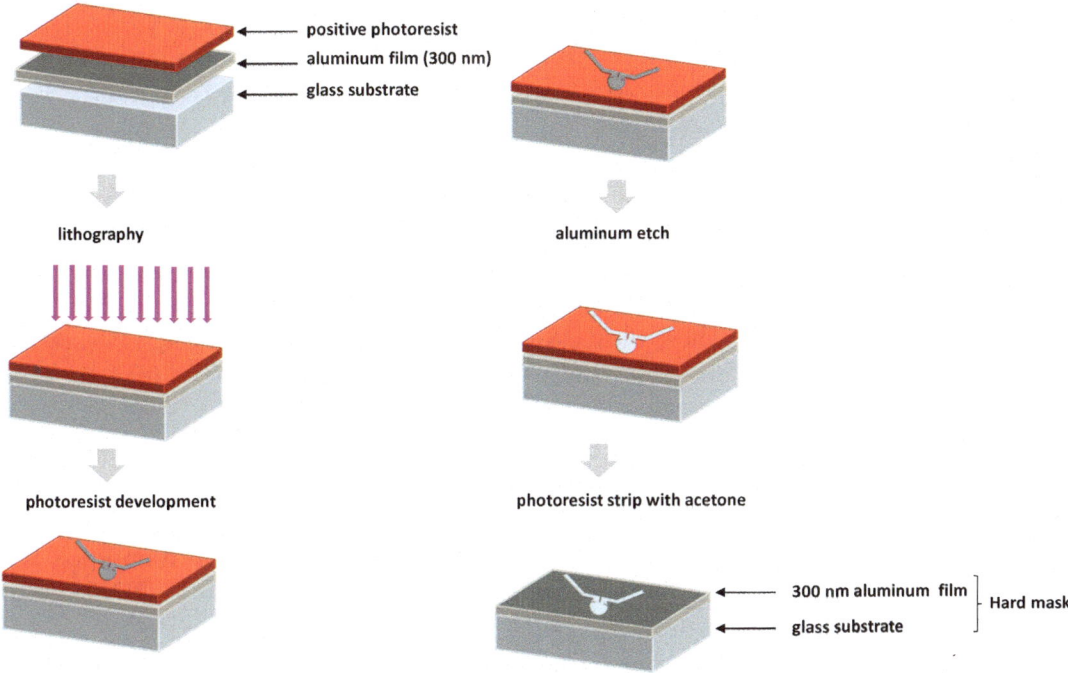

Fig. 2 Schematics of the hard mask fabrication method

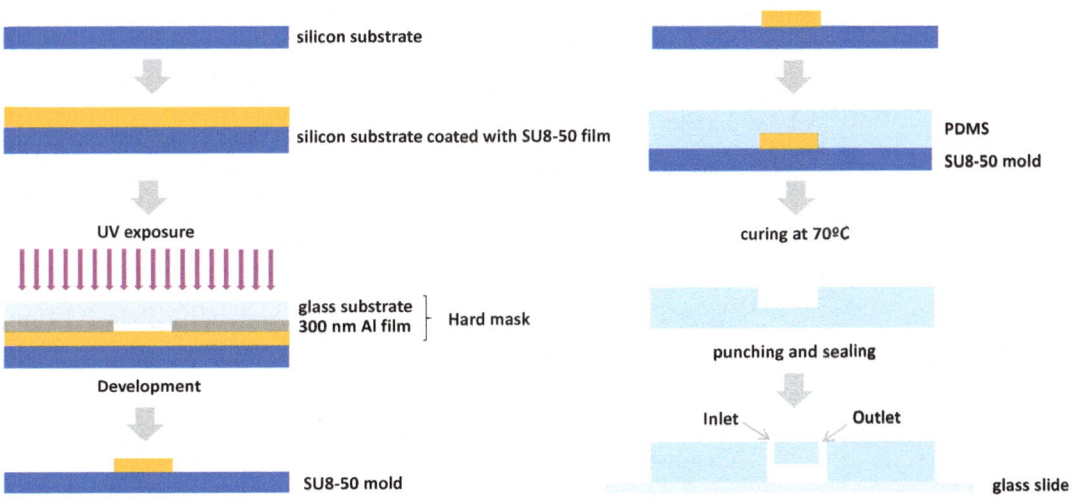

Fig. 3 Schematics of the SU8-50 mold fabrication method

12. To prepare the channels mold, transfer the hard-mask pattern to the SU8-substrate by joining the hard mask with the SU8-50-coated substrate (Fig. 3). Make sure that the aluminum-coated side of the hard mask is in close contact with the SU8-50-coated side of the silicon substrate (*see* **Note 12**).

13. Expose to UV light for 25 s (*see* **Note 13**).

14. After exposure, bake the SU8-50-coated substrate at 65 °C for 3 min, 95 °C for 5 min and cool down to room temperature.

15. Develop the substrate in a PGMEA solution for 8 min or until the channel structures are clearly defined. Rinse with isopropyl alcohol and then with dH$_2$O.

16. Dry the substrate under a stream of compressed air. At this point the channel mold is ready. A correct mold shows the 3D channel's geometry made of SU8-50 (representing the areas where the SU8-50 substrate was exposed to the UV light). In this case, the channel mold has a height of 60 μm, which corresponds to the SU8-50 film thickness.

17. Perform a final hard-bake by placing the mold in the hot-plate, at 150 °C for 15 min. Let the mold cool down to room temperature.

18. Prepare a PDMS solution by mixing PDMS monomers with the curing agent in the proportion of 10:1 (PDMS:curing agent).

19. Mix vigorously to ensure homogeneity. Remove air bubbles by placing the solution in the exicator.

20. Place the SU8 mold (with channels facing up) inside a Petri dish and gently pour down the degassed PDMS solution, avoiding the formation of air bubbles.

21. Cure the PDMS at 70 °C for 90 min in order to promote the hardening of the polymer.

22. Carefully, peel-off the solidified PDMS channels structures.

23. Use a 20 Gauge syringe needle to punch the inlets and outlets of the channel structure (*see* **Note 14**).

24. Wash standard microscopy glass slides and the PDMS structures (one glass slide for each PDMS structure) in Alconox detergent. Rinse with isopropyl alcohol and dH$_2$O. Dry them under a stream of compressed air.

25. Place the glass slides and PDMS structures on the UV-Ozone cleaner making sure that the channels structures in the PDMS are facing up. Program the UVO cleaner for 5 min of UV light plus 5 min ventilation.

26. Immediately seal the channels by aligning and pressing the PDMS structures against the glass slides. Make sure that it is the PDMS side containing the channels (hence exposed to the UV) that is pressed against the glass slide (surface side exposed to UV).

27. Allow the sealed channels to rest at least 24 h before use.

3.3 Stimulation Assays in Microtiter Plates

1. Start from a Working Cell Bank, with cell cultures that have reached 80% confluency (*see* **Note 15**).

2. Remove the culture medium from the T75 flasks with the aid of disposable micropipettes and replace it by 6 mL of sterile PBS. Agitate softly to remove any nonattached cells.

3. Discard the PBS solution and add 4 mL of a trypsin-0.05% EDTA solution.

4. Incubate for 3 min in a water bath at 37 °C.

5. Add 6 mL of complete cell culture medium and agitate softly to detach the remaining cells.

6. Transfer the 10 mL of the cell suspension to a 15 mL Falcon tube and centrifuge at $150 \times g$ for 3 min, at 20 °C.

7. Discard the supernatant and resuspend the cells in 1 mL of sterile PBS.

8. Prepare aliquots of 100 μL of 0.4% Trypan Blue. Add to each aliquot 100 μL of the cell suspension in PBS.

9. Using a microscope, focus the grid lines of the hemocytometer with a $10\times$ objective and count the viable cells, from at least three diagonal squares (*see* **Note 16**).

10. Calculate the cell concentration in cells/mL using Eq. (1):

$$\text{Cells/mL} = \frac{\text{Total viable cell number}}{\text{N}^\circ \text{ of squares}} \times \text{ dilution factor}$$

$$\times 10^{-4} \, \text{mL} \tag{1}$$

In our case, the dilution factor is 2 (100 μL of cell suspension +100 μL of Trypan Blue).

11. Dilute the PBS cell suspension in complete cell culture medium to a final concentration of 1×10^4 cells/mL.

12. Transfer 50 μL of the cell culture suspension (~500 cells) to individual wells of a sterile, cell-culture microtiter plate (*see* **Note 17**).

13. Incubate the plate overnight at 37 °C, in a 5% CO_2 atmosphere to allow for cell adhesion.

14. On the day of the assay, carefully remove the culture medium and replace it by 50 μL of the Fluo-4-Direct solution. Incubate the plate in the dark, for 30 min, at room temperature.

15. Prepare the compounds (carbachol and pirenzepine) in assay buffer. Make a stock solution considering the final concentration in the well.

16. Mount the microtiter plate on the microscope stage. Using the 10x objective, inspect and focus an area of the well where cells are correctly attached and with a homogeneous density (*see* **Note 18**).

17. Turn on the fluorescence imaging (*Ex.* 490 nm/*Em.* 520 nm) and adjust the gain, exposure time, and frame rate of the acquisition camera. Make sure to turn the fluorescent light at least 10 min before the assay (*see* **Note 19**).

18. Record the baseline fluorescence for 20 s.

19. With care, in order not to detach the adherent cells, add 25 µL of the test compound solution or assay buffer (negative control), at one point of the well, outside the field of view.

20. Record the dynamics of cells' fluorescence for a period of 180 s.

3.4 Stimulation Assays in Microfluidics

1. Start from the sealed channels obtained in Subheading 3.2, **item 27**. Insert the capillary tubes both at the inlets and outlets to allow the delivery of the required solutions (*see* **Notes 20** and **21**).

2. One day prior to the assay, fill a syringe with ethanol at 96% and connect the inlet tube to the syringe. Connect the outlet tube to a waste reservoir (e.g., Eppendorf tube).

3. Mount the syringe on the infusion pump. Use enough tube length to allow manipulation of the structures without breaking them or disconnecting the tubing.

4. Set the syringe flow rate at 5 µL/min and start flowing ethanol 96% until the liquid fills completely the channel structure. Let the channels settle overnight, at 4 °C. The purpose of this step (priming) is to provide the channels with a "wettable" surface and decrease the probability of emergence of air bubbles.

5. On the day of the assay, remove the ethanol 96% by flowing dH_2O at 5 µL/min for at least 5 min. Make sure that no air bubbles are forming inside the channel. If this happens, continue to flow water until the air bubbles are completely removed.

6. Prepare 1 mL solution of fibronectin in H_2O with the concentration of 100 µg/mL and fill a new syringe with this new solution.

7. Connect the inlet tube to the syringe containing the fibronectin solution and start flowing the solution at a flow rate of 5 µL/min until the channels are filled.

8. Carefully disconnect the inlet tube and incubate the channels for 2 h at 37 °C.

9. Remove the fibronectin solution by flowing 50 µL of dH20 at 5 µL/min.

10. Mount the channels on the microscope stage and focus the channel chamber using the 10× objective.

11. Prepare the cells as described in Subheading 3.3 up to **step 10**.

12. Dilute the PBS cell suspension in complete cell culture medium to a final concentration of 1×10^6 cells/mL. Make sure that you have at least 1 mL of this concentrated cell suspension.

13. Fill a new syringe with the cell suspension and connect it to the inlet tube.

14. Start flowing the cell suspension at a flow rate of 5 µL/min. This step must be monitored under the microscope to inspect the number of cells that are trapped in the chamber. Stop the flow when approximately 100 cells become trapped in the chamber.

15. Carefully detach the inlet tube and incubate the channels at 37 °C, under a 5% CO_2 atmosphere for 30 min to allow the cells to adhere.

16. Under the microscope, flow 20 µL of pre-warmed (37 °C), complete cell culture medium at a flow rate of 1.5 µL/min to remove non-adherent cells (*see* **Note 22**).

17. Prepare the calcium sensitive dye solution by diluting 250 µL of Fluo-4-Direct in 750 µL complete cell culture medium.

18. Fill a new syringe with the calcium sensitive dye and connect it to the inlet tube.

19. Under the microscope, flow the solution at a flow rate of 1.5 µL/min solution, making sure that the channel is completely filled and that no air bubbles are formed.

20. Disconnect the tubes carefully and incubate the channels for 30 min at 37 °C under a 5% CO_2 atmosphere followed by a 30 min incubation period at room temperature, in the dark.

21. Prepare fresh solutions of carbachol in assay buffer, with concentrations ranging from 10^{-8} to 10^{-2} M.

22. Prepare fresh solutions of a mixture of pirenzepine and carbachol in assay buffer. For each solution, fix the concentration of pirenzepine at 5 µM while varying the carbachol concentration from 10^{-7} to 10^{-3} M. Reserve at least one chamber to run assay buffer as negative control (*see* **Note 23**).

23. Mount again the channel under the microscope and focus the cell area using the 10× objective.

24. Turn on the fluorescent light setting the *Ex.* and *Em.* filters to 490 nm and 520 nm, respectively. Adjust the gain, exposure time, and frame rate of the imaging acquisition system. Make sure to turn the fluorescent light at least 10 min before the assay (*see* **Note 23**).

25. Start flowing the compound solution at 1.5 µL/min and record the cells fluorescence (*see* **Note 24**), for a period of 300 s (Fig. 4 and Electronic Supplementary Video S1).

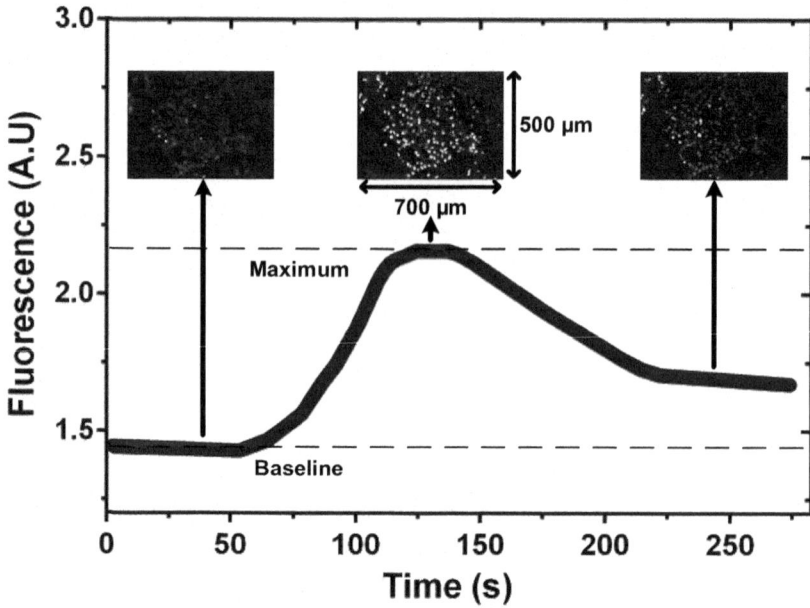

Fig. 4 Representative raw data for real-time cell response due to GPCR activation with carbachol (1 mM) in microfluidics using an initial flow rate of 1.5 μL/min. Three regions are identified: a baseline cell fluorescence prior to carbachol contacting the cell area; an increase of fluorescence reaching a maximum due to carbachol-induced GPCR activation; and a decrease in fluorescence as cells revert toward the basal level and the GPCR enters a refractory period

3.5 Data Analysis

1. Use Image J to convert the fluorescence recordings in time-dependent light intensity data.

2. Take the baseline cell intensity (S_o), defined as the average intensity recorded during the first 20–30 s of each assay.

3. Convert each time point response signal (S_i), into a normalized signal according to the expression: $(S_i - S_o)/S_o$.

4. Take the maximum of the normalized signal obtained for each concentration of the stimulating compound tested as representative of the maximum cells' response.

5. Normalize the highest cell response value in the dataset cell response vs. concentration to 100% and the lowest (negative control), to 0%.

6. Use these values to generate a normalized Hill dose response curve, in accordance to Eq. (2) using the software GraphPad Prism, where y is the % of maximum response, × the logarithmic value of a given concentration, p the Hill slope, and EC50 the effective concentration that generates the half-maximal response (Fig. 5).

$$y = \frac{100}{1 + 10^{[(LogEC50-x)\times p]}} \qquad (2)$$

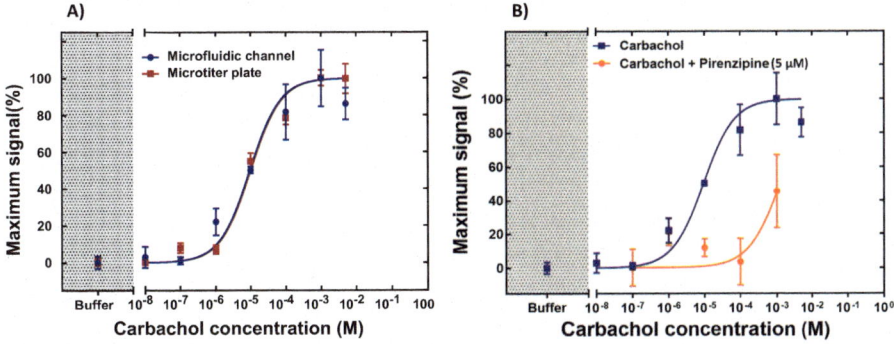

Fig. 5 Dose-response curves for GPCR activation. (**a**) Dose-response curve for the muscarinic M1 GPCR using carbachol as agonist. Representative data from both microtiter plates and microfluidics are presented, showing similar dose response patterns. (**b**) Effect of the antagonist pirenzepine in the M1 GPCR activation in microfluidics. Dose-response curves for the muscarinic M1 GPCR using carbachol-only and a mixture of carbachol and 5 μM pirenzepine

4 Notes

1. In our laboratory, microfluidics fabrication is based on soft lithography methods. The latter combines a group of techniques for replicating structures into an elastomer using stamps or molds. The process starts with the transfer of a channel design (from a CAD file), by photolithography, into a patterned Al mask (hard mask). The mask is then used to make a mold, which is a negative replica of the channel. The latter is then fabricated from the mold (or stamp) using an elastomer (PDMS in our case). In this section we describe all steps required for the fabrication of microfluidics using soft lithography including the fabrication of the photomask. However, other strategies for fabrication of microfluidics can be considered, including plastic micromachining or 3D printing.

2. Spin-coaters are used in thin-film fabrication techniques. The aim is to homogeneously spread a thin film (nm-μm film thicknesses) of a given material (in the liquid state), over a surface. In our case, we used the spin-coater to coat either glass or silicon substrates with the photoresist films of the desired thickness. Rotation speed and rotation time requires optimization as a function of thickness.

3. Photoresists are light-sensitive polymers. Positive photoresists become soluble when exposed to light and are removed by a developer solution. A positive photoresist was used to fabricate the Al hard mask.

4. Negative photoresists polymerize when exposed to light. The negative photoresist SU8-50 was used to fabricate the microfluidics mold.

5. Upon receiving the solution, we advise to perform 10 mL aliquots and store it at −20 °C.

6. Fluo-4 fluorophores form a group of Ca^{2+} indicator dyes that show enhanced fluorescence upon calcium binding. The ester forms of these dyes are cell permeable and bind iCa^{2+} with high affinity. Fluo-4-Direct is a commercial formulation that comprises a quencher to prevent background fluorescence, thus eliminating post-staining washing protocols (*see* Fig. 1).

7. The original cell bank is preserved in DMSO, stored at −80 °C.

8. HEK 293 T are adherent cells. Percentage of confluency relates to the surface coverage of adherent cells. An 80% confluency implies that 80% of the surface of the T75 flask is covered with cells.

9. Aluminum film deposition requires dedicated processes and equipment. We deposited aluminum from the gas phase using physical vapor deposition (PVD) techniques, typically available only at microfabrication facilities. However, patterned hard masks can be purchased from dedicated companies. In this case, the client designs the required microfluidic structures and orders the mask fabrication.

10. Aluminum etchant solution is extremely corrosive. Work in fume cupboards, with exhaust. Manipulate the sample with plastic forceps.

11. UV-Ozone is a surface cleaning technique. Ultra-violet (UV) light of two different wavelengths is used to generate ozone (O_3) from oxygen (O_2) and to convert O_3 into oxygen free radicals, which have a strong oxidation power. The latter reacts with organic matter on the surfaces, causing scissions at the molecular structure level. Potential organic contaminants present at the surfaces are thus removed as volatile compounds.

12. At this point SU8-50 is not polymerized, so avoid touching the substrate. Manipulate with forceps.

13. Exposure time must be optimized as function of the desired thickness and according to the power of the UV source. We used an UV hand lamp in the range of 315–400 nm, with a power of 250 W. Prolonged exposure times may result in unwanted SU8 structures (other than the desired channels), or cracks on the channel mold. Low exposure times will provide insufficient cross-linkage and the channels will be removed by lift-off during the PGMEA development step. Refer to the photoresist's instruction manual for the correct power and exposure times as a function of final thickness.

14. We used syringe needles to manually punch the inlets and outlets of the channels. This operation may damage the PDMS and release small PDMS pieces that can clog the

microchannel. To decrease the risk of clogging, place the PDMS channels facing up and punch the device outward. In this way, any PDMS piece that is formed will be pushed toward the PDMS surface.

15. Working cell banks can be maintained for several months by performing cell passages. For stimulation assays we used the cells between the second and the tenth passage. After this, a new cell Working Bank was prepared.

16. Each laboratory usually has its own procedure to count the cells. In our case, we usually count 3 diagonal squares zones. Other protocols may suggest counting additional regions of the hemocytometer.

17. The number of used wells depends on the number of assays. We usually perform calibration curves in triplicates, each with at least 5 different compound concentrations. We also reserve some wells (3–5 wells) to perform negative controls, in which cells are assayed with assay buffer.

18. Each well is analyzed separately, under the microscope. We do not advise using the entire microtiter plate because incubation times, particularly with the Fluo-4-Direct dye, would vary substantially between the first and last assays, which may induce artifacts on the expected cellular response.

19. At this point, cells present a basal fluorescence resulting from basal levels of iCa^{2+} at resting state (no stimulation) and staining with Fluo-4-Direct. It is important not to saturate the pixels intensity in order to correctly monitor the fluorescence dynamics. In our case we used an inverted Olympus microscope equipped with a 50 W mercury arc lamp. Exposure time and gain were set, respectively, to 1 s and 6 dB. The $10\times$ objective provided a field of view of 700 μm × 500 μm and the real-time images were captured at 1 frame/s.

20. Each channel is tested with different compound concentrations and disposed afterward. In view of this, prepare enough channels to perform a calibration curve and the respective negative controls.

21. We used PE capillary tubing to connect the syringe (with the desired solutions) and the microchannels. Make sure that the chosen capillaries have a diameter that tightly fits into the inlets and outlets of the channels to avoid fluid leaks.

22. From this point, cells are partially adherent to the channels due to interactions with fibronectin. However, any air bubble can seriously compromise the cell layer. When changing solutions, make sure to avoid entrance of air into the fluidic system.

23. Sequential addition of antagonists and agonists in microfluidics is not feasible. Due to the characteristic laminar flow, the antagonist is displaced by the agonist and it is therefore not effective in blocking agonist-induced activation. The specificity of the assay could nevertheless be observed by exposing the microfluidic cell culture to mixed solutions, containing varying concentrations of carbachol and a fix concentration of pirenzepine (5 μM), and comparing dose response curves in the presence and absence of pirenzepine.

24. For the microfluidics assay the acquisition settings were set to 1 s and the gain increased to 12 dB. Real-time images were captured with a rate of 1 frame/s.

References

1. Lefkowitz RJ (2007) Seven transmembrane receptors: something old, something new. Acta Physiol 190:9–19

2. Eglen RM, Bosse R, Reisine T (2007) Emerging concepts of guanine nucleotide-binding protein-coupled receptor (GPCR) function and implications for high throughput screening. Assay Drug Dev Technol 5:425–451

3. Hill SJ (2006) G-protein-coupled receptors: past, present and future. Br J Pharmacol 147: S27–S37

4. Overington JP, Al-Lazikani B, Hopkins AL (2006) Opinion—how many drug targets are there? Nat Rev Drug Discov 5:993–996

5. Hopkins AL, Groom CR (2002) The druggable genome. Nat Rev Drug Discov 1:727–730

6. Rosenbaum DM, Rasmussen SG, Kobilka BK (2009) The structure and function of G-protein-coupled receptors. Nature 459:356–363

7. Martins SAM, Trabuco JRC, Monteiro GA, Chu V, Conde JP, Prazeres DMF (2012) Towards the miniaturization of GPCR-based live-cell screening assays. Trends Biotechnol 30:566–574

8. Zhang R, Xie X (2012) Tools for GPCR drug discovery. Acta Pharmacol Sin 33:372–384

9. Eglen R, Reisine T (2011) Primary cells and stem cells in drug discovery: emerging tools for high-throughput screening. Assay Drug Dev Technol 9:108–124

10. Wu M-H, Huang S-B, Lee G-B (2010) Microfluidic cell culture systems for drug research. Lab Chip 10:939–956

11. Young EWK, Beebe DJ (2010) Fundamentals of microfluidic cell culture in controlled microenvironments. Chem Soc Rev 39:1036–1048

Chapter 20

Homology Modeling Using GPCRM Web Service

Przemysław Miszta, Szymon Niewieczerzał, Paweł Pasznik, and Sławomir Filipek

Abstract

Homology modeling methods are commonly used for quick and precise construction of a desired protein or its mutant using protein templates, which were determined by crystallography, cryo-EM, or NMR. Due to the increasing number of such structures, the obtained models are precise even in the case of small similarity between sequences of template and modeled proteins. The reason for that is a high evolutionary conservation in the structure regions responsible for keeping the function of proteins. This is also the case for G protein-coupled receptors (GPCRs), which constitute the largest family of membrane receptors with nearly 800 proteins. The GPCRM web service (https://gpcrm.biomodellab.eu/) was set up for the nearly automatic generation of high-quality structures of modeled GPCRs. The three possible paths: "High similarity," "Quick path," and "Long path" allow the user to choose between a fast but less reliable path, up to more reliable but longer procedures. In the Advanced mode the service allows for user modifications including selection of template(s) and a manual adjustment of the sequence alignment.

Key words Homology modeling, GPCRM web service, GPCRs, Multiple templates, Profile-profile alignment

1 Introduction

The GPCRM web service [1] contains many features that can be used by a broad range of researchers, students, modelers, and experimentalists. The service is friendly enough to be used by a person unexperienced in modeling procedures. The *High similarity* option additionally allows for fast generation of good models when a close homolog structure is known. Such a procedure can be used when mutations and/or deletions or insertions are introduced into the known receptor structures. Therefore, this option is useful to test changes in the receptor structure upon modifications of a known receptor. The GPCRM service includes also the possibility to model large loops as well as N- and C-termini of the receptor, a feature that is not available in other services. It is especially useful for testing allosteric drug binding when the allosteric sites are

Sofia Aires M. Martins and Duarte Miguel F. Prazeres (eds.), *G Protein-Coupled Receptor Screening Assays: Methods and Protocols*, Methods in Molecular Biology, vol. 2268, https://doi.org/10.1007/978-1-0716-1221-7_20,
© Springer Science+Business Media, LLC, part of Springer Nature 2021

located in terminal parts of the receptor. The obtained homology models can be also used to study activation processes by molecular dynamics when agonists are docked, or to investigate functional selectivity when biased ligands are bound. The *Advanced* mode, which can be used by more experienced users, allows to modify nearly every step of the modeling procedure for fine-tuning the constructed models. Apart from generating the homology models, the updated GPCRM service offers a set of useful information from the large and sortable tables. The tables contain links to Protein Data Bank (PDB) and UNIPROT databases on GPCRs as well as figures showing ligand–receptor interactions. The molecular viewers implemented in the service allow for an online examination of the obtained models. GPCRM is continually upgraded in a semiautomatic way and the number of template structures has increased from 20 in 2013 [2] to over 90, including structures of the same receptor with different ligands in various stages of activation.

In this chapter, we present the GPCRM protocol used to build a multi-template homology model with the histamine H3 receptor as an example. The common options and specificities of each of the available three paths—*High similarity*, *Quick path*, and *Long path*—are explained. In each case, ten final models of the receptor are produced for each of the scoring methods: Rosetta [3], Rosetta-MP [4], and BCL::Score [5]. However, the number of internal models, which correlates with quality of modeling, depends on the modeling path and is described in Table 1 [1]. The GPCRM service uses the profile-profile alignment [6] as one of four methods for sequence alignment and the multiple structural templates [7] during receptor construction.

2   Materials

To model the required GPCR structure it is necessary to prepare the amino acid sequence of the protein in a FASTA format. The protein sequence is automatically tested on-the-fly for correctness (unusual amino acid codes) and similarity to GPCRs (*see* **Note 1**), while placed in the required field of the form. To make the prediction more reliable, it is also possible to check whether the receptor can form disulfide bridges and prepare a list of residue pairs forming disulfide bonds.

1. Protein sequence in FASTA format (required).

2. List of residue pairs forming disulfide bonds (optional).

Table 1
Maximal number of models generated in each option and mode of GPCRM service

Name of path	Step 1 Modeller generating 1	Step 2 Modeller refining 2	Step 3 Modeller scoring 3	Rosetta mode	Step 4 Rosetta loop refining 4	Step 5 Scoring (BCL, Rosetta, Rosetta-MP) 5	Average working time 6
	Number of models generated by Modeller	Number of loop models per each model from Step 1	Total number of models		Number of loop models refined by Rosetta using 10 best models from Step 3	Number of final models	Time
Quick path	20	10	200	No	X	10	2 h
	20	1	20	Fast	$10 \times 5 = 50$	10	4 h
	20	1	20	Slow	$10 \times 20 = 200$	10	8 h
Long path	100	50	5000	No	X	10	10 h
	100	1	100	Fast	$10 \times 50 = 500$	10	2 days
	100	1	100	Slow	$10 \times 150 = 1500$	10	3 days
High similarity	20	5	100	No	X	10	20 min
	20	1	20	Yes termini	$10 \times 5 = 50$	10	1 h

The number of final models may be smaller due to removal of defective loops

3   Methods

3.1   Input Methods

1. To start working with the web service, select the appropriate calculation path. There are three paths available, which are readily visible at the starting page (Fig. 1): *Quick path*, *Long path*, and *High similarity*.

2. Each of these paths in the final version is used to create 3D models of GPCR proteins based on the entered sequence, but each of these paths implements this process in a different way (Table 1). A detailed description of the paths will be provided later but it is worth to mention the most important differences between them.

   (a) The *High similarity* (*HS*) path should be selected if the target protein has a high or medium sequence similarity to an already known protein, with a determined structure present in GPCRM website as a template. This path is very useful to introduce mutations/deletions/insertions and add the fragments missing in the crystal structure (*see* **Notes 2** and **3**). The *HS* path is also characterized by a relatively short calculation time (about **20** min—when building of termini are not selected).

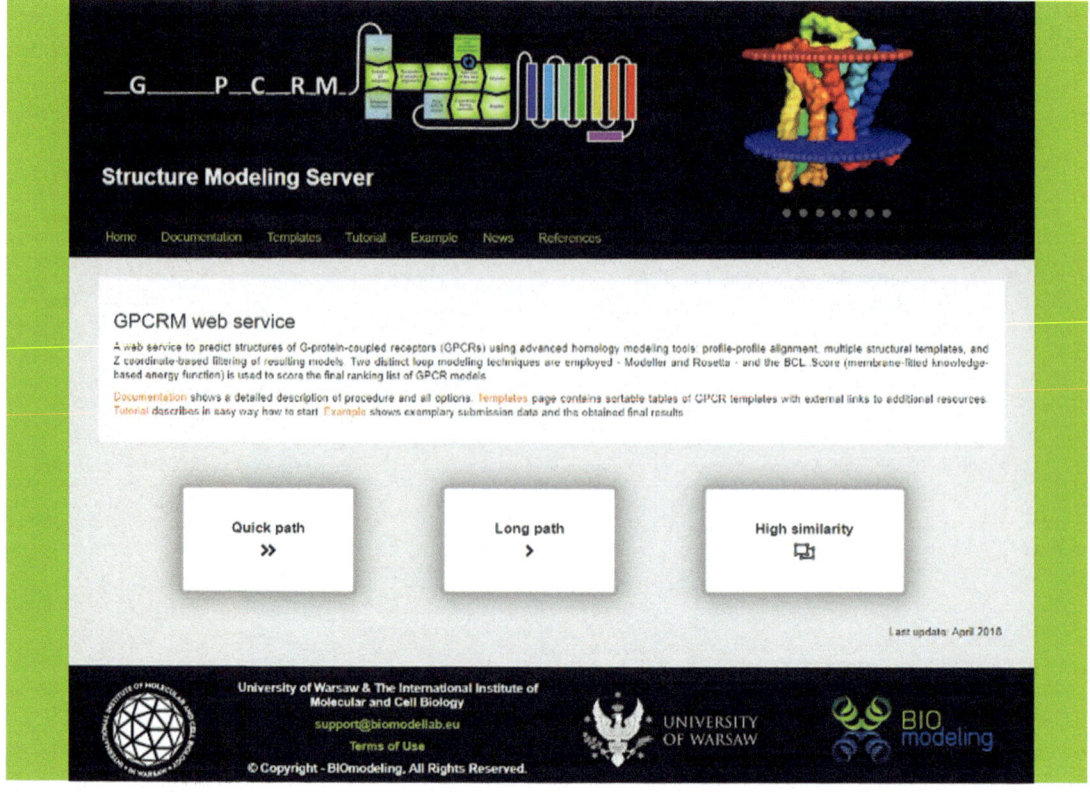

Fig. 1 The starting page of the GPCRM web service with three calculation paths

(b) The *Quick path* (*QP*) is the most common path in the service. It allows to construct the structure of a protein in a reasonable time (usually few hours). This method is recommended when the user intends to generate a large number of GPCR protein models within a fairly short time.

(c) The *Long path* (*LP*), as the name suggests, generates results after much longer times (up to several days). It offers the most extensive methods and usually leads to the best results. However, this method, due to its very long time of calculations, is recommended for individual, finally selected cases.

3. Enter the sequence of the required receptor in FASTA format, regardless of the path selection. Upload a FASTA file from disk, or type or copy/paste a protein FASTA sequence into the Query sequence field. The FASTA format is a text-based format for representing either nucleotide sequences or amino acid (protein) sequences, in which nucleotides or amino acids are represented using single-letter codes (*see* **Note 4**). It is worth noting that the uploaded sequence is automatically checked for

(a)

| Sequence status: | ⊗ The sequence seems not to be GPCR. |
| Query sequence: | GPCRM |

(b)

| Sequence status: | ⊗ The sequence is not formally correct. Check documentation. |
| Query sequence: | BOOK CHAPTER |

(c)

Username:		✎ Fill with sample data
E-mail address:		
	E-mail address is optional. You could submit your e-mail address to receive the link to the results for further usage.	
Job description:	CB2 CNR2_HUMAN Cannabinoid receptor 2	
Sequence status:	✓ The sequence is **correct** and passed GPCRs similarity test.	
Query sequence:	MEECWVTEIANGSKDGLDSNPMKDYMILSGPQKTAVAVLCTLLGLLSALENVAVLYLILSSHQLRRKPSYLFIGSLAGADFLASVVFACSFVNFHVFHGVDSKAV FLLKIGSVTMTFTASVGSLLLTAIDRYLCLRYPPSYKALLTRGRALVTLGIMWVLSALVSYLPLMGWTCCPRPCSELFPLIPNDYLLSWLLFIAFLFSGIIYTYGHVL WKAHQHVASLSGHQDRQVPGMARMRLDVRLAKTLGLVLAVLLICWFPVLALMAHSLATTLSDQVKKAFAFCSMLCLINSMVNPVIYALRSGEIRSSAHHCLAH WKKCVRGLGSEAKEEAPRSSVTETEADGKITPWPDSRDLDLSDC	

Fig. 2 Examples of introduced incorrect sequences and resulting messages from the service. (**a**) Lack of similarity to GPCRs; (**b**) incorrect amino acid letters; (**c**) sequence from sample data (CB2 receptor)

correctness and characterized as belonging (or not) to the GPCR family. When the sequence does not belong to the GPCR family, a warning immediately appears: "The sequence seems not to be GPCR" (Fig. 2a) but it does not block the job submission. When the user types characters other than those encoding amino acids, the following warning will appear: "The sequence is not formally correct. Check documentation" (Fig. 2b) and it blocks the job submission.

4. There is also a possibility to run sample data and use the *Fill with sample data button*, which will automatically fill the *Query sequence* field with the example sequence, cannabinoid receptor CB2 (Fig. 2c). When the entered protein sequence is correct, a label appears confirming the correctness of the entered sequence: "The sequence is **correct** and passed GPCRs similarity test" (*see* **Note 5**).

3.2 Entering Parameters/Selecting Options in Auto Mode

1. An option is available to improve accuracy of the alignment and model building by specifying a location of S-S bridge between the extracellular loop 2 (ECL2) and the transmembrane helix TM3 of GPCR. Do this by entering the appropriate cysteine

Fig. 3 The exemplary data of S-S bridge for CB2 receptor

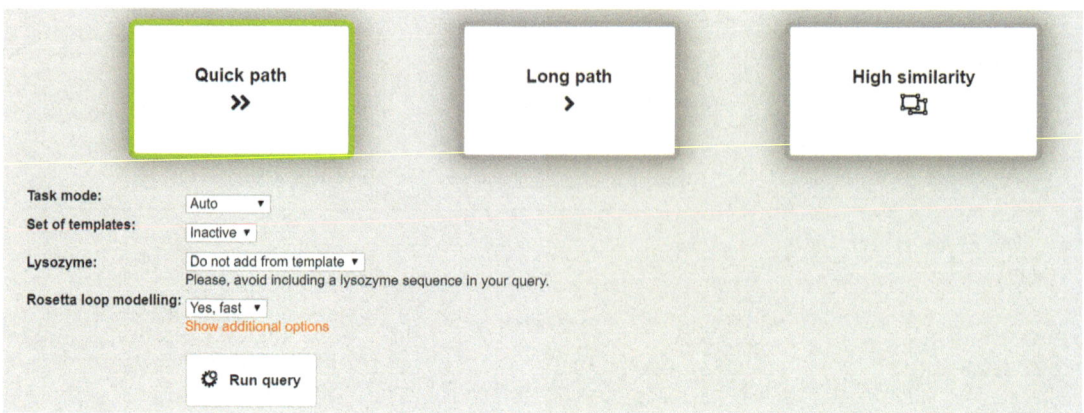

Fig. 4 Selected parameters for a typical run in *QP* and *LP* options

numbers in the empty fields of **S-S bridge**. Showing an example for CB2 receptor (Fig. 3), this field is automatically filled by cysteines 174 and 179 (*see* **Note 6**).

2. After selecting the path (*QP* or *LP*), choose either *Auto* or *Advanced* mode (the latter is not available for the *HS* path). For novice users, the simpler and more intuitive *Auto* mode is recommended. In this mode, in the field **Set of templates**, the user can choose either the active or inactive form of the receptor. The user can also decide whether 3D modeling should be based on patterns including fusion proteins such as lysozyme. By default, it is recommended to make a receptor model without a fusion protein (*see* **Note 7**). Before running calculations, the user should additionally decide how the GPCRM service should model the loops. There are 3 options:

(a) *No*—without loop modeling.

(b) *Yes fast*—with loop modeling by Rosetta program using fewer generated models.

(c) *Yes slow*—with loop modeling by Rosetta program using a larger number of generated models.

The exemplary selected parameters in *QP* option are shown in Fig. 4.

3. In the *HS* path (see Fig. 5): the parameters **Task mode, Lysozyme,** and **Rosetta loop modelling** are not available, because the default mode is *Auto* without a fusion protein and without

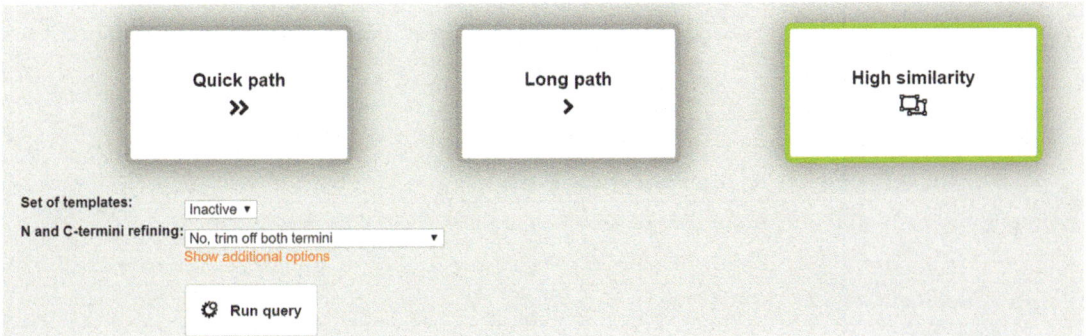

Fig. 5 Selected parameters for a typical run in *HS* path

Username:	AAA		✏ **Fill with sample data**
E-mail address:	xxxxxx@yyy.zz		
	E-mail address is optional. You could submit your e-mail address to receive the link to the results for further usage.		
Job description:	UUU|		

Fig. 6 The job management fields, optional

the possibility of generating loops using Rosetta. However, in the *HS* path there is an opportunity to decide how to model the N and C-termini of a protein. In **N and C-termini refining** selection, the user has three options to choose from:

(a) *No, trim off both termini*—creating a model without the N- and C-termini of GPCR.

(b) *No, Modeller models without refining*—creating a model with N- and C-termini using Modeller.

(c) *Rosetta refinement of termini*—creating a model with N- and C-termini of using Rosetta (*see* **Note 8**).

4. There are also three useful (but not mandatory), fields—**Username, E-mail address**, and **Job description**—that make the management of obtained results more friendly (Fig. 6). If the user provides the e-mail address, a link to the result is sent when the job is completed, while a **Job description** is useful for later management of results.

5. The results are stored on the server for about **2** weeks. After that and due to the high interest in the service, the results are deleted as the disk space is limited.

3.3 Running the Job

1. After starting the calculations using the **Run query** button ⚙ Run query , the user is redirected to the page with information about the status of the calculation (Fig. 7). The provided *Task token* is a link to the results when the Internet connection is broken or when the calculations will take a long time to complete. . If the Task mode was set to *Auto*, the

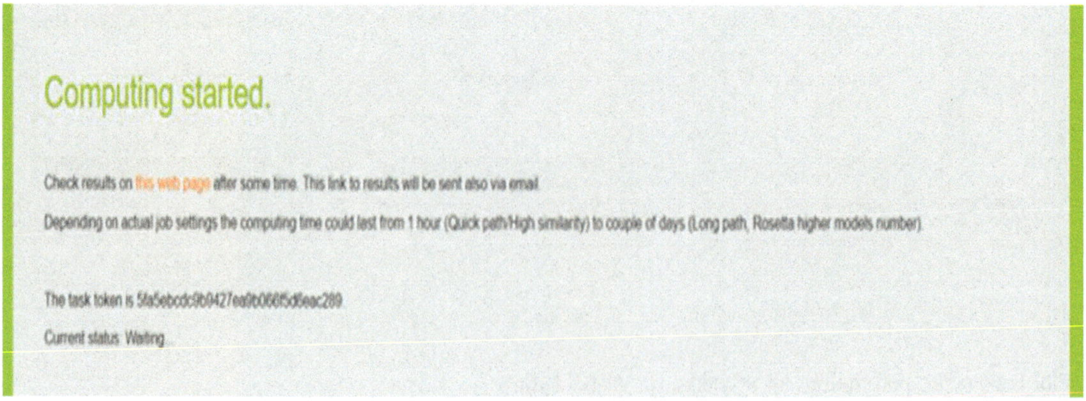

Computing started.

Check results on this web page after some time. This link to results will be sent also via email.

Depending on actual job settings the computing time could last from 1 hour (Quick path/High similarity) to couple of days (Long path, Rosetta higher models number).

The task token is 5fa5ebcdc9b9427ea9b06605d6eac289

Current status: Waiting

Fig. 7 The page informing about the status of the running job

computing of the submitted job starts. Otherwise (Task mode set to *Advanced*), after some sequences alignment computations, the user is redirected to additional forms to provide more detailed information.

2. The calculations are divided into stages: (i) searching for similar sequences and aligning the user-provided sequence with sequences of available templates; (ii) receptor modeling, using the sequence provided by the user, using the Modeller program; (iii) modeling loops using the Rosetta program (if this stage was selected by user). When the *Auto* mode is selected, all stages are carried out automatically without the possibility for any further interference at the individual stages. If the user does not provide the e-mail, the page informing about the status of the calculations, after they are performed, will automatically change to the page presenting the results.

3.4 The Advanced *Mode*

1. In the *Advanced* mode, the **Task mode** option, available only for *QP* and *LP* paths after the first stage of calculations, the user has the possibility to select the individual templates. The page with the status of calculations (Fig. 7) changes to the results page from the first stage showing information on sequence similarity and selected receptor sequences. The latter will be used as templates for further calculations. The service GPCRM automatically selects one or two templates from different families of GPCRs with the largest similarity to the target sequence. The user can leave the selection suggested by the service or choose other sequences available in the database (Fig. 8), (at least one structure) to serve as a template for further modeling. Then, the user can continue the job by clicking *Run query* button again.

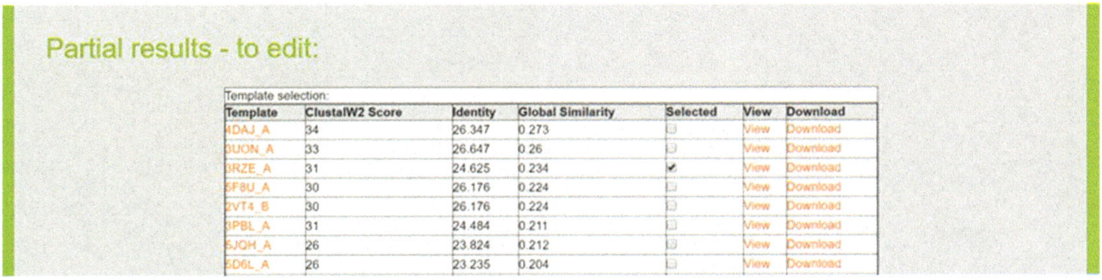
Partial results - to edit:

Template selection:

Template	ClustalW2 Score	Identity	Global Similarity	Selected	View	Download
4DAJ_A	34	26.347	0.273	☐	View	Download
3UON_A	33	26.647	0.26	☐	View	Download
3RZE_A	31	24.625	0.234	✔	View	Download
5F8U_A	30	26.176	0.224	☐	View	Download
2VT4_B	30	26.176	0.224	☐	View	Download
3PBL_A	31	24.484	0.211	☐	View	Download
5JQH_A	26	23.824	0.212	☐	View	Download
5D6L_A	26	23.235	0.204	☐	View	Download

Fig. 8 Manual selection of templates in *Advanced* mode

2. In the next stage, the user can choose the alignment method (PSA, MSA, or simple pairwise), manually improve the sequence alignment, and further change the definition of loop beginnings and ends (*see* **Note 9**), both for Modeller and Rosetta programs (usually the same values). At this point, the final stage of calculations can start.
3. A link to results of each individual stage of calculations is sent to the e-mail address provided by the user.
4. When calculations fail, the user can read the obtained partial results and make corrections in the input form.
5. In the *Advanced* mode, the user should monitor individual stages of calculations, which sometimes take several hours, and make suitable corrections/selections at each stage. In this view, this mode is not recommended for casual users. For detailed description of all available options in the service please see the *Documentation* tab in GPCRM menu.

3.5 Description of Output

Only the major results are described here. It is recommended to see the Example tab in GPCRM menu for detailed description of all tables in the Results page.

1. Regardless of the path and options selected, the results are represented in a similar way (*see* **Note 10**). Here, the results will be presented for modeling of inactive histamine H3 receptor for the user-selected *QP* path in *Advanced* mode. The first element in the results page is an information table collecting information about the completed job: task mode, user name, unique task code, dates and run time, length of the sequence, and all the selected options. Figure 9 contains the sequence of the modeled receptor.
2. The table in Fig. 10 shows the information about the sequence similarity and identity to the particular templates. Marks Yes/No indicate which template(s) was chosen for further calculations (*see* **Note 11**). In the analyzed case, the receptor with the PDB id:3RZE from histamine receptor family

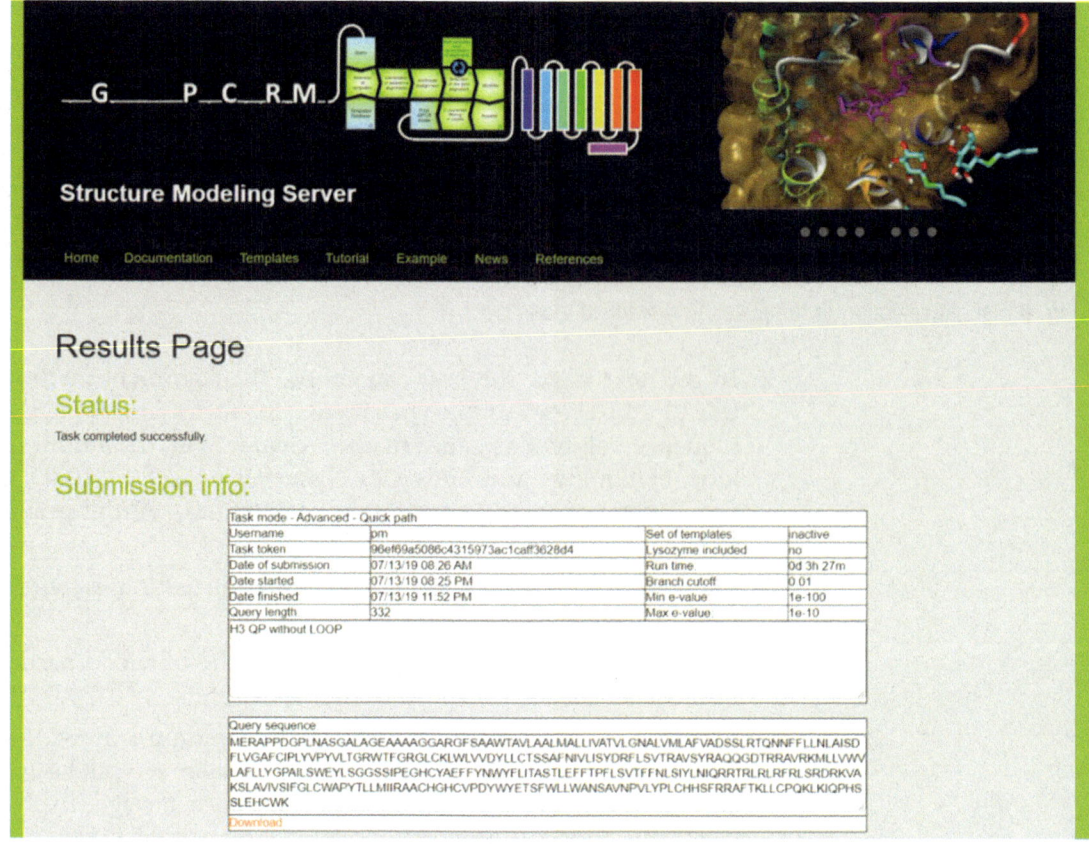

Fig. 9 The initial part of Results page showing selected options and dates

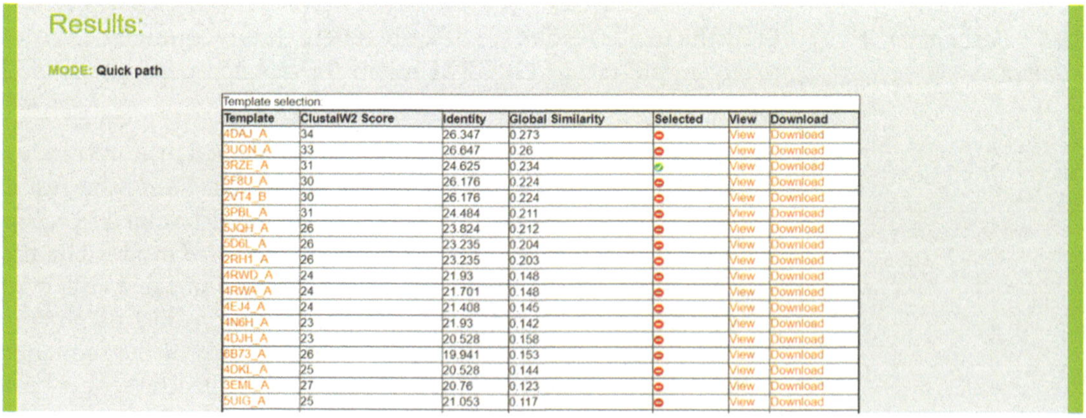

Fig. 10 Selection of templates for construction of the receptor model

(H1) was chosen even not having the highest identity and similarity. It was done in *Advanced* mode manually by the user.

3. When a template from a different family is chosen in *Auto* mode because of its high similarity to the target sequence, it

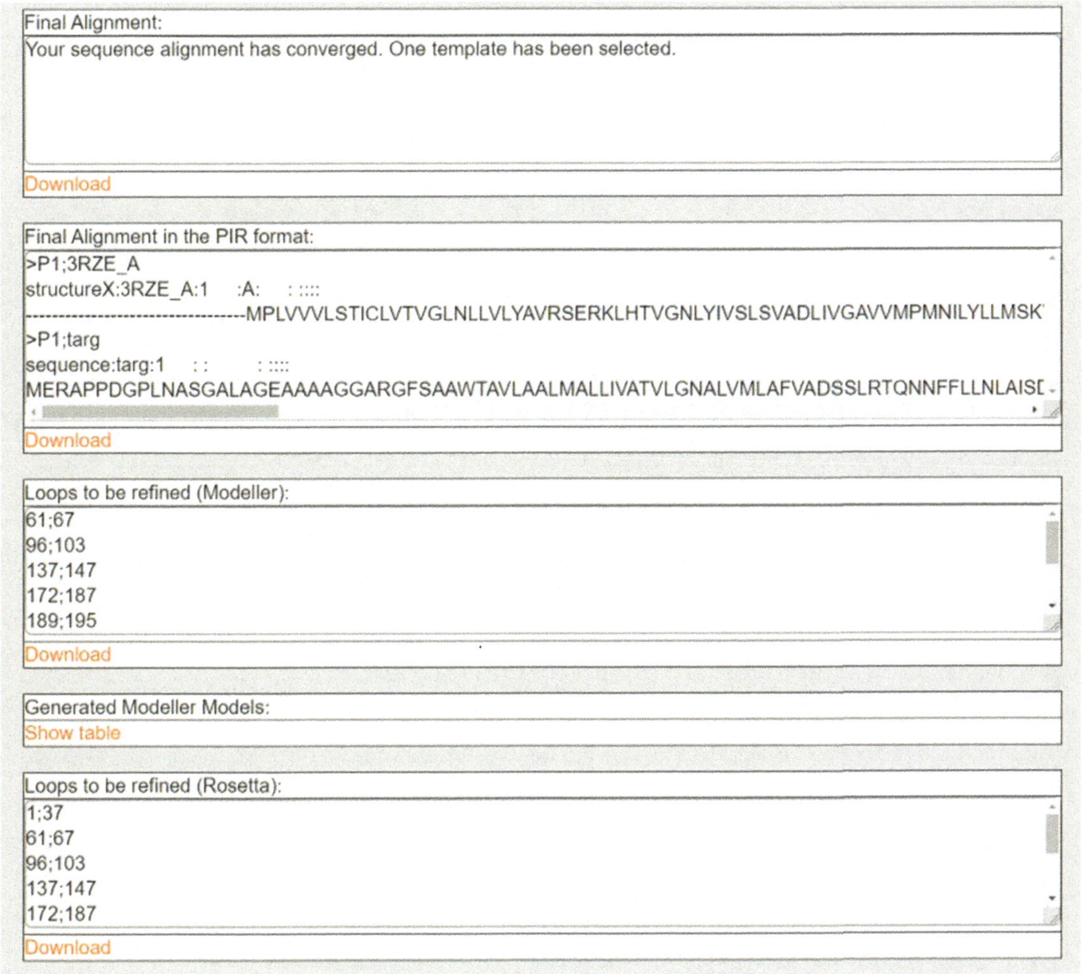

Fig. 11 The alignment of the target sequence with all selected (in this case only one) templates

is up to the user to keep the results or to run calculation in the *Advanced* mode.

4. The table in Fig. 11 provides information on the aligning process and the selected loops for modeling (not shown).

5. Results in Fig. 12 include the constructed 3D models of the target receptor. The final 30 models are provided each 10 ranked by three scoring functions: *Rosetta, Rosetta MP,* and *BCL::score*, to select accurate models for the required purposes: the structure of the binding site, the transmembrane domain, or the overall shape of the receptor, respectively. Each constructed model can be viewed in the graphics window powered by **NGL** or **JSMOL.** The models can be downloaded separately in PDB format using the *Download PDB* button.

Generated Rosetta Models (sorted by BCL score):

PDB file:	Rosetta Total Energy	RosettaMP Score	BCL::Score	Model Viewer	Download PDB
gpcrm_model_b01.pdb	865.682 (25)	289.381 (26)	-3179.89 (1)	NGL JSmol	Download
gpcrm_model_b02.pdb	949.736 (29)	333.484 (29)	-3160.63 (2)	NGL JSmol	Download
gpcrm_model_b03.pdb	871.376 (27)	297.076 (27)	-3095.3 (3)	NGL JSmol	Download
gpcrm_model_b04.pdb	599.37 (4)	69.426 (3)	-3081.92 (4)	NGL JSmol	Download
gpcrm_model_b05.pdb	1254.46 (40)	606.817 (40)	-3080.92 (5)	NGL JSmol	Download
gpcrm_model_b06.pdb	745.106 (16)	195.049 (16)	-3055.63 (6)	NGL JSmol	Download
gpcrm_model_b07.pdb	799.166 (19)	214.916 (18)	-3042.95 (7)	NGL JSmol	Download
gpcrm_model_b08.pdb	802.247 (20)	244.364 (21)	-3024.72 (8)	NGL JSmol	Download
gpcrm_model_b09.pdb	834.532 (23)	261.904 (23)	-3005.94 (9)	NGL JSmol	Download
gpcrm_model_b10.pdb	955.855 (30)	347.2 (30)	-2985.55 (10)	NGL JSmol	Download

Download table

Generated Rosetta Models (sorted by RosettaMP score):

PDB file:	Rosetta Total Energy	RosettaMP Score	BCL::Score	Model Viewer	Download PDB
gpcrm_model_mp01.pdb	551.571 (1)	34.342 (1)	-2829.03 (20)	NGL JSmol	Download
gpcrm_model_mp02.pdb	551.656 (2)	37.637 (2)	-2803.09 (23)	NGL JSmol	Download
gpcrm_model_mp03.pdb	599.37 (4)	69.426 (3)	-3081.92 (4)	NGL JSmol	Download
gpcrm_model_mp04.pdb	595.5 (3)	75.676 (4)	-2874.87 (18)	NGL JSmol	Download
gpcrm_model_mp05.pdb	639.648 (5)	110.468 (5)	-2738.24 (31)	NGL JSmol	Download
gpcrm_model_mp06.pdb	671.346 (9)	122.371 (6)	-2922.44 (14)	NGL JSmol	Download
gpcrm_model_mp07.pdb	684.159 (10)	124.521 (7)	-2909.09 (15)	NGL JSmol	Download
gpcrm_model_mp08.pdb	640.391 (6)	130.728 (8)	-2699.33 (36)	NGL JSmol	Download
gpcrm_model_mp09.pdb	698.19 (12)	148.482 (9)	-2555.8 (46)	NGL JSmol	Download
gpcrm_model_mp10.pdb	668.686 (8)	152.904 (10)	-2711.52 (33)	NGL JSmol	Download

Download table

Generated Rosetta Models (sorted by Rosetta total energy):

PDB file:	Rosetta Total Energy	RosettaMP Score	BCL::Score	Model Viewer	Download PDB
gpcrm_model_r01.pdb	551.571 (1)	34.342 (1)	-2829.03 (20)	NGL JSmol	Download
gpcrm_model_r02.pdb	551.656 (2)	37.637 (2)	-2803.09 (23)	NGL JSmol	Download
gpcrm_model_r03.pdb	595.5 (3)	75.676 (4)	-2874.87 (18)	NGL JSmol	Download
gpcrm_model_r04.pdb	599.37 (4)	69.426 (3)	-3081.92 (4)	NGL JSmol	Download
gpcrm_model_r05.pdb	639.648 (5)	110.468 (5)	-2738.24 (31)	NGL JSmol	Download
gpcrm_model_r06.pdb	640.391 (6)	130.728 (8)	-2699.33 (36)	NGL JSmol	Download
gpcrm_model_r07.pdb	665.005 (7)	215.036 (19)	-2808.92 (21)	NGL JSmol	Download
gpcrm_model_r08.pdb	668.686 (8)	152.904 (10)	-2711.52 (33)	NGL JSmol	Download
gpcrm_model_r09.pdb	671.346 (9)	122.371 (6)	-2922.44 (14)	NGL JSmol	Download
gpcrm_model_r10.pdb	684.159 (10)	124.521 (7)	-2909.09 (15)	NGL JSmol	Download

Download table

Fig. 12 The constructed models of the target receptor ranked by three scoring functions: *Rosetta, Rosetta MP,* and *BCL::score*. They are sorted separately in three tables

6. It is also possible to download all information from the Results tables using the *Download table* button. Additionally, it is possible to download files containing information about sequence alignment, and about the loop modeling in Modeller and Rosetta. The last *Download* button allows for downloading all files in a compressed form.

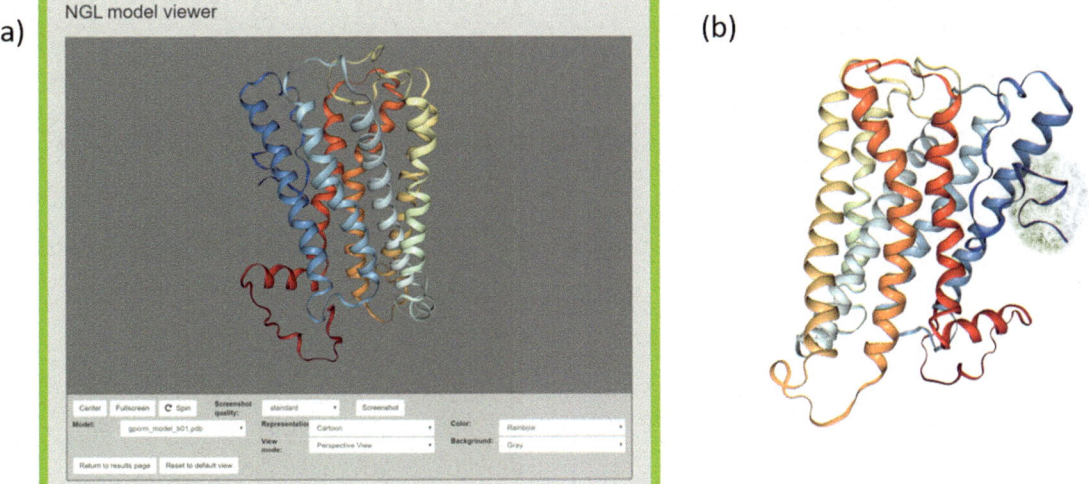

Fig. 13 Visualizations of the obtained models. (**a**) The NGL viewer showing one of constructed models. (**b**) The journal quality image of the receptor

3.6 Structural Viewers and the Table of Templates

1. The viewers installed in GPCRM, and especially NGL, offer useful display functions such as different representations, structure coloring, and background selection (Fig. 13a). The receptor models can be viewed one by one (*see* **Note 12**). The implemented *Screenshot* button can generate figures of journal quality, on white background in three different qualities: standard, high, and very high, all in PNG graphics format (Fig. 13b).

2. Finally, it is worth to mention that the *Templates* tab in GPCRM menu contains interactive and sortable data of all available templates in the service, both in inactive and active forms (Fig. 14). This database contains information related to the particular receptors such as PDB code, UNIPROT code, organism, and subfamily, with suitable links (highlighted in orange). The database also contains pictures of ligand interactions in the binding site. Such table can help to choose the most suitable receptor(s) to model the target GPCR in the *Advanced* mode.

3. Additional information on usage of the GPCRM service can be found using *Documentation* and *Tutorial* tabs in the menu. A simple, but useful graphical messages in the Tutorial appears when moving the mouse over different parts of the screen (Fig. 15), explaining the usage of particular options.

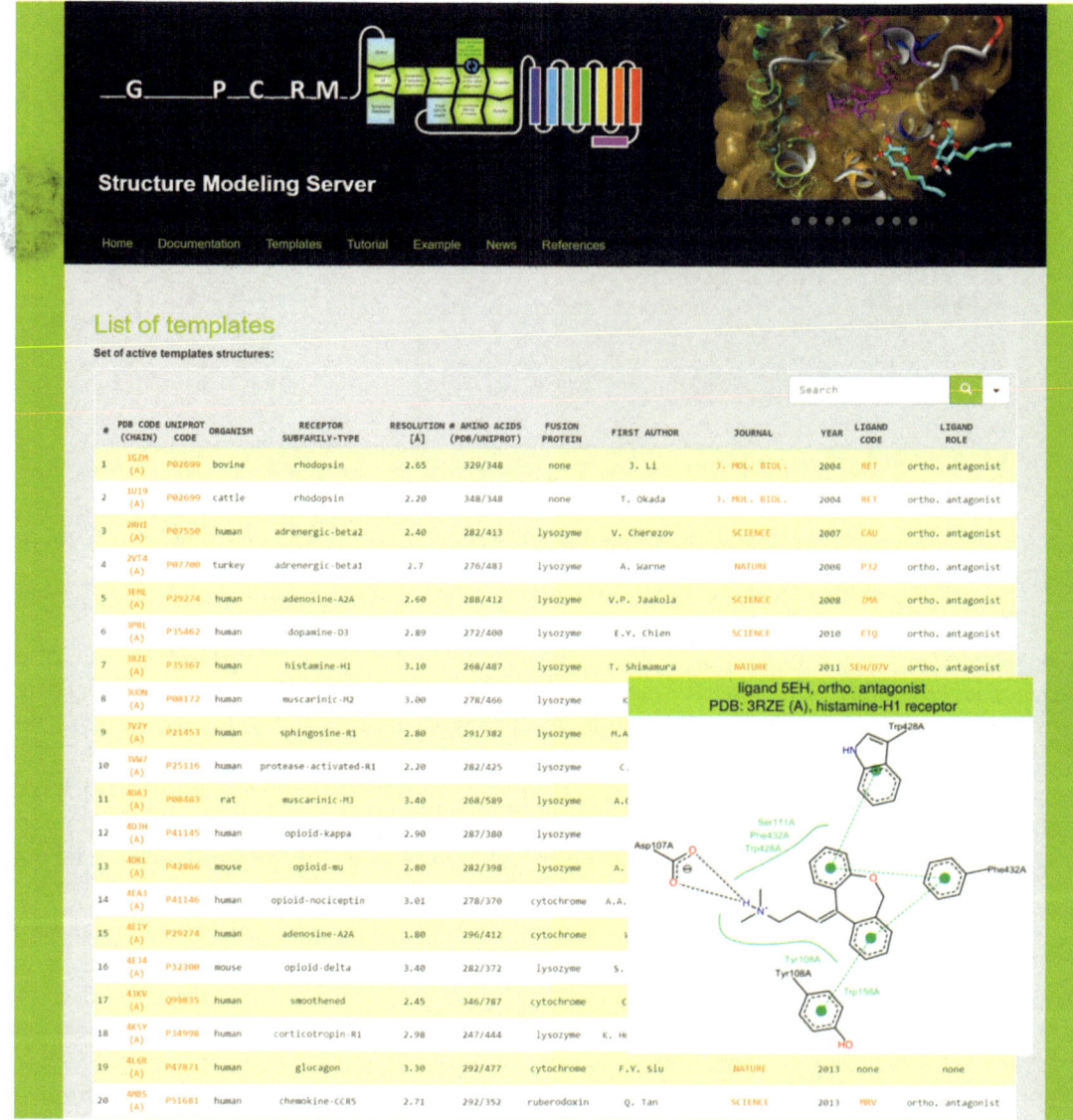

Fig. 14 The sortable table with all available templets in inactive and active forms. It contains links to other databases (PDB, UNIPROT, and to literature by DOI number) and also provides images of ligand interactions in the binding site

(a)

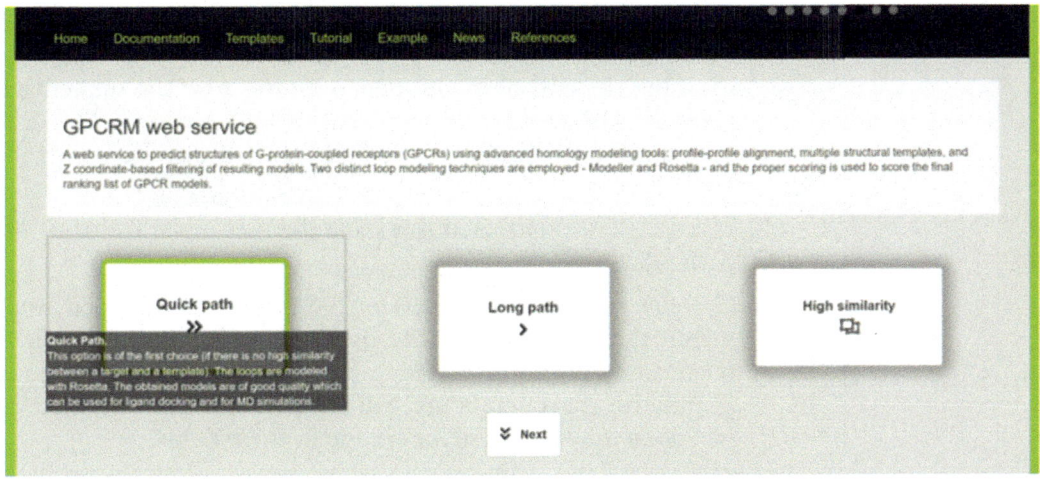

(b)

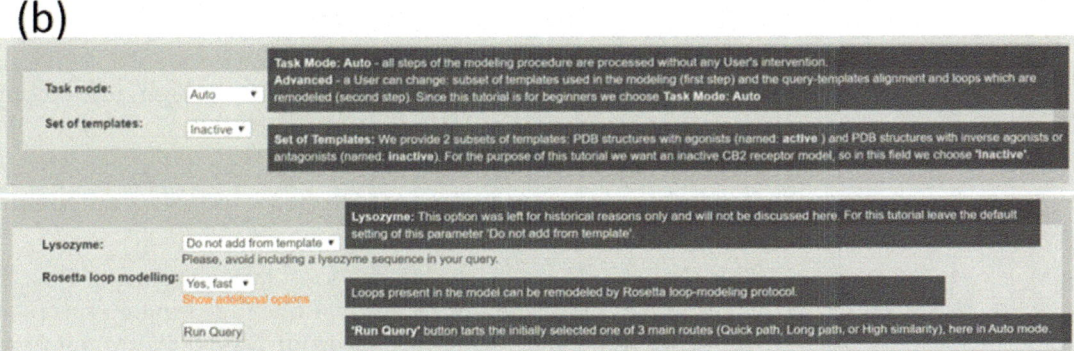

Fig. 15 The graphical messages in Tutorial explaining usage of options/modes. (**a**) Help on main paths. (**b**) Help messages on particular options

---

4 Notes

1. The initial checking for similarity to family of GPCR assumes 10% of similarity.

2. It is necessary to know in advance if there is a high-sequence similarity to one of the proteins used as a template. If the user chooses *HS* path for protein with low similarity, poor results can be obtained.

3. The *High similarity* mode is less time consuming than other modes, so it is the best mode to learn how to use GPCRM service and quickly watch results. However, it takes much longer time if N- and C-termini are selected to be refined by Rosetta (this mode includes loops over 20 amino acids too).

4. Sequences in FASTA format must contain only single letter codes of amino acids according to the IUB/IUPAC standard. No gaps ("-" character) nor ambiguous characters ("X" character) are allowed. One can introduce "*" (asterisk) character, meaning the end of the sequence. In this case the sequence is read until this character, and the rest is omitted. One can include FASTA header in the first line. It must start with the ">" character.

5. It is worth noting that at this stage the user can still change the path selection.

6. Cysteine sequential positions in "S-S bridge," if provided, must match their positions in the query sequence and are strictly related to the disulfide bridge between the extracellular loop 2 and the transmembrane helix 3. Do not provide cysteine positions for other cysteine bridge in these fields.

7. The Lysozyme option is kept for historical reasons.

8. The Rosetta refinement of termini selection requires much longer computational time. When the sequence similarity between the target protein and a template is high, it is recommended to choose the Modeler program.

9. Those loops below 20 amino acids. The longer loops are refined only by Rosetta and are selected arbitrarily if the Rosetta mode is enabled.

10. If the user does not choose the loop modeling using Rosetta program, the results of receptor modeling using Rosetta will not be generated—only results from Modeller program will remain.

11. The service does not indicate in Results whether the selection of the template(s) was automatic or manual.

12. It is currently not possible to see overlapped more than one homology models in one graphical window of NGL viewer.

Acknowledgments

This research was funded by National Science Centre, Poland, grant OPUS 2017/25/B/NZ7/02788.

References

1. Miszta P, Pasznik P, Jakowiecki J, Sztyler A, Latek D, Filipek S (2018) GPCRM: a homology modeling web service with triple membrane-fitted quality assessment of GPCR models. Nucleic Acids Res 46:W387–W395

2. Latek D, Pasznik P, Carlomagno T, Filipek S (2013) Towards improved quality of GPCR models by usage of multiple templates and profile-profile comparison. PLoS One 8:e56742

3. Rohl CA, Strauss CE, Misura KM, Baker D (2004) Protein structure prediction using Rosetta. Methods Enzymol 383:66–93

4. Alford RF, Koehler Leman J, Weitzner BD, Duran AM, Tilley DC, Elazar A, Gray JJ (2015) An integrated framework advancing membrane protein modeling and design. PLoS Comput Biol 11:e1004398

5. Woetzel N, Karakas M, Staritzbichler R, Muller R, Weiner BE, Meiler J (2012) BCL: score--knowledge based energy potentials for ranking protein models represented by idealized secondary structure elements. PLoS One 7: e49242

6. Panchenko AR (2003) Finding weak similarities between proteins by sequence profile comparison. Nucleic Acids Res 31:683–689

7. Larsson P, Wallner B, Lindahl E, Elofsson A (2008) Using multiple templates to improve quality of homology models in automated homology modeling. Protein Sci 17:990–1002

INDEX

Sofia Aires M. Martins and Duarte Miguel F. Prazeres (eds.), *G Protein-Coupled Receptor Screening Assays:
Methods and Protocols*, Methods in Molecular Biology, vol. 2268, https://doi.org/10.1007/978-1-0716-1221-7,
© Springer Science+Business Media, LLC, part of Springer Nature 2021

Printed by Printforce, the Netherlands